SOCIAL AND ECONOMIC BENEFITS OF PROTECTED AREAS

Protected areas (PAs) preserve biodiversity and ecosystems of high conservation value. In addition, these areas provide a range of benefits, both direct and indirect, to our societies and economies, i.e. so-called ecosystem services. These services include, for example, an ecosystem's ability to regulate floods and climate, purify water, secure the pollination of crops, and create opportunities for recreation, culture and tourism.

This book offers a comprehensive introduction to the socio-economic benefits of PAs and PA networks and provides step-by-step practical guidance on identifying, assessing and valuing the various ecosystem services and related benefits provided by PAs. It also aims to improve the communication of PA benefits to different stakeholders and the general public. It is shown that identifying and valuing the socio-economic benefits of PAs can be beneficial for several reasons. Demonstrating the socio-economic importance of a protected site can significantly increase political and stakeholder support for the site and resolve conflicts between different interest groups. This can lead to positive changes in policies and decision-making. Insights on PA benefits are also needed to identify a combination of actions and land use practices that best support the sustainable and equitable utilisation of these benefits, while retaining a site's conservation goals. Finally, demonstrating different benefits can help to discover alternative and sustainable sources for financing the management of PAs.

Marianne Kettunen is Senior Policy Analyst at the Institute for European Environmental Policy (IEEP) and Guest Researcher at the Finnish Environment Institute, Helsinki, Finland with dedicated experience in studies, capacity building and policy influence related to socio-economic role of biodiversity and ecosystem services and supporting the integration of these aspects into decision-making processes.

Patrick ten Brink is Senior Fellow and Head of Office at the Institute for European Environmental Policy (IEEP) in Brussels, Belgium. He is also the editor of *The Economics of Ecosystems and Biodiversity in National and International Policy Making*, developed within The Economics of Ecosystems and Biodiversity (TEEB) project, administered by the United Nations Environment Programme.

SOCIAL AND ECONOMIC BENEFITS OF PROTECTED AREAS

An assessment guide

Edited by Marianne Kettunen and Patrick ten Brink

Routledge
Taylor & Francis Group
LONDON AND NEW YORK

earthscan
from Routledge

'This book is [therefore] extremely welcome, coming at the moment when interest in assessment is higher than at any time before. The team of authors, led by Marianne Kettunen and Patrick ten Brink from the Institute for European Environmental Policy (IEEP), has assembled an impressive global evidence base and practical advice, drawing on years of experience. We urge everyone involved in protected areas conservation to benefit from its guidance and help to promote protected areas as natural solutions to many of the world's sustainability challenges.'

From the Foreword by Braulio F. de Souza Dias, Executive Secretary, Secretariat of the Convention on Biological Diversity (CBD) and Ernesto Enkerlin-Hoeflich, Chair of the World Commission on Protected Areas (WCPA), International Union for Conservation of Nature (IUCN)

'This book presents a timely and practical guide to assessing and communicating the multiple values of protected areas, whether inland or coastal wetlands, drylands, grasslands, forests or marine areas in the open oceans. Essential reading for all wishing to ensure that nature is more fully taken into account in decision making, including all those responsible for managing and maintaining the health of the over 2000 Ramsar Sites worldwide.'

Professor Nick Davidson, Deputy Secretary General, Ramsar Convention Secretariat

'This guidebook is incredibly comprehensive and useful for all practitioners attempting to present socio-economic values of PAs. It will undoubtedly lead to a whole host of new and strong PA valuation studies important for advancing the conservation agenda.'

Andrew Bovarnick, Lead Natural Resource Economist, UNDP

'The Guide provides a very welcome contribution to filling the gap that currently exists in the availability of practical tools and approaches for documenting, analysing and communicating the social and economic benefits of Protected Areas. Economists, conservation planners and policy-makers will all gain from the insights and techniques that are presented. The book is a core resource which will undoubtedly prove useful in strengthening protected area planning and management.'

Lucy Emerton, Director of Economics & Finance, Environment Management Group

'A timely, practical and inspiring guidebook that helps us to value the multiple benefits of protected areas and to communicate them better to local people and decision-makers. A must-read for all protected area managers!'

Sanna-Kaisa Juvonen, Senior Advisor for International Affairs, Metsähallitus Natural Heritage Services, Finland

'This book is an excellent, practice-oriented overview of current methodological approaches and challenges to assess the ecosystem services provided by protected areas. It gives clear indication and guidance how to better understand the potentials and shortcomings of assessing and valuing nature and how these values can be taken up by and communicated to the decision making processes.'

Alberto Arroyo Schnell, Senior Policy Advisor on Biodiversity, Andreas Baumüller, Head of Natural Resources and Land Use and Peter Torkler, EU Policy, WWF

First published 2013
by Routledge
2 Park Square, Milton Park, Abingdon, Oxon OX14 4RN

Simultaneously published in the USA and Canada
by Routledge
711 Third Avenue, New York, NY 10017

Routledge is an imprint of the Taylor & Francis Group, an Informa business

British Library Cataloguing-in-Publication Data
A catalogue record for this book is available from the British Library

Library of Congress Cataloging-in-Publication Data
Social and economic benefits of protected areas : an assessment guide / edited by Marianne Kettunen and Patrick ten Brink.
pages cm
Includes bibliographical references and index.
ISBN 978-0-415-63283-6 (hbk : alk. paper) -- ISBN 978-0-415-63284-3 (pbk : alk. paper) -- ISBN 978-0-203-09534-8 (ebk : alk. paper) 1. Protected areas--Social aspects. 2. Protected areas--Economic aspects. 3. Environmental protection--Social aspects. 4. Environmental protection--Economic aspects. 5. Environmental economics. I. Kettunen, Marianne. II. Brink, Patrick ten.
S944.5.P78S63 2013
333.72--dc23
2013004282

ISBN: 978-0-415-63283-6 (hbk)
ISBN: 978-0-415-63284-3 (pbk)
ISBN: 978-0-203-09534-8 (ebk)

Typeset in Bembo
by GreenGate Publishing Services, Tonbridge, Kent

CONTENTS

ILLUSTRATIONS

Figures

Tables

Boxes

CONTRIBUTORS

Editors and coordinating authors

Marianne Kettunen is a Senior Policy Analyst at the Institute for European Environmental Policy (IEEP), currently based as a guest researcher at the Finnish Environment Institute in Helsinki, Finland. She coordinated the assessment of socio-economic benefits of protected areas for the book *The Economics of Ecosystems and Biodiversity (TEEB) in National and International Policy Making* (Earthscan for Routledge) in 2011 and led the publication *Socio-economic Importance of Ecosystem Services in the Nordic Countries* (TemaNord). Marianne has more than ten years of experience in working on biodiversity issues at European and international level, including ample expertise in outreach, policy influence and capacity-building on the value of nature, ecosystem services and Green Economy. Marianne has an MSc in Biodiversity Research and Environmental Science from the University of Turku, Finland. Contact: mkettunen@ieep.eu

Patrick ten Brink is a senior fellow at the Institute for European Environmental Policy (IEEP) and Head of the Brussels Office, which he helped set up in 2000. He also leads IEEP's environmental economics programme. Patrick was the editor of *The Economics of Ecosystems and Biodiversity (TEEB) in National and International Policy Making* (Earthscan for Routledge) in 2011. He has more than twenty years' experience working on environmental policy and over the last ten years has developed a particular expertise in the value of biodiversity. Patrick has an MSc in Environmental and Natural Resource Economics from University College London and is on the editorial boards of the journals *European Policy and Governance* and *Ecosystem Services*. Contact: ptenbrink@ieep.eu

Contributing authors

Tomas Badura is a PhD candidate at the Centre for Social and Economic Research on the Global Environment at the University of East Anglia and holds an MSc in Theoretical Economics from the University of Amsterdam. He specialises in the economic dimension of ecosystem valuation, focusing in his current research on the methodological issues associated with the use of valuation for policy and decision-making purposes. Tomas has also worked at the Institute for European Environmental Policy (IEEP) co-authoring a number of publications related to the value of biodiversity and ecosystem services, including the economic value of the European Natura 2000 Network of protected areas.

Samuela Bassi is a policy analyst at the Grantham Research Institute on Climate Change and the Environment at the London School of Economics and Political Science and the Centre for Climate Change Economics and Policy, focused on working with topics related to green growth and climate change policy. Previously she worked at the Institute for European Environmental Policy (IEEP), working inter alia on the economic assessment of biodiversity and ecosystem services. She graduated in economics the from University of Trieste, Italy, and holds an MSc in Economics from Birkbeck College, London.

Dalia D'Amato holds an MSc in Biodiversity Survey from the University of Sussex and a Master's degree in Biodiversity and ecosystem management from Università degli Studi di Roma Tre, Rome. Dalia's academic interests have recently taken her to a number of different research institutes within Europe, including the Institute for European Environmental Policy (IEEP) and the Finnish Environmental Institute (SYKE). Dalia's expertise and interests include issues related to the socio-economic value of biodiversity and ecosystem services, links between biodiversity and climate change adaptation, and mobilising financial resources for conservation.

Nigel Dudley is a consultant working principally on issues related to protected areas and broad-scale approaches to conservation. He is Industry Fellow, School of Geography, Planning and Environmental Management at the University of Queensland and a vice-chair of the IUCN World Commission on Protected Areas. Nigel established Equilibrium Research with Sue Stolton (below) in 1991 and between them they have authored well over 200 papers, reports and books. Nigel is based in the UK, in Bristol and in the Dyfi Biosphere Reserve in Wales, but works throughout the world.

Sonja Gantioler is an expert on biodiversity and wider environmental sustainability. Currently Sonja works at the Terra Institute in Italy, focusing on issues related to the sustainability of business sectors. Previously she was part of the Institute for European Environmental Policy (IEEP) where she led several projects related to the value of biodiversity and ecosystem services, including authoring publications on the costs and benefits of the European Natura 2000 Network of protected areas and contributing to *The Economics of Ecosystems and Biodiversity (TEEB) in*

National and International Policy Making (Earthscan for Routledge). Sonja holds an MSc in Ecology and Environmental Economics from Vienna University, Vienna University of Economics and Business and Technical University Munich.

Sue Stolton established Equilibrium Research in partnership with Nigel Dudley in 1991. Equilibrium promotes positive environmental and social change by linking targeted research to field application, working with groups ranging from local communities to United Nations agencies in countries worldwide. Sue works mainly on issues relating to protected areas, in particular with respect to implementing the CBD's Programme of Work on Protected Areas. Areas of interest include management of protected areas, and in particular the assessment of management effectiveness, issues related to understanding the wider values and benefits that protected areas can provide and the development and use of the IUCN protected area management categories. Sue is a member of the IUCN World Commission on Protected Areas (WCPA) and the Commission on Environmental, Economic and Social Policy (CEESP).

FOREWORD

In the Brazilian Amazon, meticulous calculation of the economic benefits of carbon sequestration derived from the system of national parks helped to swing public opinion behind the ambitious Amazon Regional Protected Areas programme. In Mexico, rediscovery of *Zea perennis*, a wild relative of corn presumed extinct in the wild for fifty years, led to intensive research in the mountains where it was found, discovery of other economically important crop wild relatives, the creation of a biosphere reserve to ensure their survival and a national surge in conservation initiatives.

We only seem to be capable of valuing what we are about to lose. Recognition of the importance of ecosystem services is growing among governments, international institutions and the corporate sector at the very time when natural ecosystems are disappearing faster than ever. The global protected area network is a critical tool to help stem these losses. Research shows repeatedly that protected areas are effective in slowing, halting and reversing habitat loss within their borders. While as conservationists we like to think of protected areas as a small proportion of natural ecosystems, unfortunately we are forced to admit that in an increasing number of places they contain the *only* natural habitat remaining. This means that inevitably they are important not only for biodiversity but also for local people depending on nature for their livelihood.

This massively increases what is expected of protected areas. Originally set up to conserve wild species and iconic landscapes, many are now also expected to combat climate change, alleviate poverty, and supply ecosystem services that have been degraded beyond their borders such as clean water, stable soils, protection of valuable genetic resources and even day-to-day needs like herbal medicines. Such a long list of expectations puts those managing such areas under extra pressure. But on the plus side, as people recognise that protect areas do "deliver" much more than just nature conservation, the arguments for their creation and maintenance continue to increase.

A thorough social and economic understanding of these benefits can consolidate public support for conservation; and if a figure can be put on ecosystem benefits, so much the better. Economic benefits are of course not everything: we can't put a dollar sign against many of the aesthetic, spiritual and cultural values of natural ecosystems, nor the ethical duty we have to halt runaway extinction and environmental degradation. But for decision and policy makers socio-economic analysis, relying on a range of qualitative, quantitative and monetary indicators of value, still plays an extremely important role complementing the insights related to biodiversity values. The challenge for the evaluators is that any one protected area can provide many different benefits, and also bring a range of costs that need to be included in the balance.

Protected area agencies are looking for clear advice about how to identify and assess the benefits provided by their ecosystems. This book is therefore extremely welcome, coming at the moment when interest in assessment is higher than at any time before. The team of authors, led by Marianne Kettunen and Patrick ten Brink from the Institute for European Environmental Policy (IEEP), has assembled an impressive global evidence base and practical advice, drawing on years of experience. We urge everyone involved in protected areas conservation to benefit from its guidance and help to promote protected areas as natural solutions to many of the world's sustainability challenges.

Braulio F. de Souza Dias, Executive Secretary, Secretariat of the Convention on Biological Diversity (CBD) and Ernesto Enkerlin-Hoeflich, Chair of the World Commission on Protected Areas (WCPA), International Union for Conservation of Nature (IUCN)

PREFACE

> 'Everybody needs beauty as well as bread, places to play in and pray in, where nature may heal and give strength to body and soul alike.'
>
> John Muir in 'The Yosemite' (1912)

The global network of protected areas is like a web of life underlining our welfare: it will help to keep us fed and to quench our thirst, and it provides us with both tranquillity and entertainment while protecting us from many sources of harm. It is an endless source of inspiration and a living library of life holding a key to scientific knowledge and future innovations. It also maintains an open-access pharmacy and a 24-hour gym that will help to keep us healthy for years to come. However, while the awareness of the value of nature is increasing, there is still a widespread under-appreciation of the variety of social and economic benefits it provides us, especially at the practical level. Nature conservation, including investing in establishing and managing protected areas, is often considered as a hindrance rather than a benefit to increasing our welfare. Work remains to be done to increase understanding of the benefits associated with protected areas, and to demonstrate and take account of their values in concrete decision-making.

Over the past decade we have been closely engaged with the developments related to the mainstreaming of ecosystem services into biodiversity policies and the rise of socio-economic arguments as a tool to promote conservation. During these years we have become increasingly aware that highlighting the social and economic values of biodiversity can help to make a case for investing in protected areas and encouraging due measures to avoid their degradation or loss. At the same time we are conscious that the objective of protected areas – protecting rare, unique species, ecosystems, habitats and genetic resources – and objectives of achieving socio-economic benefits can sometimes work in the same direction and sometimes cause conflicts. However, we feel that the potential positive outcomes of assessing and communicating the benefits merits their systematic assessment while the risk of conflicts can often be addressed by careful planning (e.g. ensuring due zoning of protected areas) and ascertaining that any assessment of socio-economic benefits is presented in the context of biodiversity values.

We are therefore pleased to have prepared this book, with due engagement of a range of contributing authors and building on more than ten years of assessing, valuing and raising awareness on biodiversity and related socio-economic benefits. We have tried to take a cautious but positive approach as regards to valuation, including considering both monetary and non-monetary assessment of benefits. We have aimed to encourage the development of carefully crafted assessments that build on up-to-date scientific understanding and existing practical experiences, and we have aimed to do that in a manner that is accessible also to a non-academic audience. We hope that this assessment guide supports site managers and others to better understand and systematically identify, assess and communicate benefits associated with protected areas, leading to concrete results such as increasing investment in protected areas and integrating them into local land use and planning.

This book has been motivated by our genuine desire to increase the common understanding of multiple values associated with protected areas and, most importantly, to translate this understanding into concrete actions. We hope that it will be a relevant contribution to the existing knowledge base and advice available, helping to secure the future of protected areas for generations to come.

Marianne Kettunen and Patrick ten Brink

Institute for European Environmental Policy

ACKNOWLEDGEMENTS

This book has benefited significantly from the ever-increasing global body of evidence on the socio-economic importance of protected areas. We wish to thank all authors of the individual studies synthesised in the context of this book for their excellent – and often pioneering – efforts to explore, assess and raise awareness of the diverse values of protected areas across the world.

This book also builds on several core studies and initiatives. The concept for and core content of the book is based on the 'Toolkit for Assessing Socio-economic Benefits of Natura 2000' developed as a part of the project 'Financing Natura 2000: Cost estimate and benefits of Natura 2000' financed by the European Commission/ DG Environment (Contract No. 070307/2007/484403/MAR/B2). In addition, the book draws from the work carried out by the authors in the context of The Economics of Ecosystems and Biodiversity (TEEB)[1] and a number of further studies by the European Commission on costs, benefits and overall economic value of the EU Natura 2000 PA network.[2]

We are thankful for the support by the Institute for European Environmental Policy (IEEP) to our endeavour, with special thanks to David Baldock, Claire Froomberg, Graham Tucker and IEEP's board of trustees.

The encouragement and support (review, cooperation, comments, material, inspiration, moral support etc.) of the following colleagues is gratefully acknowledged: Braulio F. de Souza Dias, Sarat Babu Gidda and Markus Lehmann (Secretariat of the Convention on Biological Diversity), Prof. Ernesto Enkerlin Hoeflich and Dr Kathy MacKinnon (World Commission on Protected Areas – IUCN), Prof. Nicholas Davidson (Secretariat of Ramsar Convention on Wetlands), Stefan Leiner (European Commission, DG Environment), Elisabeth Schlaudt (IEEP), Peter Torkler (WWF Germany), Dr Rob Tinch (Environmental Futures Ltd./Iodine SPRL), Sanna-Kaisa Juvonen, Liisa Kajala, Mervi Heinonen and the rest of Metsähallitus Natural Heritage Services team, Dr Joachim Maes and Dr Grazia Zulian (Joint Research Centre of the European Commission), Dr Marius Mayer (University of Würzburg), Dr Pieter van Beukering (IVM – Institute for Environmental Studies), Lucy Emerton, Director of Economics &

Finance, Environment
Management Group, Jenny Birch (BirdLife International), Jose Benedicto Royuela (Brunel University, London) and Azucena de la Cruz (Portuguese Society for the Study of Birds).

The authors of the individual chapters of this book who heroically – and in retrospect perhaps even ever so foolishly – agreed to embark with us on this journey: Dalia, Nigel, Samuela, Sonja, Sue and Tomas, a sincere thanks for your dedication and hard work, without which this book would not exist.

On a personal note, we thank our family members and friends for their support during the process of composing this volume. Special thanks to Stefan Simis for his encouragement, patience and keeping up good spirits, and to Annikki and Hannu Kettunen for supporting their daughter to work in and experience protected areas around the world. Special thanks also to Catherine Magnant for her continued, unwavering support and Chloé and Éléa ten Brink for being an endless source of joy and inspiration.

Finally, this book is dedicated to all the protected areas around the world and to those maintaining them, safeguarding biodiversity and our well-being.

Marianne Kettunen and Patrick ten Brink (IEEP)

Notes

1 www.teebweb.org
2 http://ec.europa.eu/environment/nature/natura2000/financing/index_en.htm

ACRONYMS AND ABBREVIATIONS

CBA	Cost-benefit analysis
CBD	Convention on Biological Diversity
CVM	Contingent Valuation Method
CWR	Crop Wild Relative
ETS	Emissions Trading Scheme
FAO	Food and Agriculture Organization
FTE	Full-Time Equivalent
GEF	Global Environment Facility
GIS	Geographic Information System
IBA	Important Bird Areas
IE	Income Equivalent
IPA	Indigenous Protected Area
InVEST	Integrated Valuation of Ecosystem Services and Trade-offs
IUCN	International Union for Conservation of Nature
MA	Millennium Ecosystem Assessment
MPA	Marine Protected Area
NGO	Non-governmental organization
NP	National Park
NPP	Net primary production
NPV	Net present value
NTFP	Non-timber forest product
PA	Protected Area
PA-BAT	Protected Areas Benefits Assessment Tool
PES	Payments for Ecosystem Services
REDD	Reduced Emissions from Deforestation and Forest Degradation
RSPB	Royal Society for the Protection of Birds
TEEB	The Economics of Ecosystems and Biodiversity
TEV	Total Economic Value
UK NEA	United Kingdom National Ecosystem Assessment

UN	United Nations
UNEP	United Nations Environment Programme
UNESCO	United Nations Educational, Scientific and Cultural Organization
UNWTO	United Nations World Tourism Organization
WTP	Willingness to pay
WWF	World Wide Fund for Nature
WHO	World Health Organisation

1

INTRODUCTION, OBJECTIVES AND APPROACH

Kettunen, M. and ten Brink, P.

1.1 Introduction

Protected areas (PAs) are established to protect the world's biological diversity (ecosystems, habitat, species and genes) (Box 1.1). In addition, these areas maintain and deliver a range of benefits (direct and indirect) to societies and economies. For example, ecosystems within PAs can provide clean water, maintain healthy populations of pollinators and help to mitigate different natural hazards (floods, drought, wild fires, etc.). In addition, PAs form an important basis for maintaining human health (both physical and mental), creating opportunities for recreation and tourism, and forming cultural characteristics and values. PAs support food security by maintaining crop diversity and species with economic and/or subsistence value. They also play an important role in ecosystem-based approaches to climate change adaptation and contribute to mitigation by storing and sequestering carbon.

The understanding and appreciation of the benefits and related socio-economic values provided by PAs, however, remains limited, particularly in the context of different sectoral policies and practices affecting land use. This is especially the case at the practical level where, rather than a source of local to global benefits, PAs are often perceived as imposing costs or restrictions on communities and economies. While biodiversity values alone might be enough to guarantee support to (and resources for) the establishment and management of PAs, identifying and assessing (e.g. economically valuing) related benefits and socio-economic values can be a useful tool to enhance these efforts. Demonstrating the socio-economic importance of PAs can significantly increase political and stakeholder support for these sites by conveying a clear message on the value of biodiversity and functioning ecosystems to broader stakeholder groups. This support can further lead to positive changes in policies and decision-making. Identification and socio-economic valuation of benefits will also provide useful information for decision-making at a practical level, such

BOX 1.1

Definitions of protected areas

There are two protected area definitions, from the Convention on Biological Diversity (CBD) and the IUCN World Commission on Protected Areas: both convey a similar general message. These definitions encompass several other international classifications, such as natural World Heritage sites and biosphere reserves established by UNESCO.

CBD definition: 'A geographically defined area which is designated or regulated and managed to achieve specific conservation objectives'. (Article 2 of CBD) (CBD, 1992)

IUCN definition: 'A clearly defined geographical space, recognised, dedicated and managed, through legal or other effective means, to achieve the long-term conservation of nature with associated ecosystem services and cultural values'. (updated definition from 2007 in Dudley, 2008)

as advice about local and regional land use planning to support sustainable land use in the area. In addition, demonstrating different benefits associated with PAs – and their related beneficiaries – can help to identify alternative and sustainable sources for financing the management actions. For example, visitors' fees could contribute to covering the management or maintenance costs of PAs. Similarly, municipalities in the vicinity of a protected wetland could financially support the maintenance of the site in recognition of its water purification capacity benefitting.

The understanding of the role nature plays in underpinning human welfare is slowly increasing thanks to initiatives such as the Millennium Ecosystem Assessment (MA 2005) and *The Economics of Ecosystems and Biodiversity* – TEEB initiative (Kumar, 2010; ten Brink, 2011; Kettunen *et al.*, 2011; Bishop, 2012; Wittmer and Gundimeda, 2012). Building on these developments, the Strategic Plan for Biodiversity 2011-2020 adopted in 2010 to implement the UN Convention on Biological Diversity (CBD) outlines that 'By 2020, at the latest, biodiversity values have been integrated into national and local development and poverty reduction strategies and planning processes and are being incorporated into national accounting, as appropriate, and reporting systems' (Aichi Biodiversity Target 2 of the CBD Strategic Plan) (CBD, 2012).

The increased understanding has also led to a more concrete appreciation of how the sustainable use, protection and restoration of nature – including PAs – can be a key to more sustainable and 'greener' economic development (UNEP, 2011; ten Brink *et al.*, 2012). For example, investment in nature conservation can create a range of business opportunities and provide cost-effective solutions for different sectors. In June 2012 the UN Conference on Sustainable Development (Rio+20) agreed to consider green economy as one of the important tools available for achieving sustainable development and eradicating poverty (UNCSD, 2012), creating further political impetus for more sustainable use and management of nature's capital.

With the increased attention being focused on the benefits provided by nature and PAs increasing, there is arguably a need to provide accurate information and advice to a range of interested stakeholders on how to assess and communicate the values of PAs. Many stakeholders have limited expertise in assessing the socio-economic benefits of nature. It is important to understand that identifying, assessing and valuing such benefits should always be carefully considered, planned and carried out, based on a basic understanding of both ecology and environmental economics (TEEB, 2010). It also needs to be recognised that emphasising the role PAs play in supporting well-being should not be seen as replacing or undermining PAs' focus on biodiversity, nor should it jeopardise their set goals for conservation.

When appropriately applied, an assessment of PA-related benefits and their socio-economic value can be an influential tool for both supporting human welfare and stepping up conservation efforts. Experience shows that PAs demonstrating a high value for society enjoy much greater and more consistent political support, across the political divide, than PAs where these values are unknown or unreported (e.g. Kajala, 2013). Socio-economic assessments can also help to improve PA management efficiency by, for example, identifying areas for investment (tourism and recreation, branding and marketing of PA produces, etc.), helping to develop various forms of payments for ecosystem services, assessing compensation payments for values forgone, encouraging sustainable resource use and supporting zoning of PAs based on different management objectives.

1.2 Objectives, scope and audience

The purpose of this assessment guide is to increase the global awareness of and information on the benefits and socio-economic values of PAs and PA networks. This guide aims to do so by synthesising wide-ranging global evidence on benefits provided by PAs and providing concrete, step-wise and practice-oriented guidance on how to identify, assess and communicate the various benefits, with a specific focus on their socio-economic valuation. The guidance provided is applicable to existing PAs and it can also be used to support the designation of new PAs. In a broader policy context, this guide also aims to facilitate the integration of biodiversity, ecosystems and related services into decision-making processes at global and national levels, supporting the implementation of biodiversity goals for 2020.

This assessment guide is, first and foremost, a consolidated source of information for practitioners involved in the designation and management of PAs at national, regional and local level who are interested in exploring the socio-economic arguments for conservation (regional land use planners, local site managers, landowners and other land users, etc.). These practitioners, who are often trained in biology or wildlife management, nowadays find themselves expected to place PAs into the broader economic and societal context and respond to what must often appear to be a bewilderingly wide range of issues. It is hoped that the guide will inspire and help these practitioners identify and assess the different benefits and socio-economic 'potential' of their sites, such as possible socio-economic gains achieved by managing sites and land in a sustainable manner. This guide can also be used by a broader audience (decision-makers, researchers, academics, students, etc.) interested in the benefits and socio-economic values provided by biodiversity and ecosystem services, particularly in the context of PAs.

It is to be noted that the guide does not solely focus on carrying out a monetary assessment of the final (net) value of PAs. It rather aims to provide information on different approaches and methods available – qualitative, quantitative and monetary alike – that practitioners can use to highlight the socio-economic importance of PAs, depending on the information and resources at their disposal. It also hopes to help the practitioner put the available estimates into a proper context, and interpret and communicate them correctly. In other words, while this assessment guide helps to gain an overall picture of the benefits and socio-economic values related to PAs, it does not attempt to turn these benefits into a single aggregated monetary value (see Section 1.3 below).

Finally, it is to be noted that this guide does not aim to replace or duplicate existing information on assessing the socio-economic value of biodiversity and ecosystem services, including available guidance on economic valuation. Rather than providing a detailed account of the theory and use of specific economic valuation methodologies, the objective of this publication is to offer a systematic framework and guidance for applying these methodologies in the context of PAs, with due references to key existing literature and (generic) guidance documents.

1.3 Guiding principles

This assessment guide builds on a number of underlining guiding principles that should be kept in mind throughout its application.

Assessing the socio-economic importance and value of PAs does not aim to undermine the intrinsic value of biodiversity. The intrinsic value of biodiversity is separate from the anthropocentrically orientated consideration of PAs' benefits and socio-economic values. Regardless of the recent (policy) focus on the latter, the former still continues to be a reason enough for establishing PAs and investing money in their management (e.g. CBD, 2008; Kettunen *et al.*, 2011). However, PAs clearly also play an important role in supporting human well-being and highlighting these arguments can be justified in terms of increasing support for their conservation. There are important synergies between conserving nature for its intrinsic value and protecting the functions that underpin well-being. On the other hand, focusing only on the anthropocentric aspects can lead to overlooking the needs for biodiversity conservation. Finding the synergies while avoiding the possible risks is also part of the guidance in this book.

The identification, assessment and valuation of benefits provided by PAs can be structured and carried out in different ways and using different metrics of value (monetary and non-monetary), depending on the situation. This is also the basic premise of the global

TEEB initiative (TEEB, 2010). Simply recognising (e.g. identifying) the benefits and socio-economic values related to PAs can sometimes be sufficient to ensure their conservation and sustainable use (Step I of this guide). This may be the case, for example, where the spiritual or cultural benefits are a prominent feature of a PA. However, demonstrating benefits in economic terms is often useful for decision-making purposes, for example when trying to determine the full costs and benefits of conservation (including private and public, monetary and non-monetary benefits) (Step II of this guide). Such valuation enables one to identify and address trade-offs between different land use and management choices in a more informed manner by correcting the bias typical of much decision-making today, which tends to favour private wealth and physical capital above public wealth and natural capital (TEEB, 2010). Finally, capturing the benefits and socio-economic values of PAs involves the adoption of mechanisms that incorporate these benefits and values into PA management through different economic incentives and/or price signals (Step III of this guide). This can include payments for ecosystem services (PES) or the creation of new markets for sustainably produced goods and ecosystem services from PAs.

The benefits and socio-economic values related to PAs are manifold and cannot always be captured in monetary terms. The total benefits related to the socio-economic values of PAs consist of different components including, for example, tourism, recreational and cultural heritage value, a site's role in supporting wild pollinators and the value of wild berries and game provided by the site. In practice, only some of these benefits and values can be estimated in monetary terms (see Chapter 3) and therefore the final assessment of the overall value of a PA is always likely to be a combination of different estimates that cannot easily be merged into a single aggregated figure. Consequently, it is considered that the overall socio-economic role and importance of PAs (both for individual and multiple sites) can best – and most accurately – be assessed and expressed through a range of different qualitative, quantitative and monetary estimates, each of which is selected to reflect the value(s) of a specific benefit.

Identified benefits should be used sustainably by respecting sites' overall biodiversity goals and management plans. The socio-economic value of any benefit considered in the context of this guide should be determined on the basis of its sustainable use. This is of particular importance when considering the benefits related to the extraction of PAs' biodiversity resources; estimated values should not be calculated based on non-sustainable levels of producing crops or harvesting timber, fish, etc. In addition, the benefits considered and promoted should be compatible with the objectives and management plans of PAs. In some cases, conflicts between the two might arise. For example, an important wetland for birds could, in principle, be used for mitigating the impacts of floods (i.e. by providing a flood storage area). However, water levels might need to be kept lower than desirable for wetland habitats in order to maximise the area's flood mitigation potential, thus there might be a conflict between the specific conservation goals of the site and its potential to provide benefits for flood mitigation. Similarly, rapidly growing forest plantations are often very effective in sequestrating carbon but they are also rather biodiversity poor. In a number of cases, possible conflicts could be addressed by delineating different areas within PAs, some focused primarily on biodiversity values (core areas) and other on maintaining and sustainably using different ecosystem services (buffer areas) (see Chapter 15).

Benefits are often linked and these linkages should be understood in order not to overestimate the total value of PAs. Ecological processes, related benefits (ecosystem services and related goods) and socio-economic values are often interlinked (see Chapter 3). For example, provisioning of crops is often dependent on the availability of fresh water, pollinators, flood and erosion control, etc. Therefore, the economic value of pollination and flood and erosion control is already partly captured in the value of crops. Consequently, assessing the total socio-economic value of a PA by simply summing up the different (monetary) value estimates available can lead to overestimating the total value. This problem, called 'double counting', is further explained in Chapters 3 and 12 and it should be kept in mind when interpreting and communicating the valuation results.

1.4 Approach, data, structure and application

The approach adopted in this publication is twofold. First, the assessment guide aims to provide its reader with a basic understanding of the ecological and economic principles and key considerations behind assessing the socio-economic importance of nature (e.g. basic principles of economic valuation) (Part I). Second, building on this conceptual basis, the guide offers a structured approach and guidance for identifying, assessing and communicating benefits related to PAs – including possible methods for determining the socio-economic (e.g. monetary) value of these benefits (Part II). The guidance provided in Part II attempts to be as accessible and practice-oriented as possible, drawing on actual examples and experiences from PAs. However, since the socio-economic valuation of PAs is a relatively new field some sections of the guide might be less applied than others.

The approach builds on identifying various ecosystem services provided by PAs and assessing the benefits of and socio-economic values related to and/or arising from these services. This is because the guide does not wish to target only the tangible and most commonly understood benefits derived from PAs (sustainably harvested timber, crops and game, etc.). Instead, it seeks to draw a specific attention to the more hidden benefits and values related to PAs, such as the socio-economic significance of different ecosystem processes they support (regulation of floods, climate and water quality, etc.). These benefits and values are traditionally overlooked, thus raising awareness of their role is considered of high importance. The identification and classification of ecosystem services, related benefits and socio-economic values is based on the approaches adopted in the context of the key pioneering initiatives, including the Millennium Ecosystem Assessment (MA, 2005), TEEB (2010) and the UK National Ecosystem Assessment (UK NEA, 2011).

The book synthesises a range of existing studies assessing the benefits and/or socio-economic value of PAs around the world. While the authors of this book have made dedicated attempts to ensure the robustness and accuracy of these studies (e.g. the majority of studies are from peer-reviewed sources) it has not been feasible to carry out a detailed estimate of these aspects for each individual study. Furthermore, while most of the studies used in the book are less than ten years old – to ensure the best possible current relevance of the examples provided – the benefits provided by and/or socio-economic values associated with individual PAs might have changed. Consequently, readers are always advised to refer back to the original source and/or contact the original authors of studies before interpreting examples synthesised in this book and building on them for their own purposes.

The publication consists of three main parts:

- Part 1 'Contextual guidance' provides an overview of the benefits (ecosystem services and related goods) and associated socio-economic values provided by PAs. In addition, Part 1 gives an introduction to the general contextual framework for and principles of economic valuation in the PA context.
- Part 2 'Practical guidance' provides a structured framework and practical guidance for identifying, assessing and communicating the benefits and socio-economic values related to PAs. This part consists of three main steps that are summarised below.
  - *Step I: A (rapid) scoping assessment of possible benefits.* This first step shows how to carry out a rapid, first-stage scoping assessment of the possible benefits (ecosystem services) provided by PAs. This preliminary assessment 1) helps to obtain a general view of the full range of benefits provided by a site, including an initial assessment of their relative socio-economic importance and 2) helps to identify which benefits/ecosystem services could be selected for further in-depth analysis and valuation (Step 2). Step I is supported by a concrete example of the application of rapid assessment made available in Annex 1.
  - *Step II: Detailed socio-economic valuation of benefits, including step-wise guidance.* The purpose of this second step is to provide more specific guidance on methodologies available for estimating the socio-economic value of PA-related benefits. The introduced methodologies are used to derive estimates on the qualitative, quantitative and monetary value of different services. Step 2 focuses specifically on value estimates considered feasible to be obtained by practitioners, such as site managers and other interested stakeholders. Step 2 is supported by a detailed stand-alone, step-wise guidance for socio-economic valuation of key ecosystem services and related benefits (Annexes 2 to 6). These step-wise guidance documents are supported by concrete case study examples of using the methods in practice.
  - *Step III: Guidance on how to interpret and use the results of socio-economic assessments.* This step provides general guidance on using the results of valuation(s). In particular, Step 3 aims to assist in understanding the robustness of value estimates and using the different estimates to form an

overall picture of the total socio-economic importance and value of the site. In addition, advice on communicating the results to different relevant stakeholders and using them to support PA management is provided.

References

Bishop, J. (ed.) (2012) *The Economics of Ecosystems and Biodiversity in Business and Enterprise*, Earthscan, London

CBD (1992) *The Convention on Biological Diversity: Article 2 on the Use of Terms*, http://www.cbd.int/convention/articles/?a=cbd-02, accessed 1 November 2012

CBD (2008) *Protected Areas in Today's World: Their Values and Benefits for the Welfare of the Planet*, Technical Series no. 36, report by Secretariat of the Convention on Biological Diversity, Montreal, http://www.cbd.int/doc/publications/cbd-ts-36-en.pdf, accessed 26 September 2012

CBD (2012) *Global Biodiversity Targets for 2020 (Aichi Biodiversity Targets)*, http://www.cbd.int/sp/targets, accessed 24 September 2012

Dudley, N. (ed.) (2008) *Guidelines for Applying Protected Area Management Categories*, IUCN, Gland, Switzerland, http://data.iucn.org/dbtw-wpd/edocs/PAPS-016.pdf, accessed 1 November 2012

Kajala, L. (2013) 'TEEB Nordic case: Estimating local economic benefits of visitors' spending in protected areas', in M. Kettunen *et al. Socio-economic importance of ecosystem services in the Nordic Countries – Scoping assessment in the context of The Economics of Ecosystems and Biodiversity (TEEB)*, Nordic Council of Ministers, Copenhagen

Kettunen, M., Berghöfer, A., Brunner, A., Conner, N., Dudley, N., Ervin, J., Gidda, S. B., Mulongoy, K. J., Pabon, L. and Vakrou, A. (2011) 'Recognising the value of protected areas', in P. ten Brink (ed.) *The Economics of Ecosystems and Biodiversity in National and International Policy Making*, Earthscan, London

Kumar, P. (ed.) (2010) *The Economics of Ecosystems and Biodiversity: Ecological and Economic Foundations*, Earthscan, London

TEEB (2010) *The Economics of Ecosystems and Biodiversity: Mainstreaming the Economics of Nature: A synthesis of the approach, conclusions and recommendations of TEEB*, http://www.teebweb.org/publications/teeb-study-reports/synthesis, accessed 25 September 2012

ten Brink, P. (ed.) (2011) *The Economics of Ecosystems and Biodiversity in National and International Policy Making*, Earthscan, London

ten Brink, P., Mazza, L., Badura, T., Kettunen, M. and Withana, S. (2012) *Nature and its Role in the Transition to a Green Economy*, http://www.ieep.eu/publications/2012/10/nature-and-its-role-in-the-transition-to-a-green-economy--1157, accessed 10 June 2013

UK NEA – National Ecosystem Assessment (2011) *The UK National Ecosystem Assessment: Synthesis of the Key Findings*, UNEP-WCMC, Cambridge, http://uknea.unep-wcmc.org, accessed 18 June 2012

UNCSD (2012) Rio+20 declaration – The Future We Want, UN document A/66/L.56, http://daccess-dds-ny.un.org/doc/UNDOC/GEN/N11/476/10/PDF/N1147610.pdf?OpenElement, accessed 20 June 2013

UNEP (2011) *Towards a Green Economy – Pathways to Sustainable Development and Poverty Eradication*, a Synthesis for Policy Makers, http://www.unep.org/greeneconomy/Portals/88/documents/ger/GER_synthesis_en.pdf, accessed 18 June 2012

Wittmer, H. and Gundimeda, H. (eds) (2012) *The Economics of Ecosystems and Biodiversity in Local and Regional Policy and Management*, Earthscan, London

PART 1
Contextual guidance

2

PROTECTED AREAS

Their values and benefits

Dudley, N., Stolton, S. and Kettunen, M.

Chapter 2 forms the starting point of the guidance, providing the reader and users with a general introduction to the benefits related to PAs. It presents an overview of the ecosystem services and related goods provided by PAs, highlighting the ecological processes underpinning the benefits. The chapter also outlines some key aspects that need to be taken into consideration when assessing the socio-economic importance of PAs, including the distribution of benefits, costs of PAs and possible opportunities and risks related to the valuation of benefits.

2.1 Different benefits and values of protected areas

Protected areas (PAs), including national parks, nature reserves and other types of areas aimed at conserving nature's value, are today regarded primarily as places that society has identified and set aside to conserve *biodiversity*; this is generally interpreted to mean ensuring the survival of representative and viable populations of all the world's ecosystems and species. The recognition of the urgent need for well-managed PAs has grown significantly in recent years, driven by the awareness of the vast changes affecting many parts of the world, with natural ecosystems degraded by a mixture of development pressures, pollution and unsustainable use.

The commonly accepted IUCN definition of a PA (Chapter 1) makes it clear that nature conservation is a primary, in most cases the primary, aim of PAs: '*...to achieve the long-term conservation of nature...*' (Dudley, 2008). The message is reinforced by an accompanying principle, that: for IUCN, only those areas where the main objective is conserving nature can be considered protected areas; this can include many areas with other goals as well, at the same level, but in the case of conflict, *nature conservation will be the priority* (our emphasis). In this context 'nature' always refers to biodiversity, at genetic, species and ecosystem level, and often *also* refers to geological diversity, landform and broader natural values. 'Conservation' refers to the *in-situ* maintenance of ecosystems and natural and semi-natural habitats and of viable populations of species in their natural surroundings and, in the case of domesticated or cultivated species, in the surroundings where they have developed their distinctive properties (Dudley, 2008).

However, this isn't the whole story. The current emphasis on biological diversity was not always the primary motivation for establishing PAs and, for instance, many early national parks were promoted to conserve iconic landscapes for tourists or to preserve a nation's cultural heritage. Neither is the situation today as simple as the definition suggests. Many reserves established by local communities or indigenous peoples may include the protection of nature within a wider world view, but the traditional owners would certainly not use terms like 'biodiversity' to describe the values they seek to maintain (Borrini-Feyerabend *et al.*, 2013). More fundamentally, as described in Section 2.6, PAs often provide a range of other benefits to current and future human well-being and these may be instrumental in the decision to establish and/or manage a particular PA. These can include both use and non-use values (Chapter 3), and cover a wide range of provisioning, regulating and cultural benefits (Section 2.6). In many countries, PAs are also expected to demonstrate more immediate returns against investment including contributions to poverty alleviation programmes, tourism and other aspects of economic development. Such returns need to be compared with the costs of establishing a PA in terms of other benefits forgone. Against this context, being able to identify and understand the benefits provided by PAs, including estimating their socio-economic value, becomes increasingly important.

2.2 Basis of benefits: structure, functioning and processes in protected areas

The main benefits provided by PAs arise from the fact that PAs are predominantly set up in functioning and healthy natural ecosystems: places that are increasingly rare in some parts of the world. In heavily modified landscapes or seascapes, PAs may be the *only* places where more-or-less natural ecosystems remain. As well as protecting biodiversity, many PAs therefore also support and protect the wider benefits associated with natural ecosystems.

Ecosystem functioning involves a series of complex bio-physical processes relating to the cycling of energy, water and nutrients: energy capture from the sun; regulation of water flow (and in some ecosystems direct trapping of water from the atmosphere); soil building; and the stabilisation of slopes, river banks and seashores through the binding effects of plant roots and the buffering role of habitats like coral reefs and sand dunes. Increasingly recognised today, most healthy natural ecosystems have critically important functions within the global carbon cycle by accumulating carbon from the atmosphere, building this up in living plants, soils and the humus layer. Maintaining this function is now viewed as a critical component of efforts to slow climate change. Solar energy is channelled into plant growth, and a proportion of this energy is passed on to animals, providing a rich and renewable source of biomass and genetic material that humans use as food, materials and to support products such as medicines. The relationships between different species also provide further, unique values that spill over into human activities, such as pollination. These various interactions are complex, often poorly understood, non-linear and changeable. All can be influenced by human activities.

It should be noted that although the term 'natural ecosystem' will be used many times in this book, this does not imply that these ecosystems have been unaffected by human activity. Few, if any, parts of the world now qualify for this distinction. Even apparently natural ecosystems, like the tropical rainforests of the Amazon and Congo Basin, the great plains of southern and eastern Africa and the coral reefs of the tropical Pacific have been modified over time, deliberately or accidentally. Furthermore, many PAs are established in what are demonstrably cultural ecosystems, places where the whole fabric of ecology has been altered through millennia of interaction with a range of human societies: the Mediterranean region of Europe is a classic case in point. In these semi-natural and cultural ecosystems, some of the values (particularly some non-use values) are associated with human philosophy or artefacts, such as values relating to faith, culture and history.

Over the last decade, a series of meta-analyses have greatly increased our understanding of the links between biodiversity, ecosystem functions and processes and, consequently, the effectiveness with which ecosystem services are delivered and maintained. In some cases, like the use of wild genetic material for crop breeding and medical research, the link is obvious, but more subtle relationships also occur. An overview of such analyses by Cardinale *et al.*, (2012) concluded that biodiversity loss reduces the overall efficiency with which ecosystems collect, convert and cycle energy and other resources, with the relationship between the two being non-linear so that change accelerates as biodiversity loss increases. Research in grasslands, for instance, finds that high biodiversity natural grassland ecosystems are both more sustainable and more productive than lower diversity grasslands (Tilman *et al.*, 1996).

Of critical importance under conditions of climate change, there is also growing evidence that biodiversity confers greater resilience within ecosystems (Thompson *et al.*, 2009), large intact ecosystems are more likely to withstand climate change than disturbed and fragmented ecosystems (Noss, 2001) and ecosystems with high carbon frequently also have high biodiversity (Kapos *et al.*, 2008). PAs therefore not only supply ecosystem services, but they also provide ideal conditions for such services to be enhanced and sustained over time in the face of environmental change.

2.3 Ecosystem services and related goods provided and supported by protected areas

Building on an understanding of the processes and structure of ecosystems, the term 'ecosystem services' refers to those particular aspects of an ecosystem that are of direct or indirect benefit to humans. These services result in a range of outputs (use and non-use, material and non-material) that are of socio-economic value to people (see Figure 2.1 and elaboration in Chapter 3). Benefits from PAs then refer to ecosystem services (or the proportion of ecosystem services) and related goods contained within and/or available from PAs.

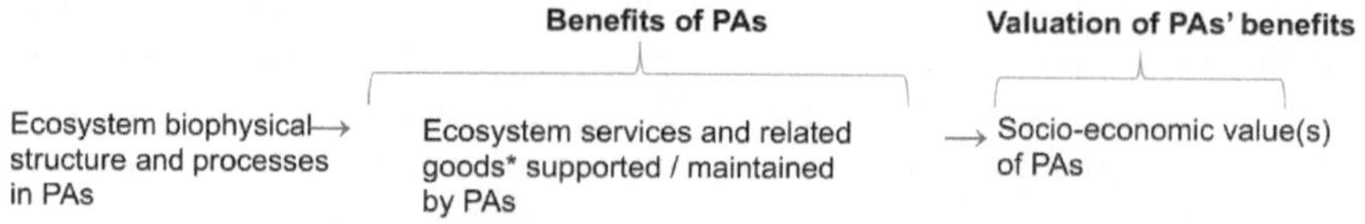

FIGURE 2.1 The relationship between ecosystem process, benefits and value

Source: building on Kettunen *et al.* (2011) and UK NEA (2011)

Ecosystem services are commonly classified as the underlying ecosystem processes (the *supporting services*), along with a range of *provisioning*, *regulating* and *cultural* services. Figure 2.2 provides an overview of these services, forming a basis for the further, more detailed classification of services and related goods in Chapters 6-9. Section 2.2 discusses the first of these (supporting services); the contribution of PAs to the other services will vary between individual PAs and will be affected by their management objectives. Provisioning services related to food, for example, may relate directly to agricultural production (as is typical in many PAs protected for their cultural landscape values) or to sustainable harvesting of wild resources within PAs (game or fisheries) for subsistence. The section below provides some illustrative examples of the ecosystem services and related goods provided and supported by PAs. Further and more detailed examples are given in Part II of the guidance.

Food provisioning e.g. hunting of wild game; use of wild food plants; fishing and/or contribution to fish stocks by protecting spawning sites and nurseries; traditional agriculture i.e. use of locally adapted crops (landraces) and/or practices and livestock grazing and fodder collection. In Lao PDR, Fish Conservation Zones, which are 'no-take' zones, are the most important co-management tool for fisheries. These zones have been established in areas selected using indigenous knowledge. Since their establishment, villagers reported significant increases in stocks of over 50 fish species (Baird, 2000). Similarly, in Vietnam approximately 5,300 people around Hon Mun Marine Protected Area (MPA) depend on the site for aquaculture and near-shore fishing with the gross fisheries value estimated at around US$15 thousand per km^2 (Pham *et al.*, 2005).

Water provisioning e.g. water quantity through groundwater renewal, maintenance of natural flows and in the case of forests such as tropical cloud forests, increase in net water flow. In Melbourne, Australia, around 80 per cent of the city's drinking water comes from uninhabited forested water catchments (around 157,000 hectares) that have been closed to the public for over 100 years with the purpose of filtering rainwater for the purposes of water supply (Melbourne Water, 2013). About 49 per cent of these old forest catchments fall within national parks. Studies of rainfall and runoff data concluded that water yield from forested catchments is related to forest age (Peel *et al.*, 2000). It was found that forest disturbance can reduce the mean annual water yield by up to 50 per cent compared to that of a mature forest, and can take as long as 150 years to recover fully. This is because evapotranspiration from older forests is lower per unit area than from younger forests.

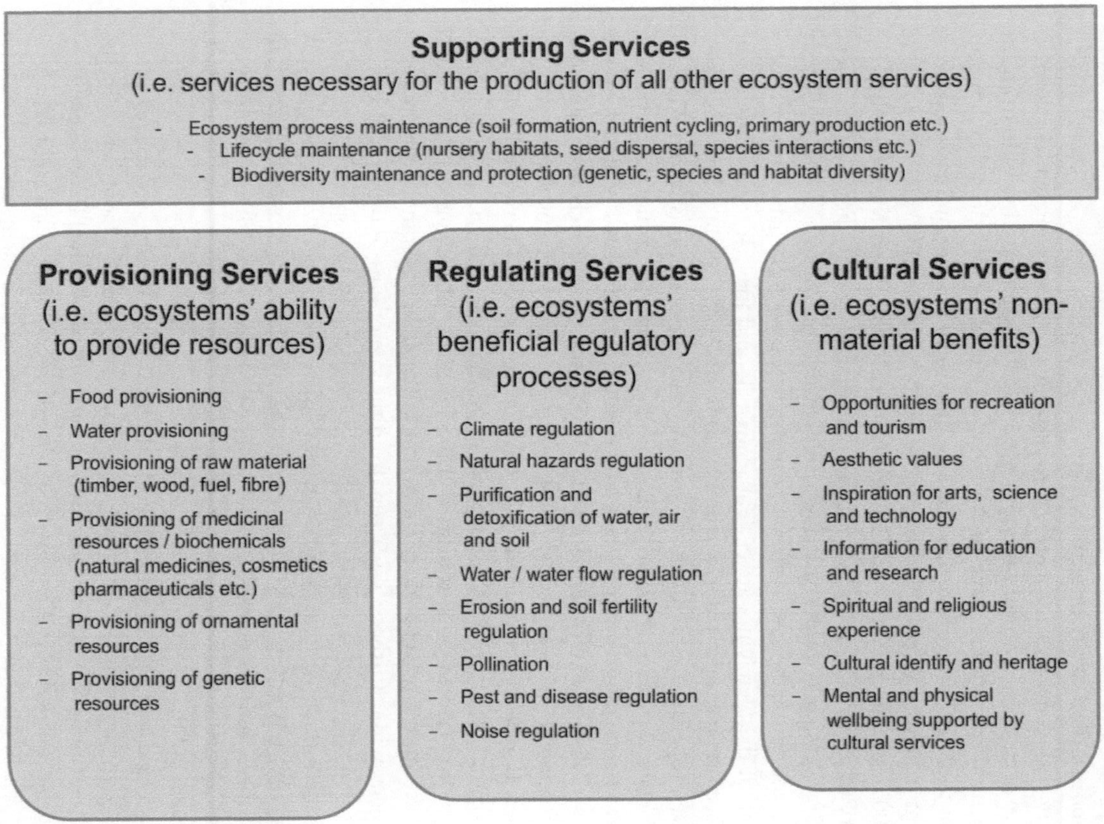

FIGURE 2.2 Commonly recognised ecosystem services and related goods from protected areas

Source: own presentation, based on MA (2005), de Groot *et al.* (2010) and UN NEA (2011)

Provisioning of raw material e.g. timber, wood, fuel and fibre. The Kayan Mentarang National Park in Indonesian Borneo is home to 16,000 Dayak people who retain rights to collect rattan (*Calamus* spp.), sang (*Licuala* spp.) and hardwoods for construction, under guidelines that are controlled by customary law (Borrini-Feyerabend *et al.*, 2004). While no quantified or monetary estimates exist on the value of these raw materials it is understood that they are of high socio-economic importance to the Dayak community.

Provisioning of medicinal resources/biochemicals e.g. natural medicines, cosmetics, pharmaceuticals. An immunosuppressant identified in a soil sample from the Hardangervidda National Park in Norway was used in the drug Sandimmun, which by 2000 was one of world's most popular medical drugs in helping prevent transplant patients rejecting new organs, with total sales of US$1.2 billion (Laird *et al.*, 2003). Studies in Carlsbad Caverns National Park, New Mexico led to the discovery of previously unknown bacteria on the walls and in the pools of several caves. By releasing enzymes that kill competitors, the bacteria compete fiercely with each other for the few available nutrients. Testing in the laboratory revealed that some of these enzymes attack leukaemia cells and may, at some point in the future, become instrumental in medical treatments (NPS, 2000), while other bacteria have properties useful for the bioremediation of pollutants (Barton, 2006).

Provisioning of ornamental resources e.g. traditional handicraft, fashion and jewellery, natural dyes and colourants/dye plants and decorative wild plants. Vallée de Mai is one of two UNESCO Natural World Heritage Sites in the Seychelles. The PA is famous for the coco de mer (*Lodoicea maldivica*), one of six palms endemic to the Seychelles. *L. maldivica* nuts are sold as souvenirs for prices of between D 190 and D 450 (around US$15 to US$35) depending on their size and symmetry. In 2009, more than 40 per cent of all tourists coming to the Seychelles visited the site and entrance fees to the site provided 76 per cent of the annual income to the management agency with a further 3.4 per cent from the sale of nuts (Rist *et al.*, 2010).

Provisioning of genetic resources e.g. crop wild relatives and tree species. The 89 ha Erebuni State Reserve in Armenia was established to protect wild wheat CWR (*Triticum* spp.) (Avagyan, 2008) and the Garrotxa Volcanic Zone Natural Park in Spain has been actively managing crop genetic diversity since 1990, when surveys identified 53 traditional varieties of eight species of fruit trees in the park (Bassols Isamat *et al.*, 2008).

Climate regulation e.g. by providing significant carbon storage and sequestration and/or by ameliorating local climate impacts. Parks Canada has researched the amount and value of the carbon stored in its network of National Parks. Total storage is estimated at 4,432 million t with a value of over US$70 billion (Kulshreshtha *et al.*, 2000). In Europe, the value of current carbon stock maintained within the EU's Natura 2000 network of PAs has been estimated to be around €600 billion to €1,130 billion (Ding, Nunes and Markandya in ten Brink *et al.*, 2011) (see Box 13.2).

Natural hazards regulation e.g. coastal mangroves, coral reefs, floodplains and forest help to buffer against natural hazards. Mantadia National Park in Madagascar protects the watershed of the Vohitra River. An analysis of the economic benefits of the park to farmers in the region found that conversion from primary forest to slash and burn agriculture can increase storm flow by as much as 4.5 times, and thus benefits from forest protection within upper watersheds reduced crop damage from floods in agricultural plots in lower basins (Kramer *et al.*, 1997).

Purification and detoxification of water, air and soil e.g. water quality through filtration. A study of the environmental services of sites managed by the United States Forest Service (many of which are classified as IUCN category V PAs) estimated that the value of their regulation of clean air provided a service worth a total US$12.5 billion (Campbell and Brown, 2012). Forested watersheds offer higher quality water than watersheds under alternative land uses, because the latter are often associated with more intensive management, less complete vegetation cover (hence more soil erosion and sediment) and more application of pollutants such as pesticides, fertiliser or toxic waste. New York, for example, has a water treatment process that is simpler than in many other American cities thanks to a package of incentives for protection and good management in the catchments upstream (Stolton and Dudley, 2010).

Water flow regulation e.g. water quantity through groundwater renewal, maintenance of natural flows and, in the case of ecosystems such as tropical montane cloud forests and *páramos*, additional water flow. For instance in Honduras, the cloud forests of La Tigra National Park (23,871 ha) provide more than 40 per cent of Tegucigalpa's annual water supply for its 850,000 inhabitants (Hamilton, 2008).

Erosion and soil fertility regulation: Assessment of the economic values of the Pantanal Conservation Complex in Brazil suggests that erosion control functions of the four PAs which make up the complex have an annual value of US$170 million (Schuyt and Brander, 2004).

Pollination e.g. by bees and other nectar-feeding insects. Bees are one of the principal pollinators of fruit trees and major staple crops, yet have declined catastrophically so that in some places farmers have to pollinate their crops manually. A study of bee populations in 40 Special Areas of Conservation in Ireland confirmed that management which prioritises the maintenance of a diversity of high-quality habitats within a broad network of PAs best facilitates bee conservation (Murray *et al.*, 2012).

Pest and disease regulation e.g. by protecting natural processes that reduce disease spread. Avoiding deforestation or restoring natural vegetation can, for example, reduce the risk of malaria and certain other diseases by slowing the spread of disease vectors (Oglethorpe *et al.*, 2008). In Indonesia, the 32,000 ha Ruteng Park on the island of Flores protects the most intact submontane and montane forests on the island. Researchers found that communities living near the PA had fewer cases of malaria and dysentery, children missed school less because of ill health, and there was less hunger associated with crop failure, than in nearby communities without intact forests (Pattanayak *et al.*, 2003). A study in the Peruvian Amazon found that the primary malaria vector, *Anopheles darlingi*, had a biting rate that was more than 278 times higher in deforested areas than in areas that were heavily forested (Vittor *et al.*, 2009).

Opportunities for recreation and tourism: Tourism is by some estimates the world's largest industry and PAs play a critical role in it. An analysis of the 26,000 Natura 2000 sites in the European Union found that they attracted an estimated 1.2–2.2 billion visitor days per annum, generating 4.5-8 million full-time equivalent jobs, with visitor expenditure of some €50-€90 billion (BIO Intelligence Service, 2011). Such benefits are distributed throughout the world but with different sites having different mixes of local, national and international visitors. For example, the tourism values of the PA system in South Korea are high; within the national park system almost 30 per cent of the budget is allocated to visitor management and associated infrastructure and there are some 38 million visitors to national parks a year. Visitation is almost entirely domestic, with only 1 per cent coming from abroad (KNPS and IUCN, 2009).

Aesthetic values e.g. beauty and scenery. Many of the PAs designated in the early twentieth century were chosen as much for their beauty and iconic scenery as for what we now call their biodiversity values. One of the most well known is the Grand Canyon in the USA, which is managed by the US Park Service to 'preserve and protect its natural and cultural resources and ecological processes, as well as its *scenic, aesthetic*, and scientific values' (emphasis added) (National Park Service, 1995).

Inspiration for the arts: The mountains of north Wales have been the inspiration for countless writers and artists long before the area was designated as the Snowdonia National Park in the 1940s. A recently constructed visitor and information centre continues the links between the landscape and the PA by including lines of poetry in its design. By Gwyn Thomas, Wales's National Poet during 2006-07, the poetry celebrates the history of the area and that of its people, and proclaims the glory of the landscape surrounding Snowdon, the highest mountain in the UK (Hourahane *et al.*, 2008).

Information for education and research i.e. formal and informal dissemination of information. The Jurassic Coast World Heritage site in the UK is creating outdoor education facilities and learning resources for local schools through harmonising the values of the World Heritage Site with the formal school curriculum for 3–19-year-olds. Nationally they coordinated the first UK UNESCO World Heritage Site Youth Summit in 2009 where over 80 young people from across the UK came together to learn and discuss the importance of World Heritage Sites (Ford, 2012). Other PAs are set up specifically for research purposes, such as the Andrews Experimental Forest in Oregon (Luoma, 1999), with over 100 research projects underway every year.

Spiritual and religious experience e.g., sacred groves, waterfalls and/or mountains. The Boabeng-Fiema Monkey sanctuary in Ghana provides an example of a sacred grove that has not only been protected by customary law, but also by modern legislature under District Council by-laws and is managed as a wildlife sanctuary. The grove is considered sacred because it supports populations of black and white colobus monkey (*Colobus vellerosus*) and mona monkey (*Cercopithecus mona*), all of which are revered and strictly protected as sons of the gods of the people of Boabeng and Fiema villages (Stolton and Dudley, 2010).

Cultural identify and heritage: Special features of land or seascapes have been identified and protected by people for millennia. They can range in size from massive landscapes like the Grand Canyon to small but special features such as waterfalls, fish pools or individual trees. The importance of Uluru-Kata Tjuta National Park, formerly called Uluru (Ayers Rock–Mount Olga) National Park, in Australia to the aboriginal people is hard to overestimate. The iconic sandstone landscape is described by them 'as a place that honours the culture of our people, preserves the fragile ecology of the land of our ancestors and upholds Tjukurpa – the Pitjantjatjara word for our history, knowledge, religion, morality and law' (UNESCO, 2012).

Finally, PA's cultural service – such as opportunities for recreation and tourism – are important contributors to *mental well-being and health*. For example, Keoladeo National Park, India, is located amidst the urban landscape of Bharatpur city. The park has designated a 2 km stretch which 'morning walkers' enjoy every day between 5 and 7 am. During summer, when daytime temperatures in Bharatpur town can reach over 45 degrees C, the number of 'morning walkers' reaches almost a thousand, providing both physical exercise and a peaceful place to relax. The park's management does not charge the usual entry fee for local people, who enjoy the vital environmental services of fresh, unpolluted air inside the park (Stolton and Dudley, 2009).

2.4 Broader benefits

The section above considered some of the individual ecosystems services that can be supplied by PAs. In addition there are more general benefits that *arise from* the ecological significance of the site but are not associated directly with a certain ecosystem service(s).

Employment by the PA, e.g. as managers, rangers or other support staff, can provide major local or regional benefits. A recent study, for example, found the 206 PAs managed by the Royal Society for the Protection of Birds (RSPB) across the UK (protecting 142,044 ha) support over 1,800 local jobs (full-time equivalent). Importantly, these jobs are often located in the more remote, rural or coastal areas of the UK, where economic opportunities tend to be fewer and less diverse (RSPB, 2011).

As well as direct benefits such as employment, a range of additional benefits can extend out to the region around the PA. For example, the estimated benefits to Finnish local economies linked to visits to national parks ranges between €0.1 and €30.6 million per year for different PAs, generating between 1 and 400 person-years of employment (Kajala, 2013).

Other tangible benefits include protecting homelands and even political stability. For example, in 2006 the Government of Bolivia declared an 'exclusively reserved, inviolable and fully protected area' inside the Madidi National Park in the Amazonian highlands (Administrative Decision No. 48/2006). The area aims to protect the home of the last peoples living in isolation in Bolivia, the presumed descendants of the Toromona people. The decision prohibits any settlement other than that of the peoples living inside the inviolable area, unwanted contact with the Toromona people and activity related to prospecting for or exploiting the area's natural resources (WRM, 2009). In neighbouring Peru, the Cordillera del Cóndor Transboundary PA was declared as part of the resolution of a boundary dispute between Ecuador and Peru. The boundary conflicts, which went back to Spanish colonial times, flared up in the 1990s and some 80 people lost their lives. The consolidation of the peace process has been cemented by both the establishment and management of PAs and the promotion of sustainable development projects for local communities (Mittermeier *et al.*, 2005).

2.5 Putting the benefits into perspective

Balancing the benefits across human populations

One key challenge related to conserving and sustainably managing PA benefits is that the socio-economic values supplied by PAs are unlikely to be evenly spread among the human population with those benefiting from PAs often not being the same as those helping to maintain the benefits. Some are very diffuse values that benefit many people, who are often unaware of the connection (climate change mitigation can benefit the whole global population, etc.), while others are immediately realisable values that benefit a small group or an individual. A tropical cloud forest can, for example, provide carbon storage, abundant clean water, stabilisation of steep slopes and other quantifiable values for the global or regional population while the income from felling timber is available only to one individual, company or a small community – who are not generally paid for the diffuse values provided by ecosystems on their land.

Consequently, identifying the beneficiaries of PAs is important (Chapters 4 and 12). It is also important to understand how the costs related to the establishment and management of a PA are distributed (Chapter 4 and 11). The resulting information can be used in developing policy responses which adequately reflect the findings of the valuation and the continued role of the PA (Kettunen *et al.*, 2011). This can be far from easy. WWF, for example, commissioned an economist to calculate the values of a tropical forest in Cameroon over twenty years ago (Ruitenbeek, 1990). The finding was that the value of the standing forest was far higher than the timber. But although these figures have never been seriously disputed, they have also failed to prevent further deforestation throughout the region, because there has been no mechanism for compensating people living in or near these forests for forgoing the immediate benefits of forest clearance in favour of the wider community and global benefits from their conservation. Unless agreed ways can be found of realising diffuse values – through Payment for Ecosystem Services (PES) schemes, for instance, (Pagiola *et al.*, 2002), valuation can become of little more than academic interest. While detailed advice on policy responses falls outside the scope of this guidance, Chapter 15 provides some key considerations on how to use the results of valuations to support more effective PA management.

Protected areas as vehicles for well-being: poverty alleviation, food security and climate stability

Although there are undoubtedly situations in which the (short-term) costs of PAs exceed their benefits to poor people, for example, by reducing their access to some natural resources (Chapter 3), they also provide a range of solutions for these disadvantaged groups. Many of the food-related benefits referred to above – such as helping to rebuild fish stocks or providing emergency supplies of wild food during times of shortage and famine – are likely to be most critically important to those who cannot afford to 'buy' their way out of hunger. More generally, by safeguarding genetic material for crop breeding or animal husbandry, PAs provide global safeguards, particularly during times of rapid environmental change.

Similarly, it is becoming increasingly recognised that PAs can play a variety of roles in addressing climate change, both through mitigation and adaptation. They help mitigate climate change by maintaining natural ecosystems and thus securing carbon already stored in vegetation, soils and peat layers and by facilitating additional carbon sequestration in such places. However, it is now recognised that climate change is already occurring and will continue for the foreseeable future, highlighting the need for adaptive strategies. The maintenance of ecosystem services supported by the global PA network will also help regions and different stakeholders (local communities, businesses, etc.) to adapt to rapidly changing environmental conditions, including more extreme weather events, changing agricultural productivity and large-scale human migration (Dudley *et al.*, 2009).

Many critics of PAs regard protection as exacerbating poverty in vulnerable communities, particularly if poverty is directly linked to income. In such cases it is easy to see that in the example used above from Cameroon, restricting the cutting and thus income from timber could easily result in reduced incomes, particularly in areas where there are few other economic opportunities. However, when the whole range of possible benefits (see Sections 2.3 and 2.4 above) are reviewed for an area, the role of the PA in contributing to poverty alleviation may become clearer.

To be able to discuss fully the role of PAs in poverty exacerbation and alleviation it is worth reviewing the definition of poverty, or rather its opposite – well-being. Five fundamental dimensions of well-being can be linked to PAs – any improvement in which should contribute to reducing poverty (Dudley *et al.*, 2008):

- Subsistence: non-economic benefits that contribute to well-being, i.e. health, nutrition, clean water and shelter
- Economic: benefits which provide the ability to earn an income, to consume and to have assets
- Cultural and spiritual: pride in community, confidence, living culture, spiritual freedom, education
- Environmental services: role in environmental stability and provision of natural resources

- Political: relating to issues of governance, participation, engagement and empowerment and thus influence in decision-making processes.

Research suggests that natural ecosystems, including those contained within PAs, are generally important in providing a safety net for the poorest members of society, such as sources of food and fuel, but are less closely or automatically linked to generating substantial economic returns. In other words, PAs are better at alleviating the impacts of poverty than at reducing poverty overall. Furthermore, attempts to use PAs as tools for poverty reduction are strongly influenced by the overall quality of governance in a country. The vicious cycle of poverty and corruption means that attempts at poverty reduction in the poorest countries often fails (Dudley *et al.*, 2010).

Land use and management choices – trade-offs and synergies between benefits

As a land use choice, PAs are a significant investment by governments, communities and private landowners, representing a long-term commitment to the protection of biodiversity and sound management of ecosystems and ecological processes. In return, PAs help to prevent the degradation of ecosystems and the designation can even increase the already valuable services and related goods provided by areas. On the other hand, PAs also result in some lost socio-economic potential in terms of natural resources 'locked up' and restricted from use. While existing evidence indicates that at the global and national level the benefits of investing in protection outweigh the associated costs (Kettunen *et al.*, 2011) dedicated measures are often needed to balance out these costs at local level (see Chapter 3 for further discussion).

Land use and management practices within PAs influence the range and availability of different benefits in practice. Ecosystem services are often interlinked and therefore measures aimed at enhancing the level of one service can have both positive and/or negative effects on others. Consequently, the consideration of potential synergies and trade-offs between different ecosystem services and related goods is important in order to decide which socio-economic benefits provided by PAs should be actively promoted through management decisions. For example, tourism and recreational use rely on the maintenance of water supply and the absence of natural hazards. Consequently, the protection of PAs' water cycle and their ability to mitigate flooding, wild fires and avalanches helps also to maintain tourism and recreation. On the other hand, focusing on maximising the production of resources from PAs, such as converting forested areas into agricultural land, would reduce an ecosystems' ability to maintain several other services such as to purify water, control erosion and regulate climate.

Synergies and trade-offs can also occur between ecosystem services and conservation of biodiversity. The restoration of degraded wetlands both supports habitat conservation and enhances PAs' capacity to purify water. On the other hand, improving carbon sequestration by reforestation with monocultures is often in conflict with conservation goals. Consequently, it is important to ensure that the promotion of the different socio-economic benefits provided by the site is in line with a site's conservation goals and it is essential to identify possible conflicts between these two aspects. In addition, there is also a need to establish the sustainable level of 'using' the identified benefits. Such thresholds include, for example, the maximum number of visitors to prevent soil compression and/or erosion and the sustainable levels of extracting biodiversity resources and water.

Because PAs are specifically designated for the protection and sustainable management of natural ecosystems they are also ideal places to understand, value, manage effectively and sustain ecosystems services and related goods. Showing that PAs provide measurable and distinct benefits beyond the hard-to-quantify romantic or ethical values of preserving wildlife can help to persuade the opponents of their worth, this way also encouraging a shift towards more sustainable and biodiversity-friendly land use and management at regional and/or national level. There are plenty of examples. The Finnish PA managing agency Metsähallitus Natural Heritage Services showed that every Euro invested in protected areas generated 10 Euros for surrounding rural regions through the side-benefits of extra visitation, thereby maintaining its funding levels during a period when the government was generally unsympathetic to conservation (Kajala, 2013). The Whangamarino Ramsar site in North Island New Zealand has an estimated average value of US$600,000 per annum at 2003 values in terms of maintaining natural flood control (Department of Conservation, 2007). National parks like Serengeti in Tanzania generate many millions of dollars in tourist revenue every year; almost certainly one of the largest sources of foreign earnings in the country.

Protected area effectiveness

It should be noted that the ability of PAs to maintain ecosystem functions and supply ecosystem services depends on a number of factors including: the effectiveness of management in achieving the stated goals of the PA; the integrity of lands surrounding the boundaries of the PA and thus the value added provided by the PA; the effectiveness of the PA in buffering land from human-induced threats; and any displacement effect the creation of the PA may have on land uses that might undermine ecosystem functions elsewhere.

Research on PA effectiveness has focused on potential benefits in terms of reductions in outright habitat loss, as well as habitat degradation. One large-scale study looked at anthropogenic threats facing 92 PAs in 22 tropical countries, and concluded that the majority of PAs are successful in protecting ecosystems. Specifically they do so by stopping land clearance, as well as preventing illegal logging, hunting, fire and domestic animal-grazing compared to unprotected areas (Bruner *et al.*, 2001). A major study from 2008 assessed the effectiveness of PAs in terms of avoided land cover clearance, using a metadata analysis that incorporated 22 countries and 49 different locations. The research concluded that PAs had lower rates of land clearing compared to their surroundings, and reduced rates within their boundaries following the initiation of protection (Nagendra. 2008). Another recent report compared multiple PA management types across four tropical areas: the Amazon, the Atlantic Coast, West Africa, and the Congo. Although the research emphasised that the degree to which PAs protect natural vegetation depends on the specific geographic context, which varied greatly among these regions, overall it found that forest cover was high inside PAs compared to surrounding areas (Joppa *et al.*, 2008). A further study in 2011, based on remotely sensed information and a matching approach that compares protected and unprotected areas of similar characteristics, found that protection reduced land conversion in 75 per cent of 147 countries analysed (Joppa and Pfaff, 2011).

This growing body of scientific evidence that PAs are effective in protecting natural vegetation and preventing land use change is important in the context of what is happening to natural ecosystems globally; for example, an analysis of 140 ecoregions found natural ecosystems were being converted at least ten times more quickly than they were being protected (Hoekstra *et al.*, 2005).

2.6 Opportunities and risks of assessing benefits

The text above notes that socio-economic valuation can provide some concrete benefits; these are listed below and elaborated in more detail in Chapter 15. But valuation also brings some risks; poorly conceived or badly executed studies can backfire by creating unrealistic expectations among stakeholders or by undervaluing benefits and thus undermining the case for PAs. Consequently, economic valuations – and monetary valuations in particular – should be one part of a broader, overall assessment of the pros and cons relating to PAs. Later in this section some of the reasons why valuation can be difficult or inappropriate in some situations are discussed.

Socio-economic valuation can help make the case for PAs by showing their wider values; specific motivations can include:

Advocacy and awareness

- Showing the relationship existing between the ecosystem services and related goods provided by PAs and human well-being, including cultural, spiritual and other social values
- Showcasing the contribution of PAs to the local and national economy and to a country's development process
- Raising awareness about the actual or potential benefits to build public support for conservation and for PAs and contribute to conflict resolution
- Mainstreaming conservation into the development or the climate change mitigation agenda at national, sub-national and local level, in development policies, plans, programmes and projects

Support to decision-making and management

- Providing information to government departments, private trusts, communities, etc. to support (political) discussions and to help in integrating PAs into national, regional and local development plans as well as other strategies and plans, such as National Biodiversity Strategies and Action Plans (NBSAPs) under the UN Convention of Biological Diversity (CBD)
- Explaining the payback from PAs to Treasury and other departments providing financial support
- Providing information that can be of concrete support to sustainable land use and management within the region and used, for example, to complement existing indicators for environmental sustainability and broaden the scope of environmental impact assessments and cost-benefits analysis (see ten Brink, 2011 and Wittmer and Gundimeda, 2012 for further information)
- Supporting the development and updating of PA management plans to ensure that, in addition to the biodiversity values, they also address relevant socio-economic benefits associated with sites. This facilitates the engagement of relevant stakeholders in the process and helps to achieve the conservation objectives.

Identifying and addressing social impacts

- Ensuring a better understanding of the values important to local communities, other stakeholders (i.e. resource users such as water companies) and PA managers
- Assessing compensation payments to local communities for values forgone
- Encouraging sustainable, evidence-based resource use

Mobilising funds

- Influencing decision-makers, the private sector and international donors to secure political commitment and increase funding for the conservation of PAs and/or to improve governance, laws and regulations
- Helping develop support mechanisms for the ecosystem services and related goods the PA provides, such as payment for environmental services schemes
- Providing evidence demonstrating that PAs can be a cost-effective means to address a number of environmental challenges (climate change, flood risks, water quality, food security, etc.) and therefore provide attractive targets for 'green' investment

Different types of valuation are suitable for the different uses above. Aggregated assessments across multiple PAs are generally useful for advocacy, mobilising funds at a regional or national level and influencing policy, but do not provide the detail needed, for example, to assess the impacts of concrete land use changes, set compensation levels or carry out detailed planning in an individual PA. In these cases, more specialised, PA-specific studies are required. It is crucial to define clearly the reasons and/or specific needs for the socio-economic valuation of PA benefits and use that information as a basis to select the most appropriate type(s) of valuation. Part II of the guidance provides more information on what methods are fit for which purposes.

But valuation also carries some risks; if these are not understood and planned for, attempts at valuation can backfire. The following section outlines some of the challenges to watch out for.

It is often difficult to identify a precise socio-economic value for particular elements: many PA values are hard to quantify – let alone monetise – in economic terms. The value of a particular species is difficult to estimate, for example, unless it is known to provide some direct economically important benefits (such as medicinal properties or measurable values for ecotourism) (Chapters 6 and 8). Existence values (see Figure 3.2 in Chapter 3) are even more elusive because different people will assign very different figures. The same is true for many other non-tangible benefits that PAs might provide: the sacred natural sites that exist in many PAs are of immense value to members of a particular faith group (Wild and McLeod, 2008) but are often unnoticed by anyone else. For example, local fishing communities supported the Madagascar government's plans to convert a small offshore island into a PA, important for nesting tropicbirds, because of the presence of sacred sites; but if these had not been specifically pointed out in the planning process, they would have been indistinguishable from the surrounding habitat. There have been numerous attempts to assign economic values to particular views or landscapes, but these should be generally treated with caution and results can differ wildly. Assigned values can be distorted by political considerations, perverse incentives or a misunderstanding of full costs. As Part II of the guidance will show, methodologies exist to put a monetary value on subsistence activities in natural or semi-natural ecosystems, such as collection of wild food. However, while useful in highlighting the importance of subsistence use, these figures do not fully capture the critical nature of wild foods in the event of a harvest failure, drought or during times of civil unrest. Suboptimal values can lead to suboptimal – or even inaccurate – valuations, which can do more harm than good.

Most valuation studies in PAs are still very narrow in scope: studies tend to focus on one or two particular issues, like growth in fish population within a marine protected area (MPA) or the value of wildlife watching in a tropical savannah. In part this is because of the complexity of socio-economic valuation, which makes total valuation studies expensive and time-consuming. While some pioneering attempts have been made (e.g. in Leuser National Park in Sumatra, Indonesia by van Beukering *et al.*, 2003; see Part II of the guidance) more comprehensive assessments, looking at the full range of ecosystem services on a site, are still required. While assessment of discrete ecosystem services or tourism earnings are often useful for managers and can help build political support for a PA, they may significantly undervalue the total socio-economic importance of services and related goods.

Values can change over time: the perceived benefits of PAs can change as a result of external developments or decisions. Consequently, assessing the socio-economic importance of PAs is not a one-off exercise, but it rather needs to form an integral part of the ongoing planning and management of sites. For example, a PA valued because it preserves an important medicinal plant may suddenly find itself politically weakened if pharmaceutical companies find something more effective or succeed in synthesizing the active ingredient in a laboratory. A forest set aside to provide clean water to a municipality may lose its economic value in the eyes of local politicians if a new purification plant is built: this debate is currently taking place in New York State. Conversely, particular products obtainable from PAs, such as wild food, fodder and some kinds of ecosystem services, may also unexpectedly increase in importance and value, for instance because harvest failure causes food shortages. Similarly, if changes upstream increase flooding, there may be a need for more native vegetation to act as a buffer and the perceptions of a riparian PA may change. The perceived value of natural vegetation for buffering against storms increased rapidly in Louisiana after Hurricane Katrina devastated large parts of New Orleans and its surroundings, in part because authorities recognised the real costs of such a storm.

New values are also emerging: socio-economic assessments undertaken a few years previously may well have missed benefits that are considered critical today. The role of natural vegetation and soils in storing and sequestering carbon is a key case in point. This hardly appeared in assessments made ten years ago but carbon valuation is now a large global business (see Chapter 7 and Annex 4). The same is broadly true for benefits related to water, both quality and in some cases quantity of water, which emerged to prominence slightly earlier. People assessing PAs need to try to look ahead at what may be important in the future although such projections can only ever be partial and incomplete.

Values differ between stakeholders: not everyone has the same perceptions of what is important; an assessment that only focuses on a limited range of stakeholders is likely only to get half the story. For example, the assessment of values and benefits associated with Küre Dağları Milli Parkı in Turkey revealed that responses from the national park managers, research scientists and local communities were markedly different; all were valid (and in fact all could be accommodated within the PA's management plan) but they represented very different perspectives about what was most significant about the area (see Chapter 15 and Box 15.3 for more information).

A risky numbers game: putting values on PAs can encourage others to do the same, by working out the comparative values of different development strategies. It may be worth a mining company offering to pay communities for their loss of water-related ecosystem services for instance, or to invest in artificial slope stabilization, if it is seen as an inevitable cost of accessing a rich seam of minerals within a PA. These debates happen. A particular risk from the perspective of PAs is that, since several socio-economic benefits are difficult to capture in economic terms, the estimated monetary value of conservation remains lower than that of land conversion. Consequently, it is important to try to ensure that qualitative and quantitative evidence on impacts receives equal attention in the decision-making process. It is also important to ensure that the valuations are as robust and fit-for-purpose as possible (see Chapter 14). If valuation has been partial in ways described above, presenting a figure that can easily be capped can paradoxically further undermine the political power of PAs.

Diverting attention from the primary role of PAs: this chapter started by identifying nature conservation as the primary function of PAs as recognised by IUCN and the CBD. Many, probably most, PAs have other important values as well. One concern raised by analysts is that a focus on measurable benefits and associated socio-economic values may put PA managers into the position of being under pressure to concentrate management efforts on the most 'valuable' outputs from a PA rather than on the core values it was set up to preserve. A focus on reducing carbon emissions from deforestation and forest degradation (REDD+ mechanism), for instance, could in theory encourage managers to promote a transition from savannah to forest in dry tropical areas, because this would supply larger carbon credits, whilst at the same time it would likely undermine populations of open rangeland species. A trade-off between management for fish of commercial or subsistence value and management of other species is a real tension in some marine and freshwater PAs. More broadly, this could (and to some extent already does) result in the focus of attention, funds and capacity being directed at those sites that can show a clear and immediate return on investment, whether this is in terms of PES schemes, large numbers of tourists or realisable assets for disaster mitigation. Less 'valuable' sites may also be critical for biodiversity but attention is likely to be reduced with a risk of lesser management and funding leading to lower levels of protection. These are real concerns that should not be underestimated, but be taken into consideration when considering the need for a socio-economic assessment of PAs.

None of these challenges outlined above are insurmountable. But all are things that can trip up the unwary and need to be taken into consideration when planning and carrying out an assessment. In general, the focus of PA socio-economic assessments should seldom if ever only be on monetary values but can usefully include a wide range of quantitative and qualitative factors, and always needs to consider the primary role of biodiversity conservation (see Part II of the guidance). Valuation studies are important and if well-planned and carried out with care can help tip the balance towards both the creation and responsible management of PAs; critically such studies can also help persuade other stakeholders of these benefits. The purpose of the following chapters is to help the PA practitioners and other interested parties to understand the opportunities and challenges associated with socio-economic valuation, providing concrete guidance on how to address them.

References

Avagyan, A. (2008) 'Crop wild relatives in Armenia: diversity, legislation and conservation issues', in N. Maxted, B. V. Ford-Lloyd, S. P. Kell, J. M. Iriondo, E. Dulloo and J. Turok (eds) *Crop Wild Relative Conservation and Use*, CABI Publishing, Wallingford

Baird, I. (2000) *Integrating Community-Based Fisheries Co-Management and Protected Areas Management in Lao PDR: Opportunities for Advancement and Obstacles to Implementation*, Evaluating Eden Series, Discussion Paper No.14, IIED, London

Barton, H. A. (2006) 'Introduction to cave microbiology: A review for the non-specialist', *Journal of Cave and Karst Studies*, 68(2): 43–54

Bassols Isamat, E., Falgarona Bosch, J., Mallarach Carrera, J-M. and Perramon Ramos, B. (2008) 'Agrobiodiversity conservation in the Garrotxa Volcanic Zone Natural Park, Spain: Experience and recommendations for future directions', in Amend T., J. Brown, A. Kothari, A. Phillips and S. Stolton (eds), *Protected Landscapes and Agrobiodiversity Values*, IUCN and GTZ, Gland Switzerland and Eschborn, Germany

BIO Intelligence Service (2011) *Estimating the economic value of the benefits provided by tourism/recreation and employment supported by Natura 2000*, final report prepared for the European Commission, http://ec.europa.eu/environment/nature/natura2000/financing/docs/Economic_Benefits_of_Natura_2000_report.pdf, accessed 15 December 2012

Borrini-Feyerabend, G., Kothari, A. and Oviedo, G. (2004) *Indigenous and Local Communities and Protected Areas. Towards equity and enhanced conservation*, IUCN/WCPA Best Practice Series No. 11, IUCN, Cambridge

Borrini-Feyerabend, G., Dudley, N., Lassen, B., Jaeger, T., Pathak, N., Phillips, A. and Sandwith, T. (2013) *Governance of Protected Areas: From Understanding to Action*, Gland, Switzerland: IUCN

Bruner, A. G., Gullison, R. E., Rice, R. E. and da Fonseca, G. A. B. (2001) 'Effectiveness of parks in protecting tropical biodiversity', *Science*, 291: 125-129

Campbell, E. T. and Brown, M. T. (2012) 'Environmental accounting of natural capital and ecosystem services for the US National Forest System', *Environment, Development and Sustainability*, 14: 691–724

Cardinale, B. J., Duffy, J. E., Gonzalez, A., Hooper, D. U., Perrings, C., Venail, P., Narwani, A., Mace, G. M., Tilman, D., Kinzig, A. P., Daily, G. C., Loreau, M., Grace, J. B., Larigauderie, A., Srivastava, D. S. and Naeem, S. (2012) 'Biodiversity loss and its impact on humanity', *Nature*, 486: 59-67

Department of Conservation (2007) *Economic Values of Whangamarino Wetland*, DoC, Auckland, New Zealand

Ding, H., Nunes, P. and Markandya, A. (2011) 'Climate regulation: carbon storage' in P. ten Brink *et al. Estimating the Overall Economic Value of the Benefits provided by the Natura 2000 Network*, final report to the European Commission, Institute for European Environmental Policy/GHK/Ecologic, Brussels, http://www.ieep.eu/publications/2012/06/estimating-the-overall-economic-value-of-the-benefits-provided-by-the-natura-2000-network, accessed 12 December 2012

Dudley, N. (ed.) (2008) *Guidelines for Applying Protected Area Management Categories*, IUCN, Gland, Switzerland

Dudley, N., Mansourian, S. Stolton, S. and Suksuwan, S. (2008) *Safety Net: Protected areas and poverty reduction*, WWF, Gland, Switzerland

Dudley, N., Mansourian, S., Stolton, S. and Suksuwan, S. (2010) 'Do protected areas contribute to poverty alleviation?' *Biodiversity*, 11(4): 5-7

Dudley, N., Stolton, S., Belokurov, A., Krueger, L., Lopoukhine, N., MacKinnon, K., Sandwith, T. and Sekhran, N. (eds) (2009) *Natural Solutions: Protected areas helping people cope with climate change* IUCN-WCPA, TNC, UNDP, WCS, The World Bank and WWF, Gland, Switzerland, Washington, DC and New York

Ford, A. (2012) *Education and Training: The Jurassic Coast, UK, in UNESCO, Managing Natural World Heritage,* UNESCO/ICCROM/ICOMOS/IUCN

Groot, R. S. de, Fisher, B., Christie, M., Aronson, J., Braat, L., Gowdy, J., Haines-Young, R., Maltby, E., Neuville, A., Polasky, S., Portela, R. and Ring, I. (2010) 'Integrating the Ecological and Economic Dimensions in Biodiversity and Ecosystem Services Valuation', in Kumar, P. (ed.) *The Economics of Ecosystems and Biodiversity (TEEB) Ecological and Economic Foundations.* Earthscan, London and Washington, DC

Hamilton, L. (2008) *Forests and water,* FAO Forestry paper 155, UN Food and Agricultural Organization, Rome

Hoekstra, J. M., Boucher, T. M., Ricketts, T. H. and Roberts, C. (2005) 'Confronting a biome crisis: global disparities in habitat loss and protection', *Ecology Letters,* 8: 23-29

Hourahane, S., Stolton, S. Falzon, C. and Dudley, N. (2008) 'Landscape aesthetics in British National Parks'. In Mallarach, J-M (ed.) *Protected Landscapes and Cultural and Spiritual Value,* IUCN, GTZ and Obra Social de Caixa Catalunya. Kasparek Verlag, Heidelberg

Joppa, L. N., Loarie, S. R. and Pimm, S. L. (2008) 'On the Protection of Protected Areas', *Proceedings of the National Academy of Sciences,* 105: 6673-6678

Joppa, L. N. and Pfaff, A. (2011) 'Global protected area impacts', *Proceedings of the Royal Society B,* 278: 1633–1638

Kajala, L. (2013) TEEB Nordic case: 'Estimating local economic benefits of visitors' spending in protected areas', in Kettunen, M., Vihervaara, P., Kinnunen, S., D'Amato, D., Badura, T., Argimon, M. and ten Brink, P. *Socio-economic importance of ecosystem services in the Nordic Countries – Scoping assessment in the context of The Economics of Ecosystems and Biodiversity* (TEEB), Nordic Council of Ministers, Copenhagen

Kapos, V., Ravilious, C., Campbell, A., Dickson, B., Gibbs, H. K., Hansen, M. C., Lysenko, I., Miles, L., Price, J., Scharlemann, J. P. W. and Trumper, K. C. (2008) *Carbon and biodiversity: a demonstration atlas,* UNEP-WCMC, Cambridge

Kramer, R. A., Richter, D. D., Pattanayak, S. and Sharma, N. P. (1997) 'Ecological and Economic Analysis of Watershed Protection in Eastern Madagascar', *Journal of Environmental Management,* 49: 277–295

KNPS - Korea National Parks System and IUCN (2009) *Korea's Protected Areas: Evaluating the Effectiveness of South Korea's protected areas system,* Seoul and Gland, Switzerland

Kulshreshtha, S. N., Lac, S., Johnston, M. and Kinar, C. (2000) *Carbon Sequestration in Protected Areas of Canada: An Economic Valuation,* Economic Framework Project, Report 549, Canadian Parks Council, Warsaw, Canada

Laird, S., Johnson, S., Wynberg, R., Lisinge, E. and Lohan, D. (2003) *Biodiversity Access and Benefit-Sharing Policies for Protected Areas: An Introduction,* United Nations University Institute of Advanced Studies, Tokyo

Luoma, J. R. (1999) *The Hidden Forest: The biography of an ecosystem,* Henry Holt and Company, New York

MA – Millennium Ecosystem Assessment (2005) *Ecosystems and Human Well-being: Biodiversity Synthesis,* World Resources Institute, Washington, DC, http://www.maweb.org/documents/document.354.aspx.pdf, accessed 1 October 2012

Melbourne Water (2013) *Information about the Melbourne water catchments,* http://www.melbournewater.com.au/content/water_storages/water_supply/water_catchments.asp, accessed 10 January 2013

Mittermeier, R. A., Kormos, C. F., Mittermeier, C. G., Gil, P. R., Sandwith, T. and Besançon, C. (2005) *Transboundary Conservation,* Cemex, Mexico

Nagendra, H. (2008) 'Do Parks Work? Impact of Protected Areas on Land Cover Clearing', *Ambio*, 37: 330-337

NPS – National Park Service (2000) *Natural Resource Year in Review: 2000*. The National Park Service, U.S. Department of the Interior

Noss, R. F. (2001) 'Beyond Kyoto: Forest management in a time of rapid climate change', *Conservation Biology*, 15: 578-591

Oglethorpe, J., Honzak, C. and Margoluis, C. (2008) *Healthy people, healthy ecosystems: A manual for integrating health and family planning into conservation projects*, World Wildlife Fund, Washington, DC

Pagiola, S., Bishop, J. and Landell-Mills, N. (eds) (2002) *Selling Forest Environmental Services: Market-based mechanisms for conservation and development*, Earthscan. London

Pham, K. N., Tran, V. H. S., Cesar, H. and Pollnac, R. (2005) *Financial sustainability of the Hon Mun Marine Protected Area: Lessons for other marine parks in Vietnam*, Poverty Reduction and Environmental Management, Institute for Environmental Studies, The Netherlands, http://www.prem-online.org/archive/8/doc/PREM%20WP%2005-13.pdf, accessed 10 January 2013

Pattanayak, S. K., Corey, C. G., Lau, Y. F. and Kramer, R. A. (2003) *Forest malaria: A microeconomic study of forest protection and child malaria in Flores, Indonesia*, Duke University, www.env.duke.edu/solutions/documents/forest-malaria.pdf, accessed 20 December 2012

Peel, M., Watson, F., Vertessy, R., Lau, A., Watson, I., Sutton, M. and Rhodes, B. (2000) *Predicting the Water Yield Impacts of Forest Disturbance in the Maroondah and Thomson Catchments using the Macaque Model Technical Report*, Report 00/14, December 2000, Cooperative Research Centre for Catchment Hydrology and Melbourne Water, Australia

Rist, L., Kaiser-Bunbury, C. N., Fleischer-Dogley, F., Edwards, P., Bunbury, N. and Ghazoul, J. (2010) 'Sustainable harvesting of coco de mer, *Lodoicea maldivica*, in the Vallée de Mai, Seychelles', *Forest Ecology and Management*, 260: 2224–2231

RSPB – The *Royal Society for the Protection of Birds* (2011) *Natural Foundations: Conservation and local employment in the UK*, RSPB, Bedfordshire

Ruitenbeek, H. J. (1990) *Economic Analysis of Tropical Forest Conservation Initiatives; Examples from West Africa*, WWF UK, Godalming

Schuyt, K. and Brander, L. (2004) *The Economic Values of the World's Wetlands*, WWF, Gland, Switzerland

Stolton, S. and Dudley, N. (2009) *Vital Sites: The contribution of protected areas to human health*, WWF International, Gland, Switzerland

Stolton, S. and Dudley, N. (2010) *Arguments for protected areas: multiple benefits for conservation and use*, Earthscan, London

ten Brink, P. (ed.) (2011) *The Economics of Ecosystems and Biodiversity in National and International Policy Making*, Earthscan, London and Washington, DC

Thompson, I., Mackey, B., McNulty, S. and Mosseler, A. (2009) *Forest Resilience, Biodiversity, and Climate Change: A synthesis of the biodiversity/resilience/stability relationship in forest ecosystems*, CBD Technical Series no. 43, Secretariat of the Convention on Biological Diversity, Montreal

Tilman, D., Wedin, D. and Knops, J. (1996) 'Productivity and sustainability influenced by biodiversity in grassland ecosystems', *Nature*, 379: 718-720

UNESCO (2012) *Managing Natural World Heritage*, UNESCO/ICCROM/ICOMOS/IUCN

UK NEA – National Ecosystem Assessment (2011) *The UK National Ecosystem Assessment: Synthesis of the Key Findings*, UNEP-WCMC, Cambridge, http://uknea.unep-wcmc.org, accessed 18 June 2012

Van Beukering, P. J. H., Cesar, H. S. J. and Janssen, M. A. (2003) 'Economic valuation of the Leuser National Park on Sumatra, Indonesia', *Ecological Economics*, 44: 43-62

Vittor, A. Y., Pan, W., Gilman, R. H., Tielsch, J., Glass, G., Shields, T., Sánchez-Lozano, W., Pinedo, V. V., Salas-Cobos, E., Flores, S. and Patz, J. A. (2009) 'Linking deforestation to malaria in the Amazon: Characterization of the breeding habitat of the principal malaria vector *Anopheles darling*', *American Journal of Tropical Medicine and Hygiene*, 81: 5-12

Wild, R. and McLeod, C. (eds) (2008) *Sacred Natural Sites: Guidelines for Protected Area Managers*, IUCN, Gland, Switzerland

Wittmer, H. and Gundimeda, H. (eds) (2012) *The Economics of Ecosystems and Biodiversity in Local and Regional Policy and Management*, EarthScan/Routledge, London and Washington, DC

WRM (2009) Bolivia: For the protection of the last isolated indigenous peoples, *World Rainforest Movement Bulletin*, 141

3

GENERAL PRINCIPLES FOR ESTIMATING THE SOCIO-ECONOMIC VALUE OF BENEFITS PROVIDED BY PROTECTED AREAS

Kettunen, M., ten Brink, P. and Bassi, S.

The purpose of Chapter 3 is to provide readers and users with an introduction to the valuation of benefits related to PAs and build on the common theory and approach for socio-economic valuation of biodiversity and ecosystem services. It presents an overview of the framework for (economic) valuation by outlining different terms and commonly used methods. The chapter focuses on explaining the logic and economic principles of valuation in clear, practical terms to help avoid some of the 'mystique' surrounding valuation methods.

Part II of the assessment guide builds on the framework and principles presented in this chapter – with a strong focus on turning them into a practical guidance for users in the context of PAs. It is hoped that the information provided by this chapter will help users to follow and reflect upon the guidance in Part II, placing it accurately in a broader context of socio-economic valuation.

3.1 Defining the socio-economic value of benefits

As outlined in Chapter 2 above, the benefits people derive from nature, including PAs, can be described as ecosystem services and they are generally distinguished under the broad headings of provisioning, regulating and cultural services and underlying ecosystem processes (supporting services) (MA, 2005).[1]

When moving onto considering the socio-economic value of ecosystem services, a further conceptual distinction can usefully be made between 1) ecosystem services as the beneficial processes and outputs of ecosystems (water retention by soils or water purification by wetlands, etc.) and 2) related goods directly enjoyed and valued by people (i.e. available clean water, pollution control) (Figure 3.1). This distinction has been found useful when assessing the value of PAs (UK NEA, 2011; Bateman *et al.*, 2011) as goods are often easier to measure and can therefore be used as indirect means (proxy) for valuing the underlying services. In this context, the term 'goods' is defined much more broadly than in traditional economics: it does not simply refer to tangible items bought and sold in markets, but it also includes a range of other (current and potential) use and non-use, material and non-material benefits – most of which do not have a market price (flood control, climate change mitigation, open-access outdoor recreation, aesthetic values and scientific knowledge, etc.) (UK NEA, 2011; Bateman *et al.*, 2011).

The socio-economic value of benefits provided by nature and PAs (i.e. ecosystem services and related goods) can be categorised in a number of ways. For the purposes of this guide, two general types of socio-economic values are identified: economic value and broader welfare value (Figure 3.1). The former refers to the estimated monetary value of PAs to different stakeholders (individuals, businesses, local communities, regions, etc.), consisting of both 'real' economic advantages (revenue, profits, avoided costs, etc.) and estimates of 'theoretical' monetary value (willingness to pay and/or accept, see 3.2 below). The latter includes wider welfare and social values to individuals and societies that are often difficult to determine in monetary terms (different health benefits, aesthetic enjoyment and cultural identity, etc.).

In general, economic and welfare values – including values not yet known, captured or realised by humans – form the total (socio-)economic value of PAs (Figure 3.2 below). This 'total value' consists of both use (direct and indirect) and non-use values associated with PAs. In addition, the ability of PAs to maintain these socio-economic values over time (so-called 'insurance' value) needs to be recognised, although the existing valuation methods fall short in providing reliable quantitative and/or monetary indicators for this value (Pascual *et al.*, 2010). In addition to these values that are based on an 'anthropocentric' appreciation, nature also has value in its own right. However, as explained in Chapter 1, this 'intrinsic value' falls outside of the socio-economic framework.

The estimated socio-economic value of different benefits (ecosystem services and related goods) is generally context-specific and varies over time and space (UK NEA, 2011; Bateman *et al.*, 2011; Kettunen *et al.*, 2011). For example, the recreational value of a PA is likely to be higher when the area is easily accessible by people (PAs located on the edge of a city, etc.) than when it is situated in a remote place. Similarly, benefits related to PAs' role in regulating the quantity and quality of water depend on whether the site is located up- or downstream from population centres. This context specificity is of importance when comparing the socio-economic role and importance of different PAs or when considering whether the values assessed for one PA can be useful in understanding and demonstrating the value of another PA (see 'value transfer' in Box 3.1).

The conceptual separation between ecosystem services and related goods is helpful because, while ecosystem services are fundamental sources for the related goods, many of these goods are generated with additional human inputs and investment; i.e. with the help of manufactured capital (machinery, etc.) and human capital (skills, etc.) (UK NEA, 2011, Bateman *et al.*, 2011). In other words, the estimated socio-economic value of a good, such as existing market values, often reflects not

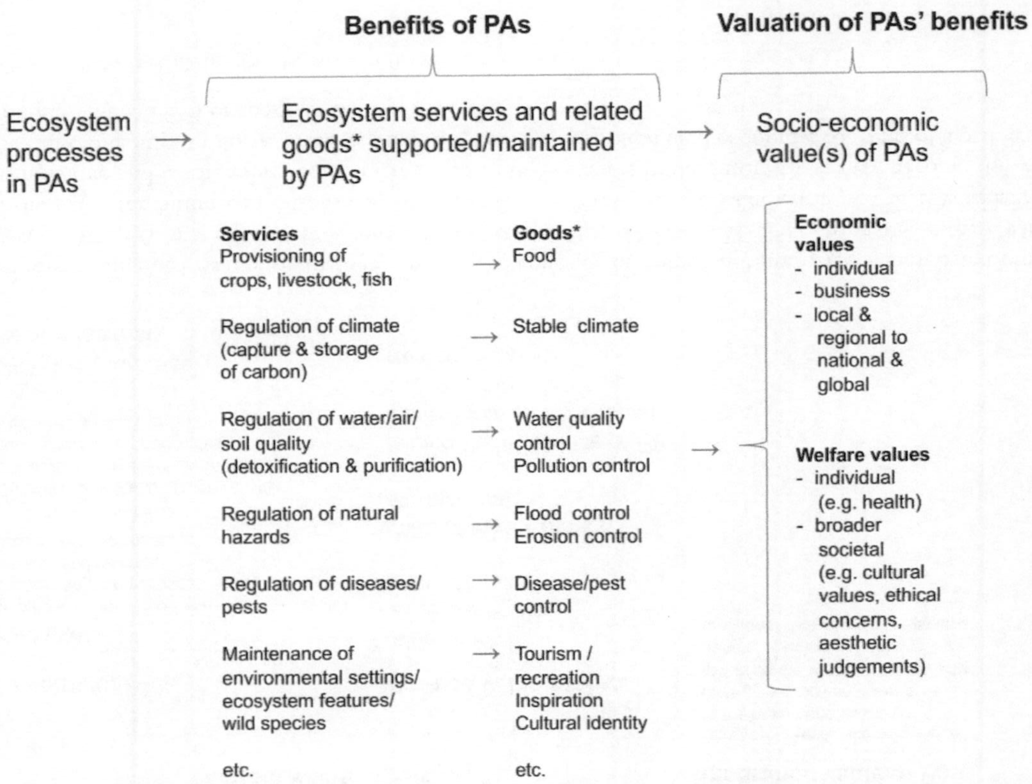

FIGURE 3.1 Conceptual classification of benefits provided by PAs (ecosystem services and related goods) and the socio-economic value of these benefits

* Good(s) includes all use and non-use, material and non-material outputs from ecosystems that have (direct) value for people (UK NEA, 2011; Bateman *et al.*, 2011).

Note: goods often also include human inputs (labour costs, etc.)

Source: own presentation, based on Kettunen *et al.* (2011) and UK NEA (2011)

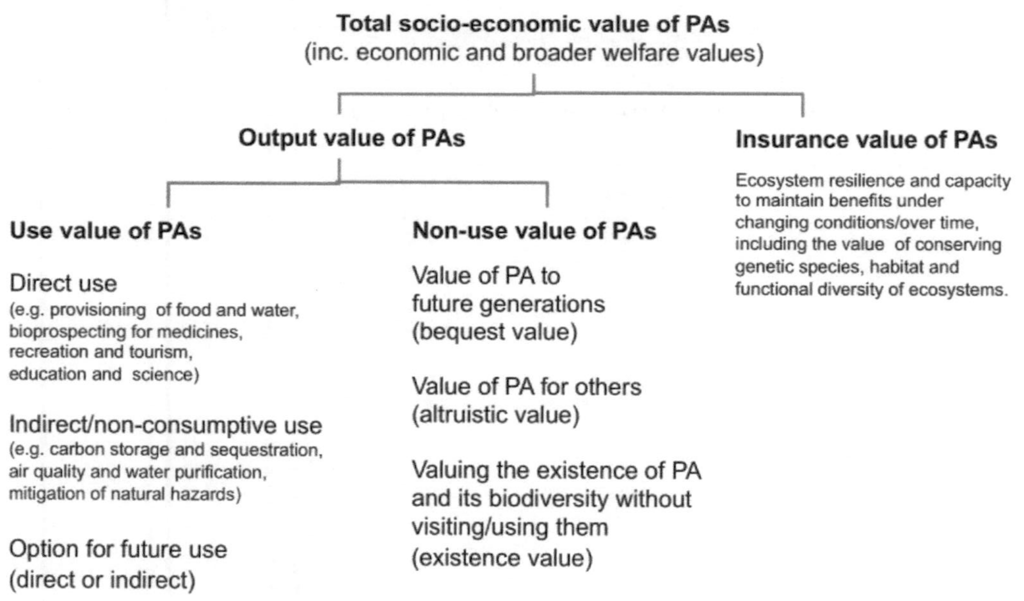

FIGURE 3.2 Total socio-economic value of PAs, conceptualised through the concepts of output and insurance values. The output value refers to the values associated with PAs in a given state whereas the insurance value refers to PAs' ability to maintain these values in the future. Consequently, the value for future generations and the altruistic value refer to the value attributed to PAs today/from the current beneficiaries' perspective and is, therefore, considered as 'non-use' – even if in the future these may become use values. Note: total socio-economic value and output value as presented here are similar to the often used economic concepts of Total System Value (TSV) and Total Economic Value (TEV) of ecosystems, respectively[2]

Source: own presentation, based on Gantioler *et al.* (2010), Pasqual *et al.* (2010), White *et al.* (2011)

only the value of the underlying ecosystem service, but also the value of associated human inputs. Consequently, in a number of cases, the (estimated) final value of a good arguably overstates the socio-economic value of the ecosystem service on its own. For example, while PAs are often ideal destinations for recreation and tourism, their appeal to people is significantly increased by marketing and the establishment of supporting infrastructure such as walking paths and camping facilities. While one could argue that the ecosystem qualities help catalyse the other investments, one could also argue that the final value of a site would not be reached without the man-made investments.

Consequently, there is a need to try to assess nature's 'share' in the final estimated socio-economic value. While this assessment guide aims to provide advice and tools for estimating this share, it also recognises that, depending on the information, resources and time available, this might not be always feasible in practice. Similarly, it is possible that the possibilities for site-specific socio-economic valuations are limited and there is a need to use existing estimates from other PAs as rough indicators of value. In such cases it is important that the shortcomings of valuation are interpreted and communicated accurately (Chapter 15).

3.2 Estimating the socio-economic value of benefits

Different indicators of value

The valuation of different socio-economic values (economic and welfare) can be done at three levels: qualitative, quantitative and monetary (e.g. White *et al.*, 2011; UK NEA, 2011; Kettunen *et al.*, 2013). Qualitative valuation focuses on non-numerical indications of value such as PAs' role in supporting local culture and identity and maintaining mental and physical health. Qualitative methods are also suitable for demonstrating social benefits associated with recreation or highlighting benefits related to security and broader well-being. Quantitative valuation focuses on numerical data, including the number of visits and jobs associated with PAs, quantity of carbon stored in and/or sequestrated by PAs, and biophysical indicators reflecting the quality of water purified by PAs. Monetary valuation focuses on capturing or reflecting the different values of PAs in monetary terms; for example, by calculating the revenue generated by visitors to PAs, estimating the amount of avoided costs when maintaining PAs' ability to purify water or mitigate flooding, or defining the value of the carbon stored and annually sequestered by the PA. In general, only a limited part of the socio-economic value of PAs can be captured through monetary indicators (Figure 3.3).

In practice, the type of approach used (i.e. qualitative, quantitative or monetary) depends on the time and resources available and type of benefit measured. Only some services are amenable to monetary analysis. Generally speaking, qualitative assessment is the least challenging and resource-intensive type of assessment, while monetary assessments are the most resource-intensive.

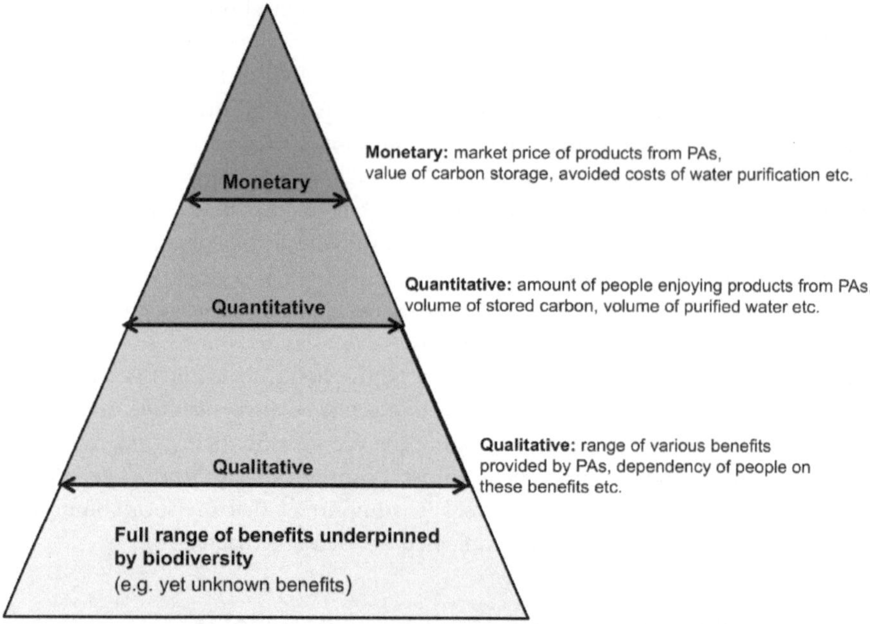

FIGURE 3.3 Benefits pyramid illustrating the availability of different indicators for the socio-economic value of PAs

Source: own presentation

It is also crucial to define clearly the reasons and/or specific needs for the socio-economic valuation of PA benefits and use that information as a basis to select the most appropriate type(s) of valuation. In a number of cases, qualitative and quantitative assessments (such as estimating the number of jobs supported by PAs or highlighting their unique cultural, historic or religious values) can be sufficient to communicate PA-related benefits to stakeholders, whereas in other cases a monetary assessment is needed to support decision-making (TEEB, 2010) (see Section 3.4 below and Chapter 5).

A range of monetary, quantitative and qualitative methods are available and currently used to estimate the socio-economic value of biodiversity, ecosystem services and related goods (Defra, 2007; Pasqual *et al.*, 2010; Bateman *et al.*, 2011; White *et al.*, 2011). Table 3.1 summarises the key methods available, whereas further guidance on the selection and application of the methods is provided in Part II of the publication.

Methods for monetary valuation

In terms of monetary valuation, the value of some PA-related benefits is either directly or indirectly apparent through markets. For example, food, medicinal and ornamental products originating from PAs are often sold to consumers and consequently their economic value can be determined through their market price.

Other market-related indicators of value might be less obvious; house prices may be higher in the vicinity of PAs which can in turn lead to higher property taxes and tax-related revenue supporting local welfare. There are also benefits that have potential, however not yet existing, market value. For example, some PAs may harbour genetic material that may provide the basis of valuable medicines in the future. Furthermore, given the great concerns related to climate change there are expectations that the stock of carbon stored in ecosystems, including ecosystems within PAs, might soon receive an explicit market value.

Market values and prices are generally useful for estimating the socio-economic value of biodiversity resources, recreation and tourism. However, they offer little help when estimating indirect and non-use benefits provided by many regulating services and related goods. In addition, the existing market prices may be distorted by subsidies or other measures affecting their 'real' value. In the case of PAs, this might refer to, for example, financial support or tax incentives given to the production of certified sustainable products. Finally, it is important to remember that the market prices of goods also integrate a range of human inputs (see Section 3.1 above). Therefore, in order to reflect more accurately the value of the ecosystem service, the value of these inputs should be deducted from market prices. Consideration can also be given to the fact that market prices may not reflect the total consumer value of the good (i.e. consumers would be willing to pay more than the current market price) and, when considering net values, the opportunity costs should also be taken into consideration (see Section 3.3 below).

There are also several benefits that are not captured by the markets but can be valued in monetary terms by using cost-based methods. For example, PAs' natural ability to purify water can be captured as the avoided costs of pre-treatment by water companies. PAs' role in water retention and mitigation of potential downstream flooding can be estimated as avoided costs to property and livelihoods (avoided costs of clean-up, shared between individuals, the insurance industry and the state, etc.). PAs' contribution to air quality in towns and cities can be translated into monetary values through estimating the quantity of captured air pollution and calculating the corresponding reduction in likely respiratory diseases and diminishing health-related expenditure (reduced costs to society and states, etc.) and losses of output (days off work, etc.).

In addition to cost-based estimates, revealed preference methods can be used to assess the monetary value of (actual) expenditure related to benefits provides by PAs, including recreation, tourism and aesthetics appeal. For instance, the costs people incur by travelling to PAs can be used as an indication for the socio-economic value of recreation (travel cost method). Similarly, statistical analysis of the house prices at different locations can be carried out to estimate the values people attribute to living in areas in close proximity to PAs (hedonic pricing). Stated preference methods can be used to develop indicative/'hypothetical' market values for a range of PA benefits. These methods aim to assess people's willingness to pay for (or willingness to accept) the provision of ecosystem services and related goods. Two distinct stated preference methods can be distinguished: contingent valuation (CV) and choice modelling (CM). In CV surveys the respondents are asked to state directly their values, whereas in CM respondents are offered a range of scenarios with differing environmental and economic parameters from which they are asked to choose (or to rank them).

While commonly used, both revealed and stated preference methods face a number of challenges – particularly with regards to the demands for data and resources and their significant reliance on a range of underlying assumptions. Revealed preference methods are relatively robust, relying on the observed choices in the actual markets. However, they require specific data and dedicated analysis, including outlining a set of plausible assumptions. Therefore, the application of the method is rather time-consuming and generally requires dedicated resources and expertise. In comparison, estimated monetary values obtained with stated preference techniques are often criticised for being reliant on survey responses rather than real market decisions. Therefore, they should be used to indicate the preferences and values of different stakeholders or the broader public, but not be interpreted as 'real' market value or without due consideration used as a basis for defining schemes for monetary compensation. Also, careful attention needs to be given to ensure appropriate design of such studies, for example to account for the fact that people tend to estimate their willingness to pay to be higher in theory than it is in practice (so-called hypothetical bias).

Finally, in some cases monetary estimates obtained from one PA can be used to develop a rough estimate of similar benefits on another PA (Pasqual *et al.*, 2010; White *et al.*, 2011). This 'benefit or value transfer' is an option to consider when resources to carry out valuation studies are scarce (Box 3.1).

BOX 3.1

Benefits or value transfer

Benefits or value transfer is a pragmatic way of dealing with information gaps and resource (time and money) constraints. The underlying thinking is that there might be sufficient commonalities between different PAs to allow insights from one area to be transferred to another. Naturally, the suitability of using benefits transfer needs to be carefully considered. In this context, it is important to assess both the ecological, spatial and socio-economic attributes of PAs and determine whether these attributes are similar enough to allow and/or justify the use of values across the sites.

The key considerations for benefit transfer when building on the values from one site (original source site) and applying them to another site (study site) are 1) the ecosystem service or related goods valued are very similar at the site where the estimates were made and the site which they are applied to, 2) affected stakeholders have very similar characteristics, 3) the original estimates are considered reliable, and 4) there is no reason to expect that the socio-economic value(s) associated with a benefit have changed since the development of the existing estimate (value of carbon markets, etc.). Note that there is some flexibility in these considerations (notably 1, 2 and 4) if and where differences between the study site and the site which the original value was taken from can be 'corrected' for. For example, knowing differences in timber type or quality between the original and new study sites can allow the unit value of timber provisioning to be adjusted. Similarly, knowledge of the relative population levels and GDP per capita can help to adjust the original value to ensure that the transferred value is representative of the new site. It is important to offer a due rationale and data trail for the adjustments made and ensure that the attempts to adjust an existing estimate do not raise more questions than they try to solve (see also Chapter 14).

When a good base of source studies is available it is also possible, by comparing all the values of these sites (meta-analysis) to derive a 'benefits production function' and use that as a basis for benefit transfer. This benefits production function is an equation that relates existing benefits values to a range of different explanatory variables (size of PA, type of site, country income levels, proximity to urban populations, etc.). By applying these variables, the benefits production function can be used to derive a corresponding local number. This may also require some adjustment in light of above points 1, 2 and 4, especially if there are differences between the study site and studies used to develop the benefits function. However, if the pool of source studies is representative then the need for adjustments becomes less critical as several critical considerations are already partly taken into account by the method itself.

Example: In relation to ecosystems' role in maintaining air quality, it has been statistically shown that the relationship between exposure to air pollution and incidence of illness is generally similar across people across the world. The dose-response relationship established in one country can therefore be applied with some confidence in another country. The main location-specific differences that need to be considered include: level of air quality (i.e. how polluted the air is locally), number of people benefiting from air quality (pollution density, proximity of PA to inhabited areas/number of visitors, etc.) and estimated value of good air quality to the public/society in question (avoided illness, reduced hospitalisation costs, loss of work output and eventual loss of life expectancy/early mortality, etc.). Some of these location-specific factors can be addressed relatively easily, for example, allowing for a quantitative assessment of the number of avoided cases of illness and early mortality.

As regard monetary estimates, an accepted benefits transfer approach between regions has been, for example, to weigh willingness to pay estimates by the relative GDP per capita. Similarly, people's estimated willingness to pay for the protection of a particular animal (charismatic species such as gorilla, panda or whale, etc.) can be transferred between regions and countries also by weighing the estimates on the basis of relative wealth.

Source and further information: Pasqual *et al.* (2010), White *et al.* (2011)

Methods for quantitative and qualitative valuation

PAs have several wider welfare (non-economic) values that cannot be easily captured in monetary terms and are therefore mainly subject to qualitative and quantitative valuation. For example, it is difficult to find relevant monetary indicators for PAs' role in supporting mental health and cultural identity or providing aesthetic pleasure and spiritual fulfilment. People may also value the possibility of using certain PAs in the future or just simply appreciate the existence of the sites and their species and habitats, even though they have no intention of visiting any of them. This appreciation may be cultural, spiritual and philanthropic or caused by a wider vision of responsibility.

A number of quantitative indicators are available to demonstrate the socio-economic value of PAs. Such indictors include the number of visitors to PAs (cultural services/goods), the number of households using water purified by PAs (regulating services/goods) and the amount of non-timber resources originating from PAs (provisioning services/goods). Similarly, assessments of health-related benefits, such as reduced levels of stress related to recreation and outdoor activities on PAs, provide concrete evidence for the broader welfare value of PAs. Such studies can, however, be rather resource-intensive to carry out. Finally, biophysical availability of different ecosystem services and related goods, such as (changes in) PAs' capacity to purify water, store and/or sequester carbon and maintain healthy populations of pollinators, can also be used as indicators of PAs' socio-economic role and importance – even though such biophysical attributes do not directly capture the economic value of specific services.

A range of qualitative methods such as public surveys, expert opinions and structured stakeholder discussions can be used to gain useful insights on the socio-economic role, importance and value of PAs. Such methods are applicable for assessing all ecosystem services and goods related to PAs and they also allow for exploring people's understanding and appreciation of ecological processes and the role of biodiversity in underpinning different benefits. Furthermore, qualitative methods (expert opinions, scientific reviews, etc.) are currently the only methods available for exploring and demonstrating the insurance value of PAs.

Other methods

PAs can also contribute to local, regional and national economies and welfare by creating job opportunities, such as opportunities related to recreation and tourism or site management, and increasing the income of local businesses both on and off-site, for example through direct expenditures of the reserve and of its employees. Furthermore, PAs can help generate public and private investment in the area through funding of research and development and/or creation of 'green' business opportunities.

As explained in Chapter 2 above, these benefits are related to the overall 'existence' of PAs and they cannot be easily attributed to a specific ecosystem service or good. Consequently, when considering the overall benefits of PAs, the economic values related to jobs, employment and investment need to be considered and communicated separately and cannot be 'summed up' with the estimated values of ecosystem services and related benefits (Chapters 11 and 13). It also needs to be clarified that, in addition to possible links with PAs, employment and regional investments are affected by a number of broader, non-PA related socio-economic factors and policies.

3.3 Estimating the costs of protected areas

Ensuring the provision of benefits from PAs requires society to incur costs, for example for site designation and ongoing protection and management. As with benefits, costs depend significantly on location, planning processes and management strategy. Costs are commonly grouped into management, human-wildlife, opportunity and displacement costs outlined below (e.g. Kettunen *et al.*, 2011; ten Brink *et al.*, 2011).

Management costs include expenditures arising from actions and measures ensuring that appropriate management leads to effective provision of benefits. They typically include one-off costs related to PA management (drafting management plans, etc.) or specific investment needs (land purchase, compensation to landowners, one-off restoration and infrastructure for access, etc.). The one-off costs are incurred only a limited number of times, typically at the early stages of designation. Management costs also include recurrent costs for ongoing management and monitoring activities. Such costs include costs related to different regular management and monitoring activities, salaries and rent of management bodies, and investment in communication. They are incurred regularly, for example, once a year or once a month, throughout the life of a PA.

Human-wildlife conflicts refer to damage caused by wildlife in areas used for human activities. Such conflicts can result in a number of costs, including both costs of prevention and costs of damage (damage to property, crop and livestock, etc.). Conserving rare or endangered species is one of the common goals of PAs, but if a site is successful at increasing the size of wildlife populations, the damages caused by certain species may also increase. For example, the protection of large predators can result in the loss of livestock, and agricultural output may be reduced due to grazing by protected herbivores like goats or elephants.

Opportunity costs refer to costs arising from opportunities forgone due to the designation of an area as a PA. The establishment of PAs might result in forgoing a number of alternative uses of the area, such as agriculture or housing development, or it might prevent access to natural resources. These restrictions can give rise to opportunity costs. Formally, such costs can be defined as the maximum return forgone associated with having to adapt to the restrictions posed by PAs, including being unable to undertake certain economic activities in or near PAs (Jacobs, 2004). Key opportunity costs include the loss of potential profits and/or employment opportunities, for example, due to prohibition of economic activities such as intensive agriculture and mining. Limitations related to the direct harvesting of

TABLE 3.1 Methodologies for socio-economic valuation of PA benefits

Type of method	*Name of method*	*Description*	*Preliminary use*
Monetary (market)	Market value/price (adjusted)	Market prices of goods originating from PAs, preferably adjusted for distortions such as subsidies and taxes	Provisioning services/goods Cultural services/goods: recreation and tourism
	Production function approach	Estimating the share of the (adjusted) market value that can be attributed to ecosystem services (i.e. deducting the value of human inputs)	Provisioning services/goods Cultural services/goods: recreation and tourism
Monetary (non-market)	Costs (damage) avoided	Costs which are avoided by not allowing ecosystems within PAs to degrade: damage costs avoided, avoided expenditure/investment costs of replacement, avoided costs for compensation of loss of benefits, avoided losses of revenue, avoided restoration or repair costs, avoided insurance costs, avoided costs to health, etc.	Regulating services/goods
	Revealed preferences	Expenditure (actual) on ecosystem services and related goods: costs related to travel to PAs (travel cost method) or property prices close to PAs (hedonic pricing)	Cultural services/goods: recreation and tourism, aesthetic appeal
	Stated preferences	Expenditure (potential) assessed via surveys exploring the 'demand' for a benefit in hypothetical markets and establishing people's willingness to pay (WTP) for maintaining/ restoring PA-related ecosystem services and goods and/or accepting changes in the provision of these benefits (contingent valuation, choice modelling)	All ecosystem services/goods Also possible to explore (in monetary terms) appreciation of economic/welfare importance of biodiversity and ecosystem processes
	Deliberative monetary valuation	A hybrid of stated preference and participatory methods (below) with the objective of using formal methods of deliberation to express values in monetary terms.	

Type of method	*Name of method*	*Description*	*Preliminary use*
Non-monetary (qualitative and quantitative)	Non-monetary indicators of economic value	Quantitative estimates for socio-economic importance and value: number of visitors to PAs, number of households using water purified by PAs, number of local communities depending on PAs for their livelihood, amount of non-timber resources originating from PAs, etc.	Provisioning services/goods Regulating services/goods Cultural services/goods: recreation and tourism
	Health-based valuation	Measuring health-related benefits of PAs such as reduced levels of stress through recreation/ outdoor activities.	Regulating services: mental and physical health
	Biophysical Indicators for status of ecosystem service	Biophysical availability of ecosystem services and related goods, including (changes in) PAs' capacity to purify water, store/sequester carbon, maintain pollinators and attract visitors, availability of biodiversity resources from PAs, etc.	Provisioning services/goods Regulating services/goods Cultural services/goods: recreation and tourism
	Focus group/ survey/review based valuations	Identification of different socio-economic value(s) of PAs based on public and/or expert opinions, surveys and discussions (Q-methodology, Delphi surveys)	All ecosystem services/goods Also possible to explore appreciation of economic/welfare importance of biodiversity and ecosystem processes
Revenue-based (monetary and quantitative)	Jobs and employment	Number of jobs and employment related to PAs (tourism/recreation and management), local/ regional revenue streams created by tourism to PAs	Broader welfare values (not directly related to ecosystem services/goods)
	Socio-economic investments (public and private)	PA-related investment to local communities/region such as PA-related research and development, PA-related business opportunities	As above

Source: own presentation, based on White *et al.* (2011), Bateman *et al.* (2011), Defra (2007)

particular species (endangered fish, game, edible plants, etc.) or to land management activities negatively affecting species' habitats can also trigger further costs in the form of forgone activities. It should be noted that the activities forgone because of the establishment of PAs can be more or less sustainable. Opportunity costs do not reflect the scale of impacts of these activities, but simply the amount of revenue that will be lost if such activities are not allowed to occur. Some (long-term) impacts may be so detrimental to the environment or to the society at large that arguably they should be not allowed even in the absence of PA. Therefore, insights on the types of activities prevented and on their impact on a broader geographical scale may be needed to interpret opportunity costs appropriately.

Finally, displacement costs are related to people being directly displaced from certain sites when these are designated as PA. While less frequent in developed countries, this has been known to be a real issue in some developing countries, leading to significant social and economic costs. Even when not resulting in complete displacement, the establishment of PAs might lead to social and welfare costs caused by losing the governance of land that has been important for subsistence (fuel wood, medicinal plants, etc.) and/or social, cultural and religious reasons.

Like benefits, costs are experienced by different stakeholders at different levels ranging from global to local (Table 3.2). Global costs take place when the management of PAs is supported at inter- and/or multinational scale, for example in the context of mobilising international funds to support PAs in vulnerable countries or when financing cross-border PA management programmes. More frequently, costs are incurred at national, regional and local levels. The former can include the direct expenditure on PAs from national budgets (management, land purchase, compensation, etc.), cost of national or regional bodies responsible for planning the management of PA networks, and forgone tax revenues (such as tax rebates allowed to some PA-related activities). Most of the costs related to PAs are, however, experienced at local level including direct management costs, costs of displacement, human-wildlife conflicts, as well as a range of opportunity costs.

Costs can be assessed in qualitative, quantitative and monetary terms. In comparison with benefits, several costs have an explicit market value and are therefore often more amenable to monetary valuation. This is particularly true for management costs which can be estimated, for example, through the costs of equipment and machinery used in managing PAs, staff employment and salary costs, and land purchase prices (see Chapter 11 for further information). Market values can also help to determine costs of human-wildlife conflicts, for instance when increased numbers of wildlife have resulted in loss of yields or cattle. Opportunity costs, however, are often more difficult to assess – and sometimes even to identify – in monetary terms. Some estimates for opportunity costs can be obtained through surveys and questionnaires that reveal the (perceived) costs related to the establishment and/or management of PAs including, for example, people's willingness to accept certain restrictions and/or pay for certain management activities. Alternatively,

TABLE 3.2 Examples of protected area costs accruing at different scales

Level	*Costs*
Global/ multinational	Management (global transfers to developing countries) Alternative development programmes (global transfers to developing countries)
National	Land purchase Coordination of management (undertaken by national bodies such as Ministry of the Environment and Environmental Agencies) Compensation for forgone activities (national budget) Compensation for human-wildlife conflicts (national budget) Opportunity costs of forgone tax revenue
Local/ regional	Restricted access to resources (commercial or subsistence) Displacement Management (private land owners, municipal lands) Opportunity costs of forgone economic activities Social costs of losing governance of land (social, cultural or religious implications) Human-wildlife conflict (damage costs and/or costs of conflict prevention)

Source: own presentation, based on Kettunen *et al.* (2011)

opportunity costs may be indirectly estimated based on the sums allocated to compensation payments (payments to compensate forgone agricultural activities, etc.). Nevertheless, several opportunity costs can only be assessed through qualitative and quantitative methods. The same applies to costs related to displacement. Further guidance on how to assess different costs related to PAs is provided in Chapter 11.

3.4 Estimating the total (net) benefits of a protected area

As described above, PAs can bring a wide range of benefits and, in order to ensure that the flow of these benefits is maintained or enhanced a number of costs are likely to occur. When assessing the overall value of PAs it may, therefore, be necessary to think in terms of 'net' benefits. Put simply, the total net benefits of a protected site in a given moment are the 'sum' of all the identified benefits minus different costs (Figure 3.4). In practice, however, costs and benefits may not be fully comparable: only a limited number of benefits can be assessed in monetary terms whereas costs are more amenable to monetary valuation. Consequently, if PA's net benefits are assessed only in monetary terms the costs may sometimes appear to outweigh benefits. However, when the socio-economic value of PA benefits is assessed in its entirety, including quantitative and qualitative estimates, the benefits can substantially outweigh costs.

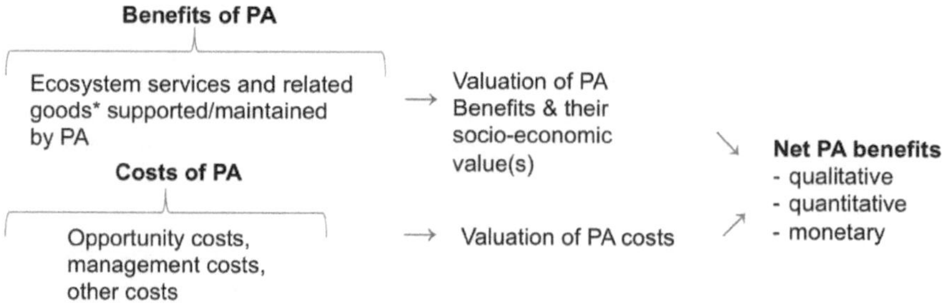

• Good(s) includes all use and non-use, material and non-material outputs from ecosystems that have (direct) value for people.

FIGURE 3.4 Conceptual classification of the total (net) value of a PA

Source: own presentation, based on Kettunen *et al.* (2011)

Two approaches can be adopted for the socio-economic valuation of total net benefits provided by PAs: 1) assessing the overall socio-economic value of PA benefits and costs in a given moment or over time, or 2) assessing the change in the socio-economic value of benefits and costs resulting from (anticipated) changes in land use within PAs. The former approach can integrate both annual flows of benefits and insights on their 'stock' values (amount of carbon stored, etc.). It is suitable for raising awareness of the benefits associated with PAs among stakeholders and can, for example, be used for advocacy to gain public support of PAs and improve the allocation of resources for PA management. The latter approach needs to be adopted when considering the possible socio-economic implications of changes in a PA's conservation objectives and/or management regime or when seeking to support the establishment of PAs with non-biodiversity related arguments (see Section 3.6 below).

Given that only some of the benefits are recognised by markets, both market and non-market valuation methods (complemented by qualitative and quantitative valuation methods) are needed to provide a comprehensive and truthful picture of the overall socio-economic importance of PAs. Consequently, it is unlikely – and indeed even unwise – to assume that the assessment of the total net benefits of a PA will lead to a single aggregated monetary estimate for its socio-economic value. The final outcome is likely to be a combination of qualitative, quantitative and monetary insights, possibly targeted towards certain decision-making processes.

PAs offer a range of ecosystem services some of which are interlinked. Consequently, the provisioning of one ecosystem service and related good may have a negative impact on another. Careful consideration of these interlinkages and their changes over time is needed when assessing the overall value of total net benefits provided by PAs. This is especially important when assessing the socio-economic (and biodiversity) impacts of change such as when considering the establishment of a new PA with its related opportunity costs or changes in a PA's management regime. There are often conflicts between maximising the provisioning of resources and maintaining the regulative functions of ecosystems. For example, forests are valuable carbon stores but wood production generally diminishes their carbon stocks. In addition, there might be conflicts between human interests (benefits) and set goals for biodiversity conservation. For example, the overuse of areas for recreation and outdoor activities might lead to the degradation of PAs' natural values (see Section 15.5).

In terms of economic assessments, the interlinkages between ecosystem services can also lead to the risk of double counting the value of different benefits. For example, maintaining populations of pollinators (regulating service) ensures the availability of several biodiversity resources such as honey, fruit and vegetables (provisioning service). Thus, when both of these benefits are included in an economic (monetary) assessment, depending on the valuation methods used, there is a risk that the same benefit will be accounted for twice. The risk of double counting is also the reason why the underlying valuable ecosystem processes (i.e. supporting services) need to be considered separately in the context of valuation (see Section 3.1 above).

Distribution of total net benefits

PAs provide benefits to different stakeholders at different scales (local, regional, national and global) (Chapter 2). Many key benefits from PAs benefit local actors the most – from sustainable resource use to disease control to local cultural or spiritual values (Kettunen *et al.*, 2011). Values related to watershed protection are of benefit both locally and regionally. Benefits related to the mitigation of climate change take place at global scale with indirect effects, such as the reduced risk of extreme weather phenomena at the local and regional level. Similarly, costs related to the establishment and management of PAs are distributed between stakeholders at different scales. The costs related to the management of PAs are mainly paid at the national or global level whereas several other PA-related costs (including human-wildlife conflicts, access to resources and other opportunity costs) occur at local level possibly undermining local support for conservation efforts. Consequently, the identified net benefits of PAs consist of a range of socio-economic values experienced by a range of stakeholders.

While the overall economic and welfare benefits of PAs are often positive, the local balance sheet might end up negative (Kettunen *et al.*, 2011). A comprehensive assessment of PA benefits and their socio-economic value should also consider the distribution of these benefits and related costs between stakeholders. Such considerations are often the key for the application of results in practice, for example establishing effective and fair mechanisms for funding PA management.

It should be noted that both the range of benefits and socio-economic values people associate with a given benefit change over time. For example, the carbon prices – often used as an indicator for the value of carbon sequestration and storage – are expected to rise in the future. The integration of a time perspective into the assessment of total net PA benefits may be essential, however the extent to which this is needed will depend on the context and purpose of the assessment and the questions it seeks to answer.

3.5 Estimating the added value of protected area designation

Some benefits associated with existing or planned PAs are likely to have continued to exist with or without a formal designation of a site as a PA. Therefore, PAs' total net benefits and their socio-economic value, as explained above, can be divided into two components: 1) added value of designation (i.e. benefits and values related to PA status such as increased ecotourism interest, benefits and values of avoided degradation and increased benefits, values due to appropriate management and investment, and minus the associated costs) and 2) value of benefits maintained even without designation (Kettunen *et al.*, 2011) (Figure 3.5).

The added value of PA designation depends on the specific characteristics of designation (rules and regulations related to conservation, etc.), applied management measures and, importantly, what would have happened in their absence (Kettunen *et al.*, 2011). If non-designation and the lack of associated measures were to lead to ecosystem degradation and loss of services and related goods, then designation and effective protection would obviously offer significant added value and, consequently, several identified benefits could be attributed to the PA status of an area. Conversely, if there is only a low threat of degradation, PA designation offers fewer additional benefits. Also, as outlined above, costs of management, investment, implementation and control can also increase with PA designation in order to meet conservation objectives. There will also be a range of opportunity costs, particularly to local communities.

In practice, it can be difficult to distinguish the added value of designation from a PA's total ecosystem value, especially over time. Consequently, it is not assumed that the practitioners assessing the benefits and socio-economic values of existing PAs would always be able to separate the added value of designation from the value of the site, especially in quantitative and monetary terms. The attempts to distinguish between the socio-economic value of a site with or without designation may not always be sensible, especially in the case of PAs that have been in place for a long time. The added value of designation can, however, be important when considering the establishment of new PAs. Whether an assessment should focus on the overall benefits and socio-economic value of a PA or specifically on the added value of designation is something that should be decided and clarified at the very early stages of the process (Chapter 5).

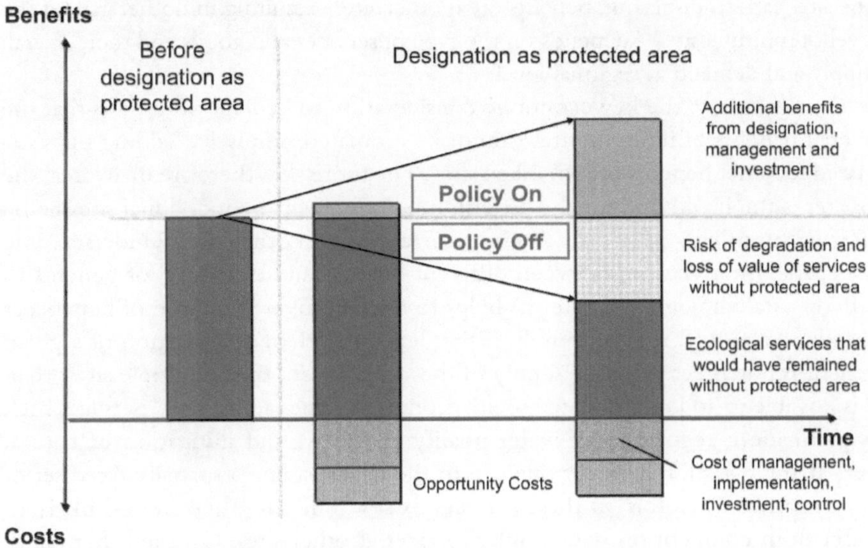

FIGURE 3.5 Overall benefits and socio-economic value of a PA vs. added value of PA designation (an illustration). 'Policy on' illustrates the foreseen development when the decision (policy action) is taken to establish a PA whereas 'policy off' refers to the development without designation

Source: own presentation, based on Kettunen *et al.* (2011)

3.6 Estimating benefits from multiple sites

In most cases, the socio-economic value of benefits provided by an individual site, including both their current value and the possible increased value following further investment in conservation, is complementary or additive to the value of benefits from other sites. In other words, promoting the role of benefits at one site does not diminish or cancel out the importance and socio-economic value of similar benefits at neighbouring sites. For example, increasing the carbon storage on one site is unlikely to affect the socio-economic value of carbon storage at another site.

However, in some cases investments aimed at enhancing the value of socio-economic benefits stemming from a given site may lead to competitive impacts and the decline of socio-economic value of other sites. For example, development and branding of 'PA friendly' products might diminish the market share of similar products produced on other sites in the same region. Similarly, encouraging ecotourism and recreation at one site may lead to a reduction in visitors at a neighbouring site. On the other hand, increasing tourism and recreational opportunities at multiple sites within the same region might attract more tourism to the area when this is seen as an opportunity to enjoy several sites during one visit. Whether the promotion of and/or investment in ecosystem services and related goods at one site leads to displacing benefits on another site – resulting in no net gain at the level of multiple sites – depends on the type of services and goods and their overall supply and demand at regional level.

Consequently, the key economic consideration to be kept in mind is that the overall benefits of multiple sites cannot be estimated simply by 'adding up' estimated total net benefits of individual sites. The more sites there are in an area, the greater is the likelihood for the provisioning of benefits by individual sites being affected (positively or negatively) by interaction with other sites. Understanding the spatial interrelations between different benefits and the 'flow' of benefits to different stakeholders (Chapter 2) helps to form an overall picture of benefits at broader scale. As a rule of thumb, if the demand and/or appreciation of a given benefit is not affected by the supply of the same benefit from multiple sites, then it is possible to add up estimated socio-economic values across sites. Services such as pollination, regulation of water quality and flow, and mitigation of natural hazard are less 'mobile' from one site to the next, having a spatially fixed set of beneficiaries. Therefore, if these services exist at one site, they are less likely to suffer from competition with similar services at other sites. On the other hand, provisioning and cultural services and related goods are enjoyed at broader level (regional, national and global) and therefore competition may occur between different sites (see Chapter 13 for further information).

Notes

1 The supporting services have also been termed 'intermediate services' and the other services 'final ecosystem services' to help clarify the causal hierarchy of services and avoid double counting of the benefits (UK NEA 2011). Also, in the context of monetary valuation the category 'habitat services' has been used to capture the value of those supporting services that have more direct socio-economic benefits, including maintenance of genetic diversity and life cycle for migratory species (Kumar, 2010).

2 It should be noted that there are differing opinions whether 'option value' should be considered as a true component of 'use value' (defined as TEV) (Freeman, 1993 as in Pascual *et al.*, 2010). Option value, referring to values not yet known, captured or realised by humans, forms an important consideration in the context of PAs. Therefore, for the purposes of this book it has been included in the conceptual presentation of the total socio-economic value of PAs.

References

Bateman, I., Mace, G., Fezzi, C., Atkinson, G. and Turner, K. (2011) 'Economic Analysis for Ecosystem Service Assessments', *Environmental and Resource Economics*, 48: 177–218

Defra – Department for the Environment, Food and Rural Affairs (2007) *Introductory Guidance for Valuing Ecosystem Services*, Defra, London, https://www.gov.uk/government/publications/an-introductory-guide-to-valuing-ecosystem-services, accessed 9 June 2013

Freeman, A. M. (1993) *The Management of Environmental and Resources Values*, Resources for the Future Press, Baltimore, MD

Gantioler, S., Rayment, M., Bassi, S., Kettunen, M., McConville, A., Landgrebe, R., Gerdes, H. and ten Brink, P. (2010) *Costs and Socio-Economic Benefits associated with the*

Natura 2000 Network, final report to the European Commission/DG Environment, Institute for European Environmental Policy/GHK/ Ecologic, Brussels, http://ec.europa.eu/environment/nature/natura2000/financing/docs/natura2000_costs_benefits.pdf, accessed 2 October 2012

Jacobs (2004) *Environment Group Research Report: An Economic Assessment of the Costs and Benefits of Natura 2000 Sites in Scotland*, 2004 Final Report, the Scottish Government, http://www.scotland.gov.uk/Resource/Doc/47251/0014580.pdf, accessed 19 October 2012

Kettunen, M., Berghöfer, A., Brunner, A., Conner, N., Dudley, N., Ervin, J., Gidda, S. B., Mulongoy, K. J., Pabon, L. and Vakrou, A. (2011) 'Recognising the value of protected areas', in P. ten Brink (ed.) *The Economics of Ecosystems and Biodiversity in National and International Policy Making*, Earthscan, London and Washington, DC

Kettunen, M., Vihervaara, P., Kinnunen, S., D'Amato, D., Badura, T., Argimon, M. and ten Brink, P. (2013) *Socio-economic importance of ecosystem services in the Nordic Countries – Synthesis in the context of The Economics of Ecosystems and Biodiversity (TEEB)*, Nordic Council of Ministers, Copenhagen

MA – Millennium Ecosystem Assessment (2005) *Ecosystems and Human Well-being: Biodiversity Synthesis*, World Resources Institute, Washington, DC, http://www.maweb.org/documents/document.354.aspx.pdf, accessed 1 October 2012

Pascual, U., Muradian, R., Brander, L., Gomez-Baggethun, E., Martin-Lopez, B., Verma, M., Armsworth, P., Christie, M., Cornelissen, H., Eppink, F., Farley, J., Loomis, J., Pearson, L., Perrings, C. and Polasky, S. (2010) 'The Economics of Valuing Ecosystem Services and Biodiversity', in P. Kumar (ed.) *The Economics of Ecosystems and Biodiversity (TEEB) Ecological and Economic Foundations*, Earthscan, London and Washington, DC

TEEB (2010) *The Economics of Ecosystems and Biodiversity: Mainstreaming the Economics of Nature: A synthesis of the approach, conclusions and recommendations of TEEB*, http://www.teebweb.org/publications/teeb-study-reports/synthesis, accessed 9 June 2013

ten Brink, P., Badura, T., Bassi, S., Daly, E., Dickie, I., Ding, H., Gantioler, S., Gerdes, H., Kettunen, M., Lago, M., Lang, S., Markandya, A., Nunes, P. A. L. D., Pieterse, M., Rayment, M. and Tinch, R. (2011) *Estimating the Overall Economic Value of the Benefits provided by the Natura 2000 Network*, Final Report to the European Commission, Institute for European Environmental Policy/GHK/Ecologic, Brussels, http://www.ieep.eu/assets/954/Economic_Benefits_of_Natura_2000_Network_Main_Report.pdf, accessed 21 October 2012

UK NEA – National Ecosystem Assessment (2011) *The UK National Ecosystem Assessment: Synthesis of the Key Findings*, UNEP-WCMC, Cambridge, http://uknea.unep-wcmc.org, accessed 18 June 2012

White, S., ten Brink, P. and Simmons, B. (2011) 'Recognising the Value of Biodiversity: New Approaches to Policy Assessment', in P. ten Brink (ed.) *The Economics of Ecosystems and Biodiversity in National and International Policy Making*, Earthscan, London and Washington, DC

PART 2

Practical guidance

STEP I

A scoping assessment of possible benefits

The purpose of Step I of the practical guidance (Chapters 4 and 5) is to offer a simple, yet systematic, framework for carrying out a rapid, first-cut scoping assessment of the possible benefits (ecosystem services and related goods) provided by PAs.

The purpose of this scoping assessment framework is twofold. First, it helps to obtain a general view of the full range of benefits provided by a PA, including an initial assessment of their relative importance (Chapter 4). It also gives guidance on how to derive and interpret these first-stage results, including how to communicate them to relevant stakeholders. Second, the assessment framework helps to identify which benefits could be selected for further in-depth analysis, following advice under Step II of this guide (Chapter 5).

The framework is designed to be applied at the site level and is suitable for identifying benefits provided by both existing PAs and areas considered to be designated as PAs. Advice is also provided for situations where focus on multiple sites is required.

Finally, the advice given under Step I builds mainly on generic considerations and theoretical examples applicable to all types of PAs. A concrete 'real-life' example on the use of the scoping assessment has been provided in Annex 1. Existing examples of the assessed socio-economic value of benefits associated with different types of PAs, including how these values have been estimated, are integrated into Step II and the accompanying Annexes of the guide.

4

SCOPING ASSESSMENT OF BENEFITS PROVIDED BY PROTECTED AREAS

Kettunen, M. and ten Brink, P.

It is recommended that every socio-economic valuation of PAs should start with a short scoping assessment aimed at gaining an overall picture of the full range of benefits (ecosystem services and related goods) provided by a site. While only some of these benefits might be selected for further valuation, a comprehensive scoping assessment helps ensure that all benefits are identified and accounted for. A scoping assessment can help ensure that no benefits are neglected, for example, due to the lack of quantitative or monetary information. It can also help to draw attention to benefits that might become more important in the future, for example due to climate change or resource scarcity.

A scoping assessment also allows the valuation of PAs to be placed in a broader socio-economic context, helping to ensure correct use and targeting of results. It helps to form a picture of the distribution of benefits (and costs) between different stakeholders at local, regional, national and global level. It also allows one to draw preliminary conclusions on the possible distribution of associated costs. As a result, a scoping assessment supports the identification of stakeholders that should be engaged in the process, helping to ensure that the outcomes of valuation successfully contribute to the management and/or establishment of a PA – both within the site and at the broader landscape level.

Finally, a scoping assessment will help to consider the opportunities for and feasibility of further, more detailed (monetary) socio-economic valuation, with due attention to possible risks (Chapters 2 and 5).

4.1 How to carry out a scoping assessment

A scoping assessment can be carried out based on two approaches: 1) a review and analysis of existing information, including literature, expert knowledge, statistical information and plans and maps outlining the land use in the area, and/or 2) stakeholder questionnaires, interviews and workshops. These approaches are not exclusive but can – and arguably should – be used to complement one another. In practice, the approach and/or combination of approaches adopted depends on the time and resources available and the scope of foreseen valuation.

Review and analysis of existing information

A review of existing information is the most basic and least resource-intensive method to identify the benefits of PAs and gather available information on their socio-economic value.

In addition to synthesising existing evidence on site's socio-economic value, a literature review should also integrate information on studies related to the ecological characteristics of the site. The latter helps to understand the biophysical basis underlying the provisioning of ecosystem services and also supports the identification of possible future benefits, such as the ecosystem services likely to gain more importance due to climate change. Finally, efforts should be made to identify the key beneficiaries of different services and related goods (see below).

Relevant sources of information include:

- scientific studies (ecologic and economic), whether in peer-reviewed journals, university studies or NGO assessments
- assessments of PAs status, functioning and trends (local and regional) – including data and statistics from PA management authorities
- data on resource use within and around PAs (e.g. water quality and consumption, use of non-timber forest resources), including maps to clarify the proximity and connectivity of the PA and the surrounding communities
- data on the socio-economic activities within and around PAs (e.g. sources of livelihood, dependency on non-market resources), including PAs possible role in supporting these activities
- data on environmental risks within and around PAs (e.g. frequency and risks of flooding, drought, wild fires and erosion), including risks maps and data on the impacts of incidents
- data on possibilities for substituting ecosystem services and related goods provided by PAs (e.g. vicinity of alternative sources for water, available resources, technology and know-how for substituting services)
- data on possible stakeholders benefiting from ecosystem services and related goods from PAs, including maps outlining the spatial distribution inhabitants, population centres and businesses in the area.

While providing a useful starting point for scoping assessment, the review and analysis of existing information might be limited by the availability of existing information. Consequently, a scoping assessment carried out based on the review of existing information still risks underestimating benefits that have not yet been identified or for which no quantitative/monetary information is available.

Expert/stakeholder questionnaires, interviews and workshops

A targeted questionnaire to and/or interviews of key experts could help to further identify and gain information on benefits associated with PAs. The use of questionnaires and interviews could also be extended to include a number of key stakeholders.

With sufficient time and resources available, dedicated expert and/or stakeholder workshops are considered to be a valuable – and rather cost-effective – way to gain an overall picture of the benefits provided by a PA. In addition to supporting the planned socio-economic valuation, stakeholder workshops can raise awareness of different benefits and also help to openly discuss the distribution of costs related to these benefits. Consequently, they can support the engagement of relevant stakeholders and local communities, enabling their participation in the management and/or establishment of a PA.

While the outcomes of questionnaires, interviews and workshops are greatly dependent on people's personal views and level of appreciation and/or information, they can be useful in complementing the results of a literature review. The outcomes are, however, largely dependent on the attendants and/or target audience. Therefore, due attention should be given to ensure a comprehensive representation of all relevant stakeholders.

Key stakeholders to be addressed include:

- PA managers
- Landowners and land users within PAs
- Representatives of local and/or regional governments
- Representatives of local communities (within/in the vicinity of PAs)
- Representatives of local service providers (water companies, hydroelectricity providers, etc.)
- Representatives of relevant local businesses (nature tourism businesses, water and soft drink producers, etc.)
- Representatives of other PA users
- Relevant experts and researchers
- Local NGOs.

Finally, in several parts of the world household activities, including the related use of resources, are divided between genders. For example, collecting water and fuel wood are predominantly conducted by women, whereas fishing and hunting are carried out by men (e.g. Chong, 2005). In such cases it is important to assure the due participation of both genders in the assessment.

Possible resources to support scoping assessment

A number of existing tools have been identified that can be used to support the development of scoping assessments:

PA benefit assessment tool (PA-BAT): PA-BAT by WWF (Stolton and Dudley, 2009) provides a standardised format for documenting and assessing multiple values (biodiversity and socio-economic) of PAs, including differentiating the varying perceptions of values and benefits amongst stakeholders. It is designed to help to carry out a multi-stakeholder assessment of the importance or potential importance of different values and benefits in PAs. The PA-BAT aims to help PA managers consider all the benefits (current and potential) that could arise from the area they manage, both to aid understanding about the importance of an area and to help ensure management protects this wide variety of values. The PA-BAT is essentially a set of datasheets that collect basic information about the types of benefits (e.g. permissible activities in relation to resource use and ecosystem services); who they are important to; and qualitative information about their level of importance; their relationship to the protected area and the times of year in which they are important. As such the PA-BAT can provide a useful tool for framing and documenting the results of a scoping assessment. Please see Box 15.3 in Chapter 15 for a concrete example of the use of the PA-BAT.

Social assessment of conservation initiatives: a review of rapid methodologies: The report provides a review of the different approaches available to assess the social impacts of PAs and, on this basis, outlines a suggested framework for rapid impacts assessments (Schreckenberg *et al.*, 2010). While aimed at monitoring and measuring PA's social impacts, the document is also a useful resource for planning broader socio-economic assessments, including providing information on the design of methodologies for (rapid) participatory assessments.

Guidance and reference documents for identifying the value of protected areas: A number of existing synthesis and analysis on socio-economic benefits of PAs can be used to further support the identification of PA benefits and costs. Such reference documents include, for example, arguments for protected areas' multiple benefits by Stolton and Dudley (2010), integrating the socio-economic importance of PAs into international, national, regional and local policies by Kettunen *et al.* (2011) and Berghöfer *et al.* (2011), quick guide for assessing protected area benefits by Pabon-Zamora *et al.* (2008) and several synthesis and guidance documents of PA benefits by the Convention on Biological Diversity and Ramsar Convention on Wetlands (e.g. de Groot *et al.*, 2006; Mulongoy and Gidda, 2008; CBD, 2008; Ervin *et al.*, 2010, Russi *et al.*, 2012).

4.2 Identification and rapid assessment of benefits

Identification of benefits

The possible benefits provided by PAs depend on a variety of ecological and socio-economic characteristics ranging from species composition and ecosystem functioning to the distribution of human settlements and sources of livelihood in the region (Chapter 2). While these characteristics are very site-specific, a number of general considerations can help to determine whether a certain benefit (e.g.

ecosystem service and related good) is of relevance in the given context. It is important to ensure that all possible benefits – with or without market value – are accounted for. In addition, the possibility of certain benefits gaining importance in the future should be systematically considered.

Table 4.1 below provides a short checklist supporting the rapid identification of benefits provided by a site. It also considers the likely availability of, or feasibility for obtaining, different types of value estimates (qualitative, quantitative and monetary), supporting the decisions taken on further socio-economic valuation (Chapter 5).

The checklist outlined in Table 4.1 can also be used when identifying benefits provided by an area that is in the process of being designated as a PA. However, in this context, it is also important to reflect on the foreseen impacts of designation on the availability and supply of benefits, i.e. to consider whether and to what extent designation brings added socio-economic value to the area (Section 3.5). The designation of an area as a PA might play an important role in protecting the future availability and sustainable use of ecosystem services and related goods; for example, by preventing the conversion of an area into a built environment, intensive land use in the area, or degradation through pollution. It might also help increase tourism and visitor flows, creating revenue and investment to the area. In such cases, the designation of an area as a PA has clear added socio-economic value.

It is also possible that, while crucial for protecting the biodiversity values of the site, a PA status has no significant foreseen impacts on the availability of ecosystem services and related goods. This is the case, for example, with sites at very remote locations. In such scenarios, biodiversity conservation remains the primary added value of a designation. Despite the lack of added socio-economic value, a PA designation is unlikely to cause identified benefits to disappear. The identification and socio-economic valuation of these sustained benefits remains important, helping to demonstrate that the designation of an area as a PA does not by any means nullify its socio-economic importance.

Identification of beneficiaries

Based on the underlying anthropocentric focus, ecosystem services and related goods provided by PAs are, in effect, defined by people using or enjoying these benefits (Chapter 2). Benefits provided by PAs differ in terms of their distribution; different ecosystem services and related goods benefit different stakeholders. The beneficiaries of a given ecosystem service and/or good can be local (e.g. farmers, foresters, fishers, local communities, local businesses), regional (e.g. consumers of fresh water within a watershed), national (e.g. national food and health sectors) and global (e.g. international tourists, consumers of an internationally marketed product). One benefit can also be of value to several different stakeholders, further increasing its overall importance.

TABLE 4.1 Checklist for preliminary identification of existing and future benefits provided by PAs

Types of benefits	*This benefit is of socio-economic importance at the site when …*	*Values likely to be available/feasible to assess*
Ecosystem services and related goods		
Provisioning (sustainable): resources Food, fibre, energy/fuel, natural medicine, ornaments, biochemicals and pharmaceuticals, etc.	The site is a source for provisioning services and the related goods are sold on the market (e.g. non-timber products, game, and sustainably produced agricultural products).	Monetary
	The site is a source for provisioning services and the related goods are used locally for subsistence (but not sold on market).	Quantitative
	The site could be a (sustainable) source for provisioning services and related goods but this potential is not used at the moment.	Quantitative and qualitative (future potential)
	The site plays a role in food security by hosting a population of local/endemic breed or variety of crop/livestock or it plays an important role in maintaining (genetically) healthy populations of species (e.g. game).	Monetary Quantitative
	The site's role in maintaining food security (as above) is likely to increase in the future due the effects of climate change, habitat degradation, fragmentation, etc.	Quantitative and qualitative (future potential)
Provisioning (sustainable): water	The site is/has in its area an important water reservoir used by local/regional stakeholders (e.g. municipalities, industries).	Monetary
	The site could be/has in its area an important water reservoir but at the moment this potential is not used. The situation could change, however, depending on future developments (e.g. climate change, increase in demand due to demography or economic development).	Quantitative and qualitative (future potential)

Types of benefits	*This benefit is of socio-economic importance at the site when …*	*Values likely to be available/feasible to assess*
Regulating: climate	The site is important for carbon storage, has a high existing stock of carbon stored, and/or has high carbon sequestration capacity and/or plays an important role in regulating local/regional microclimate(s).	Monetary Quantitative
	The site's role in climate regulation could be enhanced via appropriate management (e.g. restoration) and the achievement of its conservation goals.	Quantitative and qualitative (future potential)
Regulating: natural hazards Regulation of floods, wild fires, droughts, avalanches, etc.	The site has an important function regulating/mitigating/preventing natural hazard(s) in the area. This function benefits stakeholders at local, regional and/or national level.	Monetary Quantitative
	The site does not yet play an active role in regulation of natural hazards in the area. The situation could change, however, depending on the future developments. Improvement of site's ecological status via appropriate conservation and management measures (e.g. restoration) might increase site's capacity to provide these benefits. Impacts of climate change, development of local/regional land use, development of businesses and industry sectors, etc. could increase the 'demand' for such regulative functions.	Quantitative and qualitative (future potential)
Regulating: water flow and soil fertility	The site (vegetative cover) plays an important role in regulating the drainage and stabilisation of water flows (non-flood related) in the area, this way helping to maintain the water table at a sustainable level. The regulation of water flow can also help to prevent erosion and maintain soil fertility in the area. Site's vegetative cover can also mitigate against erosion caused by wind, both within the site and in the areas surrounding the site.	Monetary Quantitative
	The site does not yet play an active role in regulating water flow or soil fertility in the area. The situation could change, however, depending on the future developments such as land use and the impacts of climate change.	Quantitative and qualitative (future potential)

Continued

Types of benefits	*This benefit is of socio-economic importance at the site when …*	*Values likely to be available/feasible to assess*
Regulating: purification of air, water and soil	The site has an important function in purifying water/air and neutralising waste in the area (e.g. wetland sites). This function benefits stakeholders at local and/or regional level. (e.g. municipalities' water purification plants, businesses dependent on clean water, etc.)	Monetary Quantitative
	The site does not yet play an active role in purifying water/air and neutralising waste in the area. The situation could change, however, depending on the future developments (see above).	Quantitative and qualitative (future potential)
Regulating: pollination, disease/pest control	The site and/or appropriate management of the site (e.g. extensive use) has an important role in maintaining natural pollinators, seed dispersal agents and/or pest and disease control agents in the area. This function clearly benefits stakeholders at (local) level.	Monetary Quantitative
	The site does not yet play an active role in pollination and disease/pest control in the area. The situation could change, however, depending on the future developments (see above).	Quantitative and qualitative (future potential)
Cultural: recreation and tourism Hiking, camping, nature walks, jogging, skiing, canoeing, rafting, recreational fishing, wildlife watching, etc.	The site is actively used for tourism and recreation and these activities create revenue via entrance fees, accommodation, food, souvenir shops, paid excursions and outdoor activities, permits for sustainable recreational fishing, higher property values around site's recreational possibilities, etc.	Monetary Quantitative
	The site is actively used for tourism and recreation but these activities do not necessarily create revenue (i.e. the access to the site is free and there are no businesses related to visiting the site)	Quantitative Qualitative
	The site is not used for tourism and recreation but there could be potential for these activities in the future (e.g. given due investment and promotion)	Quantitative and qualitative (future potential)

Types of benefits	*This benefit is of socio-economic importance at the site when …*	*Values likely to be available/feasible to assess*
Cultural: broader cultural benefits Aesthetic, inspiration, spiritual/religious, education and research, cultural identity, cultural heritage, etc.	The site has broader cultural importance and this plays a role in creating actual revenue, e.g. attracting investments and financial support to culture and affecting property prices.	Monetary Quantitative
	The site is an important outdoor destination and plays a role in maintaining people's mental and physical health.	Quantitative Monetary
	The site has broader cultural importance but this value cannot be connected to revenue.	Quantitative and qualitative
	The site has low cultural importance - its profile could be raised.	Quantitative and qualitative (future potential)
Broader local/regional socio-economic benefits Jobs and employment, regional investment, etc.	The site helps to create jobs and employment and increase local/regional revenue streams via tourism, recreation and/or management activities.	Monetary Quantitative
	Site can be linked with investments to local communities/region, e.g. research, 'green' business development.	
Long-term resilience/insurance	The site hosts a high number of species and habitats, resulting in a high rate of ecological interactions at the site. The complexity and diversity of these ecological interactions is considered fundamental for the characteristic functioning of site's ecosystem(s), thus securing the provisioning of its services.	Qualitative
	The site plays an important role at broader local and regional level by supporting healthy and diverse populations of species (e.g. genetic diversity) and/or maintaining ecological processes such as water cycle.	

Source: own presentation, based on Kettunen *et al.* (2009)

While ecosystem services and related goods are derived from (or produced by) nature, several stakeholders play a role in ensuring their continued existence and supply. For example, local communities, landowners and land users within a PA help to maintain different regulating services by adopting sustainable management practices. It is important to recognise who is responsible for supporting the maintenance of benefits: this helps to identify and assess related costs, allowing a more accurate estimation of PAs' net benefits (see 4.3 below).

Consequently, systematic identification of stakeholders benefiting from and helping to maintain the 'flow' of ecosystem services and goods from a PA, as along the lines illustrated in Table 4.2, is an integral part of valuation. It helps to gain a more comprehensive overview of the socio-economic importance of a PA, including demonstrating this importance to a wider audience. The identification of beneficiaries also supports reflection and application of the valuation results in a broader context (Chapter 16). It also helps to select the most appropriate methods for valuation: values to individuals are amenable to monetary valuation whereas qualitative assessments are required to capture the values for broader society (Chapter 3).

4.3 Joint consideration of different benefits

A scoping assessment is unlikely to produce information that would allow detailed quantitative and monetary comparisons of identified benefits. However, a preliminary co-consideration of different benefits might be feasible based on the analysis of existing literature and, if possible, supported by the insights of experts and stakeholders.

A joint preliminary assessment of benefits could be carried out in a qualitative manner or it could be developed into a simple quantitative assessment. In the latter case, the identified benefits can be indicatively ranked to reflect their estimated and/or perceived socio-economic importance (e.g. on a scale of 1-5) with the results visualised via a basic radar chart (Table 4.3). Such indicative quantification of benefits could be implemented at any level, ranging from individual expert(s) to ranking by a range of stakeholders in the context of a dedicated workshop (Section 4.1 above). A concrete 'real-life' example of such a quantitative assessment has been provided in Annex 1. The approach has also been piloted in a number of case studies (Bulgalho, 2009; Cruz and Benedicto; 2009; Kazakova and Pop, 2009; Pabian and Jaroszewicz, 2009; Tinch, 2009).

A number of considerations should be kept in mind when quantifying the socio-economic importance of benefits. The overall socio-economic importance of ecosystem services and related goods is related not only to the 'volume' of benefit (e.g. amount of water purified by a PA) but also the number of beneficiaries depending on it (e.g. number of people using water from a PA). In general, the socio-economic significance of a given benefit increases with the number of people benefiting from it. The degree of dependency of stakeholders on different benefits should also be considered. For example, resources from a PA (non-timber forest

TABLE 4.2 Hypothetical examples of identifying beneficiaries directly or indirectly benefiting from PAs

Who helps to maintain benefits?	Who benefits?	What is the scale of benefit?
Example 1: PA as a source for pharmaceutical innovations and products		
PA staff/managers *Communities within PA* *Landowners* *Farmers* *Foresters*	*Individuals*: access to new/improved drugs and biochemical products	Local Regional National Global
	Local communities: payments/compensation for bioprospecting or shared revenue from a new commercially used compound	Local
	Businesses and industries: access to genetic resources and revenue from sold drug/biochemical/ cosmetic products	Local Regional National Global
	Local, regional and national governments: reduced health costs due to improved drugs, increased food security due to improved biochemicals	Local Regional National
	Global community: improved global health and related welfare	Global

Who helps to maintain benefits?	Who benefits?	What is the scale of benefit?
Example 2: PA playing an important role in regulating flow and quality of water in the area		
PA staff/managers *Communities within PA* *Landowners* *Farmers* *Foresters*	*Individuals:* access to clean water, reduced risk of flooding/water scarcity	Local Regional
	Local communities: access to clean water, reduced risk of flooding/water scarcity, possible payments to support land use practices that maintain PA's natural capacity to regulate water	Local Regional
	Businesses and industries: secured regular water supply for industry/business needs, reduced costs of water purification, reduced costs of water regulation for electricity and water companies	Local Regional National Global
	Local, regional and national governments: increased water security, decreased risk of flooding and droughts, reduced health costs related to lack of water and/or low water quality	Local Regional National
	Global community: improved global water and health security and (a contribution to) the mitigation of water-related global environmental risks	Global

Source: own presentation, based on Kettunen *et al.* (2009)

products, fish, game, etc.) might have no market value but they are an important source of livelihood for local communities. Similarly, a water reservoir supported by a PA might be the sole source of fresh water within the area. Substituting such benefits (e.g. finding alternative sources of livelihood) is likely to be difficult, further increasing their relative socio-economic importance. The income levels of relevant stakeholders may also affect the estimated (monetary) value of a PA: higher income households are likely to have a higher ability and/or willingness to pay for maintaining a nature-based clean water supply or access to recreational activities. Finally, the greater the value of assets at risk (houses, roads, factories), the greater the economic value of contributions of a PA to the mitigation of natural hazards.

Given the distribution of different benefits between different stakeholders and scales (above), it is advisable to estimate the importance of benefits both at local/regional and national/global levels (Table 4.3 and Figure 4.1). For example, benefits related to the regulation of climate change 'add up' at national and global level, whereas their direct local and regional importance is relatively low. There might also be a need to distinguish between the current and future socio-economic importance of benefits. This is the case when the assessment of benefits is related to the establishment of a PA or anticipated changes in land use or management practices within a PA.

A preliminary co-consideration and comparison of benefits – although primarily based on subjective interpretations – helps to identify benefits for further socio-economic valuation. It is also likely to increase stakeholder understanding and awareness, facilitating discussions in the context of valuation.

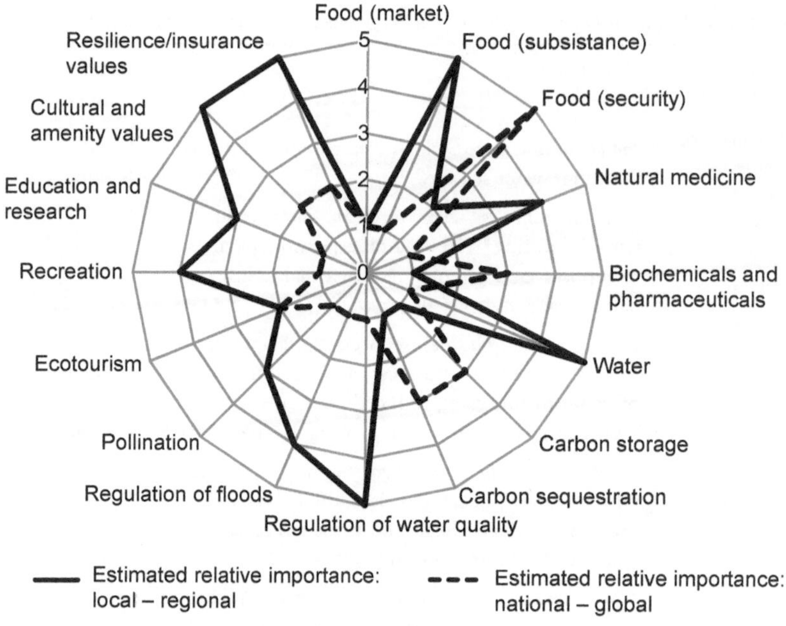

FIGURE 4.1 Illustration of the benefits provided by a hypothetical PA in Table 4.3

Source: own presentation, based on Kettunen *et al.* (2009)

TABLE 4.3 Hypothetical example of a preliminary comparison of the benefits provided by a PA by using the following quantitative scale: 0 = service is not relevant at the site, 1 = service is of very limited significance, 2 = service is of limited significance 3 = service is of moderate significance, 4 = service is of high significance and 5 = service is of very high significance. A concrete 'real-life' example of such a quantitative assessment has been provided in Annex 1

Identified benefit	*Estimated importance (on a scale of 1–5):*		*Description*
	Local – regional	*National – global*	
Food (market)	1	1	Several local communities depend on the resources provided by PA for subsistence and food security. In addition, the site hosts a wild variety of fruit/crop supporting genetic diversity and contributing to global food security (genetic 'insurance'). Currently no products are sold in the local, national or global markets.
Food (subsistence)	5	1	
Food (security)	2	5	
Natural medicine	4	1	The site is an important source of natural remedies for local communities. It has also been identified as a location for possible future bioprospecting by pharmaceutical industry.
Biochemicals and pharmaceuticals	1	3	
Water	5	1	Forest and wetland ecosystems protected by the site play an important role in the hydrology of the area, e.g. providing water to a lake situated within the site.
C storage	1	3	Forest ecosystems protected by the site function both as carbon storage and sink. However, site's contribution to the national and global carbon balance is limited by its relatively small size. In general, benefits related to the regulation of climate change mainly "add up" at national and global level, whereas their direct local and regional importance is relatively low.
C sequestration	1	3	
Regulation of water quality	5	1	Forest and wetland ecosystems protected by the site function as a natural filter around the lake situated within the site. Water from the lake is used by several local communities around the site.
Regulation of floods	4	1	Forest and wetland ecosystems protected by the site regulate water flows during the rainy season, stabilising the water table in the lake and protecting shoreline inhabitants from flooding.
Pollination	3	1	The site hosts several insect/animal pollinated plants (dependent on wild pollinators) used by local communities, e.g. a variety of wild banana.

Continued

Identified benefit	*Estimated importance (on a scale of 1–5):*		*Description*
	Local – regional	*National – global*	
Ecotourism	2	2	The site receives some international backpackers; however, it is not (yet) a popular tourist destination.
Recreation	4	1	On the other hand, the site is frequently visited for recreational purposes by urban residents a driving distance from the area.
Education and research	3	1	The site is also visited regularly by local school groups; however, it does not currently host any research facilities or activities (e.g. international research projects).
Cultural and amenity values	5	2	Finally, the site protects one of the few patches of semi-natural forests in the area and it is therefore considered to be of high cultural importance. However, such forest habitats are still relatively common on a national scale.
Resilience/ insurance values	5	2	The site has high local and regional insurance value, given its role in maintaining several regulating services and water and food security in the area. The latter is also of national and even global importance, given its crop diversity.

Source: own presentation

4.4 Consideration of benefits in the context of multiple protected areas

A scoping assessment can also be carried out at the level of multiple sites. Such assessment helps to obtain an overall picture of PA benefits at a broader level, including considering possible changes in their availability in a regional and/or national context (Box 4.1). A multi-site scoping assessment also provides a means to compare similarities and differences in the provisioning of benefits across sites, this way guiding the focus of more detailed assessments. For example, focusing resources on estimating the socio-economic value of benefits that are common to several sites can create synergies, allow the use of estimated benefits across sites (see Chapter 3) and results in possible costs savings. Scoping assessments at the level of multiple sites can also help to determine synergies across sites which can be exploited later both in management planning and capacity-building (Chapter 15). For example, by identifying all the PAs important for maintaining water supply within a region allows effective sharing of best practices and targeted capacity-building.

BOX 4.1

Ecosystem services provided by Important Bird Areas in Nepal

In Nepal, the national network of 27 Important Bird Areas (IBAs) covers over 2.6 million ha of the country. Overall, more than 80 per cent of the IBA network overlaps with the PA network with a total of 14 IBAs (2 million ha) situated entirely within PAs. Almost all of the remaining 0.6 million ha of IBAs are Community Forests managed by Forest User Groups.

The results of a scoping assessment, based on an expert consultation workshop, indicated that the network of IBAs delivers a number of important ecosystem services. For example, water provision is an important benefit associated with all the IBAs, identified as the most important service for 90 per cent of them. Harvested wild goods, including food, fibre, natural medicines and fuel wood for cooking and heating, recreation/tourism and global climate regulation were also recorded as being of high importance across the majority of sites. Local climate and air quality regulation, erosion control, spiritual/religious experience, reducing the impact of weather events and improving water quality were delivered by most sites, though these benefits were generally considered of lesser importance.

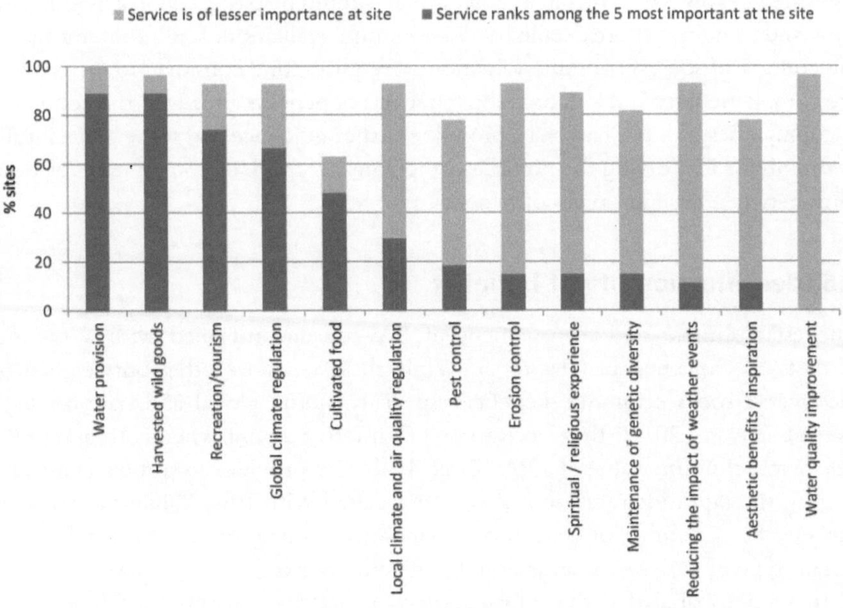

FIGURE 4.2 Quantitative assessment of ecosystem services provided by Important Bird Areas in Nepal, based on expert information

The results of the rapid assessment also indicated that the supply of several ecosystem services and related goods was decreasing. The availability of harvested wild goods (food, fibre, fuel wood and natural medicines) was estimated to decrease at over half of the IBAs where they were identified to be important. Overall, only around one third of sites showed an increase in the availability of these goods. There was a reported increase in the supply of cultivated foods at some sites although this is likely to be at the expense of other services. Ecosystems capacity to supply water and support the regulation of global climate was believed to be stable or increasing across most of the sites. Finally, recreation and tourism was reportedly on the increase at 95 per cent of IBAs subject to these activities.

Source: BCN and DNPWC (2012)

As explained in Chapter 3, in most of the cases the socio-economic value of benefits provided by an individual PA will be complementary, or additive, to the value of benefits from other sites. However, the promotion of and investment in ecosystem services at a site may lead to competitive impacts and the decline of the socio-economic value of the same service on other sites. Quantitative and monetary analysis of multiple sites is generally considered beyond the means of a scoping assessment. However, a qualitative assessment should help to provide a useful overview and to inform more detailed socio-economic valuation. It is important that, if the context of socio-economic valuation so requires, due consideration is given to the complementary and/or competitive aspects of benefits already in the context of a scoping assessment. Chapter 13 provides further guidance and some general rules of thumb for identifying the possible complementary and/or competitive relationships between different types of benefits.

4.5 Identification of net benefits

The establishment and management of PAs are accompanied with a number of costs, as explained in Chapter 3. While these costs are often outweighed by the overall socio-economic benefits, especially at the global and national level (Kettunen *et al.*, 2011), they need to be taken into account when considering the net socio-economic value of a PA. Table 4.4 below provides a short checklist supporting the rapid identification of costs associated with PAs. While the costs are site-specific, a number of general considerations can help to determine whether a certain type of cost is of relevance in the given context.

It is improbable that the information provided by a scoping assessment will allow for any detailed quantitative and monetary comparisons of benefits and costs. Consequently, the consideration of site's net benefits is likely to build on a qualitative analysis of the information available. Nevertheless, such qualitative analysis

TABLE 4.4 Checklist for preliminary identification of existing and future costs related to PAs

Types of costs	*This cost is relevant at the site when …*	*Values likely to be available/feasible to assess*
Management costs	Ongoing management activities are required to maintain or reach site's conservation objectives. For example, extensive grazing or mowing is needed to maintain open grasslands.	Monetary
	Ongoing on-site staff are required to ensure the implementation of site's conservation objectives (e.g. visitor centre staff, guides, guards to prevent poaching).	Monetary Quantitative
	One-off investment (e.g. restoration, land purchase) is required to reach site's conservation objectives (e.g. reforestation, restoration of site's natural hydrology, restocking of depleted populations of species).	Monetary
Human-wildlife conflict	There is a risk of damage caused by human-wildlife conflicts in the area. These include, for example, predation of cattle and other domesticated animals, crop raiding by birds, ungulates, primates, etc., property damage by large mammals, and raiding of nets and fish farms by sea birds.	Monetary (costs of damage) Quantitative
	There is a need for infrastructure to prevent human-wildlife conflicts (e.g. preventive fences).	Monetary (costs of investment)
	Payments to stakeholders are established to compensate the damage caused by wildlife.	Monetary (costs of payments)
Displacement	The management and/or establishment of PAs requires displacement of people from the area to reduce and/or remove human pressures on biodiversity.	Quantitative Qualitative
	Payments are established to compensate the costs of displacement.	Monetary
	The management and/or establishment of PAs results in forgone profits to individuals and businesses (e.g. due to limited access).	Monetary Quantitative
	The management and/or establishment of PAs results in forgone tax and/or land tenure revenues and forgone revenues from state-owned enterprises (e.g. forestry enterprises).	Monetary Quantitative

Continued

Types of costs	*This cost is relevant at the site when …*	*Values likely to be available/feasible to assess*
Opportunity costs	The management and/or establishment of PAs results in foreone benefits to local communities (e.g. revenue from renting the area for businesses).	Quantitative Qualitative
	The management and/or establishment of PAs reduces or prevents access to economically or culturally important resources. Such restrictions include, for example, loss of agricultural land, loss of grazing pasture, loss of access to fishing resources and restrictions to use/hunt endangered species of economic or cultural importance, loss of access to resources of high day-to-day subsistence value.	Monetary Quantitative
	Payments to stakeholders are established to compensate for the loss of access.	Monetary

Source: own presentation, based on Kettunen *et al.* (2011)

TABLE 4.5 Hypothetical example of a preliminary qualitative assessment supporting the consideration of net benefits of a PA on a scale of low – moderate – significant – highly significant, based on the consideration of benefits and costs accruing at different levels (e.g. the estimated scale of their value). Benefits with no or only a marginal socio-economic importance at the site are excluded. The example builds on the previous example in Table 4.3 in Section 4.3

	Identified benefits	*Estimated scale of socio-economic value*	*Identified costs*	*Estimated scale of socio-economic costs*
Local – regional	• Food (subsistence)	Highly significant	• Management costs: visitor infrastructure	Moderate
	• Natural medicines	Highly significant	• Management costs: guards	Significant
	• Water	Significant	• Human-wildlife conflict: costs of damage due to crop raiding	Moderate
	• Regulation of water quality	Highly significant		
	• Regulation of floods	Significant	• Opportunity costs/loss of access to natural resources: some restrictions related to grazing and hunting within PA	Moderate
	• Pollination	Moderate		
	• Recreation	Significant		
	• Education and research	Moderate		
	• Cultural and amenity values	Highly significant		
	• Resilience/insurance values	Highly significant		
National – global	• Food (security)	Highly significant	• Management costs: 'restocking' of endangered bird species	Significant
	• Biochemicals and pharmaceuticals	Moderate		
	• Mitigation of climate change (C storage and C sequestration)	Moderate	• Management costs: land purchase	Moderate
			• Opportunity costs: compensation for loss of access to grazing land	Moderate
	• Global nature-based tourism	Moderate		

Source: own presentation

helps to obtain a preliminary overview of the possible scale and extent of the site's net socio-economic value, including the distribution of benefits and costs between different stakeholders.

A qualitative assessment of net benefits, along the lines of Table 4.3, supports the overall conclusions of a scoping assessment. It also allows preliminary assumptions to be made over the general outcomes of a detailed socio-economic valuation. For example, the illustrative example outlined in Table 4.5 suggests that the range of benefits provided by a PA is higher than the associated costs. Similarly, the estimated scale of the socio-economic value of these benefits seems to outweigh the scale of costs. This indicates that there are likely to be net benefits both at local/regional and national/global levels.

References

Berghöfer, A., Dudley, N. and Förster, J. (2011) 'Ecosystem Services and Protected Areas' in H. Wittmer and H. Gundimeda (eds) The Economics of Ecosystems and Biodiversity in Local and Regional Policy and Management, EarthScan/Routledge, London and Washington, DC

Bugalho, M. (2009) *Assessing Socio-economic Benefits of Natura 2000 – a Case Study on the ecosystem service provided by the Natural Park of Vale do Guadiana (Portugal)*, report for the European Commission, http://ec.europa.eu/environment/nature/natura2000/financing/docs/guadiana_case_study.pdf, accessed 9 June 2013

CBD (2008) *Protected Areas in Today's World: Their Values and Benefits for the Welfare of the Planet*, CBD Technical Series No. 36, Secretariat of the Convention on Biological Diversity, Montreal

Chong, J. (2005) *Valuing the Role of Wetlands in Livelihoods: Constraints and Opportunities for Community Fisheries and Wetland Management in Stoeng Treng Ramsar Site, Cambodia*, IUCN Water, Nature and Economics Technical Paper No. 3, IUCN — The World Conservation Union, Ecosystems and Livelihoods Group Asia, Colombo

Cruz, A. de la and Benedicto, J. (2009) *Assessing Socio-economic Benefits of Natura 2000 – a Case Study on the Ecosystem Service Provided by SPA Pico da Vara/Ribeira do Guilherme*, report for the European Commission, http://ec.europa.eu/environment/nature/natura2000/financing/docs/azores_case_study.pdf, accessed 12 November 2012

Ervin, J., Mulongoy, K. J., Lawrence, K., Game, E., Sheppard, D., Bridgewater, P., Bennett, G., Gidda, S. B. and Bos, P. (2010) *Making Protected Areas Relevant: A guide to integrating protected areas into wider landscapes, seascapes and sectoral plans and strategies*, CBD Technical Series No. 44, Secretariat of the Convention on Biological Diversity, Montreal

Groot, R. S. de, Stuip, M. A. M., Finlayson, C. M. and Davidson, N. (2006) *Valuing wetlands: guidance for valuing the benefits derived from wetland ecosystem services*, Ramsar Technical Report No. 3/CBD Technical Series No. 27, Ramsar Convention Secretariat, Gland, Switzerland and Secretariat of the Convention on Biological Diversity, Montreal, Canada

Kazakova, Y. and Pop, E. (2009) *Assessing Socio-economic Benefits of Natura 2000 – a Case Study on the ecosystem services provided by Oaş-Gutâi Plateau and Igniş site, Maramures, Romania*, report for the European Commission, http://ec.europa.eu/environment/nature/natura2000/financing/docs/maramures_case_study.pdf, accessed 16 January 2013

Kettunen, M., Bassi, S., Gantioler, S. and ten Brink, P. (2009) *Assessing Socio-economic Benefits of Natura 2000 – a Toolkit for Practitioners* (September 2009 Edition), Institute for European

Environmental Policy (IEEP), Brussels, Belgium, http://ec.europa.eu/environment/nature/natura2000/financing/docs/benefits_toolkit.pdf, accessed 16 January 2013

Kettunen, M., Berghöfer, A., Brunner, A., Conner, N., Dudley, N., Ervin, J., Gidda, S. B., Mulongoy, K. J., Pabon, L. and Vakrou, A. (2011) 'Recognising the value of protected areas', in P. ten Brink (ed.) *The Economics of Ecosystems and Biodiversity in National and International Policy Making*, Earthscan, London and Washington, DC

Mulongoy, K. J. and Gidda, S. B. (2008) *The Value of Nature: Ecological, Economic, Cultural and Social Benefits of Protected Areas*, Secretariat of the Convention on Biological Diversity, Montreal, http://www.cbd.int/doc/publications/cbd-value-nature-en.pdf, accessed 12 December 2012

Pabian, O. and Jaroszewicz, B. (2009) *Assessing Socio-economic Benefits of Natura 2000 – a Case Study on the ecosystem service provided by Białowieża Forest, Poland*, report for the European Commission, http://ec.europa.eu/environment/nature/natura2000/financing/docs/bialowiaza_case_study.pdf, accessed 9 June 2013

Pabon-Zamora, L., Bezaury, J., Leon, F., Gill, L., Stolton, S., Grover, A., Mitchell, S. and Dudley, N. (2008) *Nature's Value: Assessing protected area benefits*, Quick Guide Series, The Nature Conservancy, Arlington, VA

Russi, D., ten Brink, P., Farmer, A., Badura, T., Coates, D., Förster, J., Kumar, R. and Davidson, N. (2012) The Economics of Ecosystems and Biodiversity for Water and Wetlands, Ramsar Convention Secretariat, Gland, Switzerland

Schreckenberg, K., Camargo, I., Withnall, K., Corrigan, C., Franks, P., Roe, D., Scherl, L. M. and Richardson, V. (2010) *Social Assessment of Conservation Initiatives: A review of rapid methodologies*, Natural Resource Issues (No 22), IIED, London, http://www.careclimatechange.org/files/reports/SAPA_IIED_Social_Assessment.pdf, accessed 1 December 2012

Stolton, S. and Dudley, N. (2009) *The Protected Areas Benefits Assessment Tool*, WWF, Gland, Switzerland, http://wwf.panda.org/what_we_do/how_we_work/conservation/forests/publications/?174401/PABAT, accessed 1 December 2012

Stolton, S. and Dudley, N. (eds) (2010) *Arguments for Protected Areas – Multiple Benefits for Conservation and Use*, EarthScan, London

Tinch, R. (2009) *Assessing Socio-economic Benefits of Natura 2000 – a Case Study on the ecosystem service provided by the Sustainable Catchment Management Programme*, report for the European Commission, http://ec.europa.eu/environment/nature/natura2000/financing/docs/scamp_case_study.pdf, accessed 9 June 2013

5

DECIDING WHICH BENEFITS TO ANALYSE IN MORE DETAIL

Kettunen, M. and ten Brink, P.

Assessing the socio-economic value of PA benefits is generally best done at site level. Many benefits are site-specific: both the range of ecosystem services and related goods provided by a PA and the costs associated with PA management can vary significantly between sites. The context in which valuation decisions are made is also specific to a certain PA or a case. Consequently, it is important that the scope of and methods for detailed socio-economic valuation, i.e. what benefits to focus on and how, are determined based on due consideration of the context and site-specific characteristics – including the results of a scoping assessment. A number of key questions supporting these decisions are listed below.

5.1 Identifying the most important benefits

A scoping assessment provides a useful tool for identifying the most important benefits provided by a PA, currently and possibly in the future. As a rule of thumb, the benefits highlighted by this assessment (e.g. benefits that receive the highest ranking under Section 4.3) are recommended to be considered for a more detailed socio-economic valuation. Further consideration should, however, be given to ensure that the outcomes of a scoping assessment, especially when based on stakeholder workshops, accurately capture all relevant site-specific characteristics. For example, there might be a need to focus the socio-economic valuation also on a number of benefits the importance of which is considered to increase in the future, such as benefits whose significance increases due to climate change.

The scoping assessment also provides an initial indication of what type of value estimates (monetary, quantitative and qualitative, including spatial insights) might be available and/or possible to obtain. This information is likely to affect the scope and focus of further socio-economic valuation, especially when the timescale and resources for valuation are limited.

In general, monetary estimates are commonly available and/or easier to attain for benefits with existing markets. Such benefits include provisioning services and related goods and benefits related to tourism and recreation. In addition, non-market monetary information on the costs of losing certain benefits often exists and can be used to illustrate the socio-economic importance of regulating services. On the other hand, monetary estimates based on the valuation of revealed and stated preferences (Chapter 3) are less common. Such estimates are based on extensive stakeholder questionnaires and are therefore very resource and time intensive to obtain. Finally, basic information on PA-related employment and investment is also often available. However, more detailed studies, such as the assessment of a PA's role in generating indirect employment and income in the area, require in-depth analysis and related expertise.

The resources and timescale available are the basic practical constraints limiting the scope of socio-economic valuation. Table 5.1 below summarises the general time and resource-related characteristics of different valuation methods. It also gives an indication of the level of expertise required to carry out valuation, indicating whether there is a need for contracting external assistance to support the valuation.

Finally, information on the available and/or feasibly obtainable data will also help to estimate the likely robustness and reliability of the final results, with possible further implications on the choice of scope. See Chapter 14 for further consideration of aspects affecting the robustness of the valuation results.

5.2 Identifying the purpose of socio-economic valuation

The most appropriate approach and methods for socio-economic valuation depends on the decision-making context (Chapter 1). If the purpose of an assessment is an academic one, then the rigours of academia need to be followed with the results presented with due consideration of the academic norms (presentation of statistical measures of confidence). If the purpose is a tool to guide practical decision-making then the key driver for determining the scope of the assessment is whether its foreseen results will be fit for purpose for the given decision-making context.

Quantitative and monetary assessments are often required when it is necessary to evaluate or demonstrate the benefits of a PA in relation to alternative land use practices, for example when designating new PAs or considering changes in land use practices within existing PAs (Hansjürgens *et al.*, 2011; ten Brink, 2011). In these cases, the valuation approach and methods need to be able to provide concrete quantitative and/or monetary evidence of the trade-offs between different benefits resulting from foreseen land use and management choices. Finally, the development of concrete mechanisms and incentives (such as payments for ecosystem services (PES) or creating new markets for sustainably produced goods and ecosystem services) requires the application of – and expertise in – detailed market-based methods (see also Chapter 15).

In general, in situations where stakeholder consensus on the value of benefits is strong and the science is clear, it may be relatively straightforward to demonstrate

TABLE 5.1 Resource and time implications of and expertise required for different socio-economic valuation methods

Type of method	*Name of method*	*Resource implications*	*Required level of knowledge*
Monetary (market)	Market value/price (adjusted)	Non-adjusted estimates are relatively easy to obtain; Adjusted estimates require additional time and resources	The development of adjusted estimates requires basic understanding of economic valuation
	Production function	Resource-intensive and time-consuming	Basic understanding of economic valuation required
Monetary (non-market)	Costs (damage) avoided	Relatively easy to obtain	Basic understanding of economic valuation required
	Revealed preferences	Resource-intensive Time-consuming	Detailed understanding of economic valuation methods required
	Stated preferences		
Non-monetary (qualitative and quantitative)	Non-monetary indicators of economic value	Relatively easy to obtain	Basic understanding of socio-economic valuation required
	Health-based valuation	Some (indirect) indicators of value easy to obtain; Dedicated studies require time and resources	Basic understanding of socio-economic valuation required; Dedicated studies require specific scientific knowledge
	Biophysical indicators for status of ecosystem services	As above	As above
	Focus group/ survey/ review based valuations	Resource-intensive and time-consuming	Detailed understanding of participatory approaches and valuation methods required

Type of method	*Name of method*	*Resource implications*	*Required level of knowledge*
Revenue-based (monetary and quantitative)	Jobs and employment	Direct employment and revenue easy to estimate; Estimating indirect impacts on employment and revenue requires time and dedicated resources (e.g. modelling)	Estimating indirect impacts on employment and revenue requires specific economic expertise
	Socio-economic investments (public and private)	Relatively easy to obtain	Basic understanding of economic valuation required

values in monetary terms and capture them in markets (TEEB, 2010). This applies most obviously to benefits with existing markets, but can equally be applied to the amount of carbon storage or the supply of clean water. On the other hand, in more complex situations involving multiple benefits and/or plurality of ethical or cultural convictions, monetary valuations may be less reliable or unsuitable. In such cases, quantitative and qualitative assessments of value may be more appropriate. In practice, it is foreseen that an assessment of the benefits and costs related to a PA will need to integrate qualitative, quantitative and monetary insights across multiple criteria.

Finally, socio-economic valuation of the benefits provided by a PA (monetary valuation in particular) might pose a number of risks as outlined in Section 2.6. These risks should be carefully considered before proceeding to carry out detailed socio-economic valuations of PAs.

5.3 Possible resources to support detailed socio-economic assessment

A number of existing tools have been identified that, based on the experience of the authors of this guidance, can be used to support the development of detailed socio-economic assessments of PAs:

Toolkit for ecosystem services site-based assessments (TESSA) – Version 1 developed by Cambridge Conservation Initiative (CCI and BirdLife International, 2011; Peh *et al.*, 2012) provides guidance on how the assessment of ecosystem services can help to determine whether conservation delivers greater benefits than conversion of land to other uses hence more reliably informing decision-making. While not specifically aimed at PAs, the tool can be of help when considering how to systematically integrate changes in land use into the socio-economic assessment of PAs and it can be particularly useful in framing socio-economic assessments of planned PAs. The CCI Toolkit for Ecosystem Services Site-based Assessment (TESSA) guides non-specialist users through methods for the rapid site-scale assessment of ecosystem services. It enables quantification of selected services and related goods under different states of the site (after conversion to agriculture or restoration of degraded forest), resulting in an analysis of net consequences of ecosystem change. The toolkit relies on a participatory approach enabling understanding of the local context to be included in the analysis while at the same time increasing ownership, legitimacy, understanding and updates of results. The tool has been piloted at a wide range of sites around the globe, including Important Bird Areas in Nepal (BCN and DNPWC, 2012), and across a range of habitats including temperate and tropical forests and wetlands providing concrete examples of how rapid assessments can provide useful information and support more detailed assessments of ecosystem services. See Box 4.1 above for an example of the use of this tool.

Guidance and reference documents for assessing the value of ecosystem services: A number of documents aimed at providing practical and user-oriented guidance on how to assess the economic value of ecosystem services are available. While not all are specifically focused on PAs, these documents provide useful information on the broader assessments frameworks and methods available (as outlined in Chapter 3). Such useful guidance and/or reference documents include, for example, an environmental economics toolkit for valuing small islands by van Beukering *et al.* (2007), introductory guide to valuing ecosystem services by Defra (2007) and toolkit for assessing the (gross) benefits of EU's Natura 2000 PA sites (Kettunen *et al.*, 2009). In addition, an assessment of the costs and benefits of the European Union (EU) -wide PA network (Natura 2000 network) by ten Brink *et al.*, (2011) provides information on the possibilities to aggregate benefits and costs at the level of multiple sites and/or PA networks (see Chapters 12 and 13)

InVEST tool – Integrated Valuation of Ecosystem Services and Trade-offs: InVEST is a freely downloadable suite of ecosystem service models being developed by the Natural Capital Project to provide information on where ecosystem services are provided and how they will be affected by alternative plans and policies (Natural Capital Project, 2012). InVEST can help to acquire information on the biophysical flow of ecosystem services on a landscape level, also in the context of PAs. InVEST is designed to help local, regional and national decision-makers incorporate ecosystem services into a range of policy and planning contexts for terrestrial, freshwater and marine ecosystems, including spatial planning, strategic environmental assessments and environmental impact assessments. InVEST models are based on production function (see Chapter 3) that define how an ecosystem's structure and function affects the flows and values of ecosystem services. The models account for both the biophysical availability of services and the location and activities of people who benefit from those services. Since data is often scarce, the models are still relatively simple and best suited for identifying general patterns in the provision and value of ecosystem services. With validation, the models can also provide useful estimates of the magnitude and value of services provided. Practical examples on the use of InVEST can be found in Chapter 15 (Box 15.5) and in Barano *et al.* (2012), Goldman *et al.* (2010) Goldstein *et al.* (2010) and Yukuan *et al.* (2010).

References

Barano, T., McKenzie, E., Bhagabati, N., Conte, M., Ennaanay, D., Hadian, O., Olwero, N., Tallis, H., Wolny, S. and Ng, G. (2010) *Integrating Ecosystem Services into Spatial Planning in Sumatra, Indonesia,* TEEBcase, http://www.teebweb.org/wp-content/uploads/2013/01/Integrating-ecosystem-services-into-spatial-planning-in-Sumatra-Indonesia.pdf, accessed 9 June 2013

BCN and DNPWC (2012) *Conserving biodiversity and delivering ecosystem services at Important Bird Areas in Nepal,* Bird Conservation Nepal (BNC), Department of National Parks and Wildlife Conservation and BirdLife International (DNPWC), Kathmandu and Cambridge, http://www.conservation.cam.ac.uk/sites/default/files/file-attachments/NepalEcosystems ServicesLowRes.pdf, accessed 1 December 2012

CCI and BirdLife International (2011) *Measuring and monitoring ecosystem services at the site scale.* Cambridge Conservation Initiative and BirdLife International, Cambridge, http://

www.conservation.cam.ac.uk/sites/default/files/file-attachments/Ecosystem%20 Services%20Toolkit%20booklet_0.pdf, accessed 1 December 2012

Defra (2007) *An introductory guide to valuing ecosystem services*, Department for Environment, Food and Rural Affairs, London, http://www.defra.gov.uk/publications/files/pb12852-eco-valuing-071205.pdf, accessed 9 June 2013

Goldman, R. L., Benitez, S., Calvache, A., Davidson, S., Ennaanay, D., McKenzie, E. and Tallis, H. (2010) *Water Funds for conservation of ecosystem services in watersheds, Colombia*, TEEBcase, http://www.teebweb.org/wp-content/uploads/2013/01/Water-Funds-for-conservation-of-ecosystem-services-in-watersheds-Colombia.pdf?, accessed 9 June 2013

Goldstein, J. H., Caldarone, G., Colvin, C., Ka`eo Duarte, T., Ennaanay, D., Fronda, K., Hannahs, N., McKenzie, E., Mendoza, G., Smith, K., Wolny, S., Woodside, U. and Daily, G. C. (2010) *Integrating ecosystem services into land-use planning in Hawai'i, USA*, TEEBcase http://www.teebweb.org/wp-content/uploads/2013/01/Integrating-ecosystem-services-into-land-use-planning-in-Hawaii-USA..pdf, accessed 13 June 2013

Hansjürgens, B., Kettunen, M., Schröter-Schlaack, C., White, S. and Wittmer, H. (2011) 'Framework and Guiding Principles for the Policy Response', in in ten Brink (ed.) *The Economics of Ecosystems and Biodiversity in National and International Policy Making*, Earthscan, London and Washington, DC

Kettunen, M., Bassi, S., Gantioler, S. and ten Brink, P. (2009) *Assessing Socio-economic Benefits of Natura 2000 – a Toolkit for Practitioners* (September 2009 Edition), Institute for European Environmental Policy (IEEP), Brussels, Belgium, http://ec.europa.eu/environment/nature/natura2000/financing/docs/benefits_toolkit.pdf, accessed 16 January 2013

Natural Capital Project (2012) *InVEST tool*, http://www.naturalcapitalproject.org, accessed 1 December 2012

Peh, K. S.-H., Balmford, A. P., Bradbury, R. B., Brown, C., Butchart, S. H. M., Hughes, F. M. R., Stattersfield, A. J., Thomas, D. H. L., Walpole, M. and Birch, J. C. (2012) *Toolkit for ecosystem service site-based assessments (TESSA)*, Version 1, Cambridge Conservation Initiative, Cambridge

ten Brink, P. (ed.) (2011) 'Rewarding Benefits Through Payments and Markets', in ten Brink (ed.) *The Economics of Ecosystems and Biodiversity in National and International Policy Making*, Earthscan, London and Washington, DC

ten Brink, P., Badura, T., Bassi, S., Daly, E., Dickie, I., Ding, H., Gantioler, S., Gerdes, H., Hart, K., Kettunen, M., Lago, M., Lang, S., Markandya, A., Mazza, L., Nunes, P. A. L. D., Pieterse, M., Rayment, M. and Tinch, R. (2011) *Estimating the Overall Economic Value of the Benefits provided by the Natura 2000 Network*, Report to the European Commission, Institute for European Environmental Policy/GHK/Ecologic, Brussels, http://ec.europa.eu/environment/nature/natura2000/financing/docs/Economic_Benefits_of_Natura_2000.pdf, accessed 12 December 2012

van Beukering, P., Brander, L., Tompkins, E. and McKenzie, E. (2007) *Valuing the Environment in Small Islands – An Environmental Economics Toolkit*, Joint Nature Conservation Committee – JNCC, UK, http://jncc.defra.gov.uk/page-4065#download, accessed 1 December 2012

Yukuan, W., Bin, F., Colvin, C., Ennaanay, D., McKenzie, E. and Min, C. (2010) *Mapping conservation areas for ecosystem services in land-use planning, China*, TEEBcase, http://www.teebweb.org/wp-content/uploads/2013/01/Mapping-conservation-areas-for-ecosystem-services-in-land-use-planning-China.pdf, accessed 9 June 2013

ANNEX 1

Scoping assessment of benefits provided by protected areas – an example of application

D'Amato, D., Kettunen, M., de la Cruz, A., Benedicto Royuela, J. and Gil, A.

The case study of Pico da Vara/Ribeira do Guilherme (Azores, Portugal) by Cruz and Benedicto (2009) demonstrates how a scoping assessment of benefits provided by a protected area (PA) (outlined in Chapters 4 and 5) can support the process of socio-economic valuation. The scoping assessment can be used as a tool to identify relevant benefits, estimate the scale of their socio-economic importance and to recognise the stakeholders involved in maintaining the benefits. Furthermore, it can be used to prioritise benefits for further, more detailed socio-economic valuation; e.g. quantitative and monetary assessments (Step II of the guidance, Chapters 6-11).

1 Case study site: Pico da Vara/Ribeira do Guilherme (Azores, Portugal)

Covering an area of over 6,000 ha, Pico da Vara/Ribeira do Guilherme PA comprises the last remaining large area of the altitudinal laurel forest on São Miguel Island (Azores archipelago, Portugal). The area hosts a number of species that are endemic to the Azores and/or rare at European level. Consequently, the site is also a part of the European-wide network of protected areas called the Natura 2000 Network. Pico da Vara/Ribeira do Guilherme is a good example of the socio-economic importance of PAs: located in the area of Nordeste and Povoação on the Island of São Miguel – two of the most rural communities in the area with 6,700 inhabitants (1,300 families) and 5,200 inhabitants (1,000 families), respectively – it provides important benefits to the local, national and global communities – as discussed further below.

2 Results of the scoping assessment

The ecosystem services and related goods supplied by Pico da Vara/Ribeira do Guilherme were identified by experts based on a literature review and ranked according to their importance on a scale of from one to five (1 = low importance, 5 = high importance, see Chapter 4), taking into consideration both the current and potential future socio-economic relevance (e.g. potentially increasing economic importance).

Provisioning services and related goods: The identified provisioning services included the provisioning of fresh water, food, and ornamental and genetic resources (Figure A1.1). Despite high rainfall, seasonal water scarcity is a critical issue in the Azores and for this reason Pico da Vara's role in supporting water supply was considered of highest importance. Laurel forest vegetation cover allows for replenishment of the aquifer, important for ensuring the availability of water in dry periods and for reducing the cost of water supply for public suppliers and for the local community. As for food and ornamental resources; while both of these ecosystem services have been ranked relatively low, they were considered to have a greater importance to the local economy in the future (Table 1). Harvesting the Azorean Blueberry (*Vaccinium cylindraceum*), used traditionally for cooking, might represent a potential economic benefit if adequately promoted by local institutions. Similarly, there was an increasing demand for ornamental plants from the Azorean laurel forest. Finally, the area hosts some endemic and rare species of potential use in agriculture or bioprospecting. These benefits are considered of high importance (option value) to the global community (Table A1.1).

Regulating services and related goods: Regulating services – including climate change regulation, water regulation and water purification – are all considered of high importance. Considering the global benefit provided by carbon storage in the PA climate change regulation was ranked as of high importance. These benefits are also likely to increase by further improving the conservation practices on the site. Water regulation by vegetation cover prevents and mitigates floods and landslide events – a severe threat on São Miguel Island. This is one of the most important services provided by the PA, both for reducing costs of damages borne by public entities and for the safety of the community. Water purification is also an essential service for the communities of Nordeste and Povoação. Currently, several areas within the PA are heavily overgrazed by domestic animals causing the degradation of water quality. This leads to high family expenses for bottled water (average costs €46.5/family/year). Improved management of the PA is foreseen to lead to a reduction of cattle in the area, enabling the recovery of endemic ecosystems and their natural ability to purify water. This will further contribute to the reduction of the consumption of bottled water.

Cultural services and related goods: Cultural services were considered of medium to high importance as they produce benefits both on the local and global scale. National and international tourism is an increasing component of the local economy in the Azores (Table A1.1). Facilities and infrastructure for recreational activities (such as birdwatching, walking and horse riding) can be improved to result in higher socio-economic value. Consequently, the importance of recreation and tourism in the PA was ranked rather high. The attractive landscape of Pico da Vara/Ribeira do Guilherme represents an amenity value for locals depending on its conservation to encourage sustainable use of the area. The PA has also a remarkable scientific and educational value, attracting the national and international research community's attention for its peculiar ecological characteristics.

Supporting services: Identified supporting services comprise of the maintenance of biological diversity and species' lifecycles. The ecological interactions of the laurel forest within the PA underpin a number of other services. In particular, high insect diversity plays a key role in maintaining the production and ecological balance within the ecosystem via pollination and control of species dynamics.

Wider socio-economic benefits: Pico da Vara/Ribeira do Guilherme creates direct and indirect employment for the local and national population. In particular, a range of jobs (PA managers and project staff members) has been created by the investment of the European Union (EU) to the conservation of the site in the context of the EU Natura 2000 Network of protected areas.

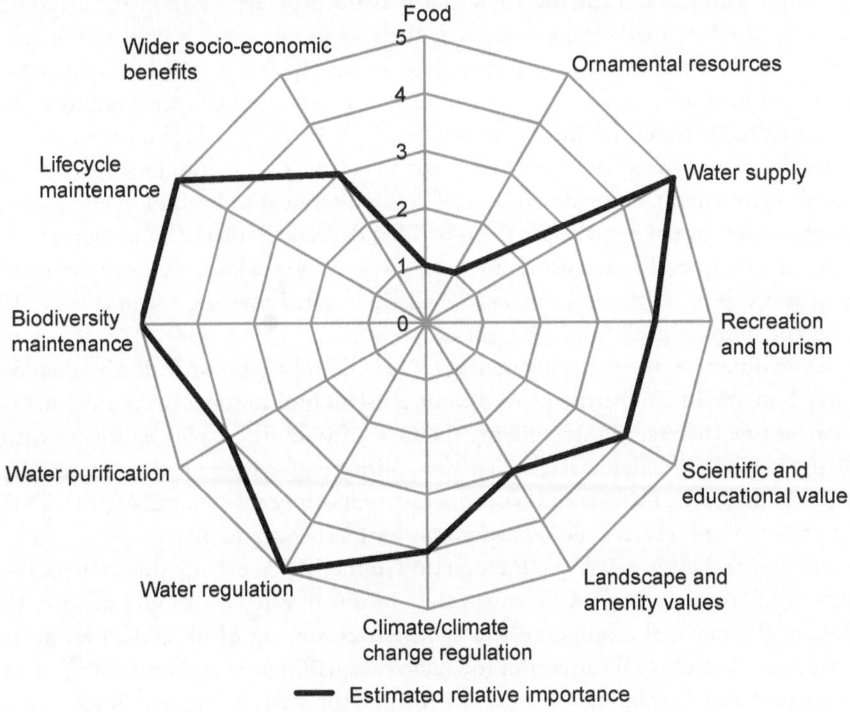

FIGURE A1.1 Socio-economic benefits provided by PA of Pico da Vara/Ribeira do Guilherme, ranked according to their perceived importance on a scale of 1-5 (1 = low importance, 5 = high importance, see Chapter 4)

3 Detailed socio-economic assessment of identified key benefits

The benefits identified as the most important through the scoping assessment were then selected for detailed socio-economic valuation (Table A1.1). The socio-economic value of water regulation (mitigation of flooding and landslides) and water supply were estimated as €20,000,000/year (potential avoided costs of damage) and €604,997/year (market price of water). Water purification was estimated to be €110,556/year based on the replacement cost method. Tourism, and scientific and education were an estimated value of €76,500/year and €2,958/year, respectively, based on market price and travel cost methods. Landscape value, amenity value, and the existence value of endemic species had an estimated overall worth of €3,000,000 for the Povoação area, based on the assessed willingness to pay to preserve these ecological features.

4 Conclusions

A scoping assessment can be an accurate and cost-effective tool to provide preliminary information about the most relevant benefits provided by the PA, including choosing the best methodology to assess their socio-economic value. In the case of Pico da Vara/Ribeira do Guilherme, water supply and purification and natural hazard mitigation via water regulation (flood and landslides prevention) were estimated to be the most important benefits by helping to reduce costs to public entities and improving the well-being and safety of local communities. The PA was also estimated to provide a range of cultural services to both local and global stakeholders. Several benefits provided by Pico da Vara/Ribeira do Guilherme PA were difficult to evaluate in economic terms (e.g. carbon storage and sequestration) or quantify (e.g. supporting services). This does not, however, lessen the socio-economic importance of these benefits.

As an outcome, the socio-economic assessment of benefits provided by Pico da Vara/ Ribeira do Guilherme proved to be a useful tool to support the PA's management (see Chapter 15). Identifying the socio-economic benefits associated with the site helped to (re)define its management priorities, focusing more on preserving functionality of the habitats and consequently supporting the sustainable provisioning of ecosystem services and related goods for local communities.

Finally, as the Pico da Vara/Ribeira do Guilherme case study shows, a socio-economic assessment of PA benefits rarely results in one economic estimate. In most of the cases, the outcome is a synthesis consisting of different indicators of value and levels of information (qualitative, quantitative and monetary). The case study also highlights the need to further the basic ecological knowledge of local ecosystems in order to improve the accuracy of assessments (e.g. to avoid the risks associated with extrapolating data from other case studies, see Chapter 3).

TABLE A1.1 Estimated quantitative and monetary values of the socio-economic benefits provided by Pico da Vara/Ribeira do Guilherme PA

Identified benefit	*Scale*	*Estimated value*	*Method/approach used*	*Comments*
Food	Local	Scoping assessment did not indicate high current socio-economic importance	–	Potential economic importance in the future if measures taken to promote the goods
Ornamental resources	Local	Scoping assessment did not indicate high current socio-economic importance	–	Potential economic importance in the future if measures taken to promote the goods
Water supply	Local	€604,997/year	Market price method (water price per sector x water consumption per sector)	The importance of this service is foreseen to increase in the future due to increasing water scarcity
Recreation and tourism	Local/ international	€76,500/year	Travel cost method (journey price + hotel price) + tourist expenditure	Potential increasing economic importance in the future if further measures taken to promote recreation and tourism
Educational value	Local	10 school visits/year; €2,958 (total)/year	Travel cost method (journey price)	Potential increasing economic importance in the future if further measures taken to promote the service
Scientific value	National/ international	8 graduates/year; 0.25 PhD/year; 1 MSc thesis/year; 10 scientific papers written	Quantitative assessment based on available statistics	
Artistic value	Local	1 photograph exhibition; 2 local painters	Quantitative assessment based on available statistics	Potential increasing economic importance in the future if further measures taken to promote the service

Continued

Identified benefit	*Scale*	*Estimated value*	*Method/approach used*	*Comments*
Landscape/amenity value and existence value of endemic species	Local/ global	€500 to €800 per person for a total of €3,000,000 for the Povoação region alone	Willingness to pay (surveys to local inhabitants) based on hypothetical taxes, free donations and/or entrance fees	The estimate is conservative, since it represents only part of the total population
Carbon storage	Global	465,364 tC/year (vegetation) + 223,667 tC/year (peat)	Quantitative assessment of carbon content stored by the PA estimated by data transfer from studies on similar vegetation	Monetary valuation of annual carbon sequestration not possible due to absence of information on annual carbon balance
Water regulation (flood and landslides prevention)	Local	Costs of damage €20,000,000 in 1997	Avoided costs of damage, avoided economic losses, avoided costs of repair	This estimate based on the costs of previous event therefore it should be considered only as a proxy/ indicative of the scale of value.
Water purification	Local	€46.5 family/year for a total of €110,556 /year	Replacement cost (bottled water expenditure per family times number of families)	The importance of this service is foreseen to increase in the future due to increasing water scarcity
Biodiversity and lifecycle maintenance, evolutionary processes	Global	Difficult to quantify/monetise	–	Important option value, including for the global community
Other socio-economic values	Local/ national	Direct employment of 21.6 full-time jobs (mean) between 2003 and 2008; Indirect employment of 4–6 full-time jobs/year; €371,360/year spent by the PA; €49,699/year spent by the PA employees	Available statistics on direct/ indirect employment and employees spending	Dependent on external funding to support the management of the site (e.g. funding from the European Union in the context of the European-wide Natura 2000 Network)

Source: Cruz and Benedicto (2009); Cruz *et al.* (2011)

References

For more information, including more detailed use of different valuation methods, please see the complete study:

Cruz, A. de la and Benedicto, J. (2009) *Assessing Socio-economic Benefits of Natura 2000 – a Case Study on the Ecosystem Services Provided by SPA Pico da Vara/Ribeira do Guilherme*, report for the European Commission, http://ec.europa.eu/environment/nature/natura2000/financing/docs/azores_case_study.pdf, accessed 12 November 2012

Cruz, A. de la, Benedicto, J. and Gil, A. (2011) 'Socio-economic Benefits of Natura 2000 in Azores Islands – a Case Study approach on Ecosystem Services provided by a Special Protected Area', *Journal of Coastal Research* 64: 1955–1959, http://www.ics2011.pl/artic/SP64_1955-1959_A.%20Cruz.pdf, accessed 12 November 2012

STEP II

Detailed methodological guidance to estimate socio-economic value of benefits

The purpose of Step II of this practical guidance (Chapters 6 to 13) is to provide specific information and advice on methodologies available for estimating the socio-economic value of PA-related benefits. The introduced methodologies are used to derive estimates of the qualitative, quantitative and monetary value of different services. Step II focuses specifically on value estimates considered feasible to obtain by practitioners, such as site managers and other interested stakeholders.

Chapters 6 to 8 focus on assessing the value of provisioning, regulating and cultural services and related goods whereas Chapter 9 draws attention to the value of the biodiversity and ecological processes underlying these services. Chapter 10 provides advice on how to reflect and estimate the wider social and economic benefits of PAs (employment and poverty reduction). Chapter 11 informs the reader on how to assess the costs related to the establishment and management of PAs while Chapters 12 and 13 provide guidance on how to assess the net benefits of PAs at the level of individual and multiple sites. Finally, Step II is supported by detailed walk-through examples of some key ecosystem services and related goods (Annexes 1 to 6).

6

PROVISIONING SERVICES AND RELATED GOODS

Kettunen, M. and D'Amato, D.

6.1 Overview of benefits

All over the world, PAs supply a wide range of provisioning services and related goods benefiting the local communities, the wider society and the economy. Benefits of provisioning services and related goods refer to resources provided and/or maintained by PAs (Chapter 2). These include, for example, the provisioning of food, water, raw material and genetic resources (Table 6.1). The benefits can occur – or be enjoyed – either within or outside PAs. In the latter case, a PA plays an important role in contributing to the provisioning of resources even though the actual harvesting of these resources takes place outside the site. This could be the case, for example, when PAs function as an important refuge or breeding place for fish or game species.

PA managers and/or other land users (farmers, foresters, landowners, local communities, etc.) play a key role in maintaining provisioning services, either by implementing direct conservation measures or by securing the sustainable management of sites through extensive, environmentally sound land use practices. Both of these activities can in turn help to maintain long-term provisioning of resources within or outside the site. The possible beneficiaries of biodiversity resources are various, ranging from local to regional and national to international, depending on the markets for different goods. For example, several goods might be produced and/or caught for subsistence use whereas some goods might be traded at different markets. The examples include, for example, fish caught at marine protected areas and sold at local market, hunting within PAs to supply game meat for restaurants within the region and harvesting of medicinal plants for the international pharmaceutical industry (Table 6.2).

TABLE 6.1 Provisioning services and related goods

Ecosystem service	*Examples of related goods*
Provisioning of food	Cereals, livestock, fish (marine and inland), game, fruit, berries, mushrooms, edible wild plants, spices, etc.
Provisioning of clean water	Drinking water, water for household use, water for industrial use, etc.
Provisioning of raw material	Timber, wood, fuel wood, energy crops, fodder and forage for livestock (hay, straw, lichen), material for construction (rattan, reed), material for clothing (cotton, linen, wool), cork, etc.
Provisioning of medical resources and biochemicals	Medicinal products (natural), cosmetics, biochemicals and pharmaceuticals, non-medicinal biochemicals, pharmaceutical models and test organisms, etc.
Provisioning of ornamental resources	Traditional handicraft, fashion and jewellery, natural dyes and colourants/dye plants, decorative wild plants, decorative animals (aquarium fish from wild), etc.
Provisioning of genetic resources	Wild and/or traditional variants and races for crop, fruit and livestock, genetic resources for pharmaceutical industry, etc.

Source: own presentation, based on MA (2005), de Groot *et al.* (2010), UK NEA (2011), Kettunen *et al.* (2013)

It is important to note that in several cases, especially in the context of agro-ecosystems, the provisioning of resources is heavily influenced by human activities and inputs, including establishing dedicated areas for cultivation and increasing productivity though irrigation and the use of fertilisers (Elmqvist *et al.*, 2010; EASAC, 2009). Consequently, while ecosystem processes such as pollination, pest regulation and regulation of erosion are necessary for the production of resources (see Chapters 7 and 9) the estimated socio-economic value of these resources (market value of cereals, livestock and cultivated timber, etc.) does not accurately reflect the ecosystem's natural capacity to provide provisioning services. PAs are rarely subject to intensive land use practices, such as large scale agriculture and forestry, and therefore the level of human input in resources they provide is more limited that in agro-ecosystems. For some PAs a certain level of human activity might even be required to maintain biodiversity, for example when conserving extensively farmed pastures and grasslands. Furthermore, the majority of provisioning services and related goods provided by PAs, such as marine and inland fisheries, game, wild timber, genetic resources and water, are almost entirely dependent on the ecosystems' own ability to maintain the service. Therefore, resources originating from sustainably managed PAs can, to a large extent, be attributed to naturally functioning ecosystems.

As for water provisioning, the total quantity of water available is determined by both biotic (vegetation cover, species composition, etc.) and abiotic (topography, temperature, etc.) attributes of PAs. It is generally not possible to specify what proportion of fresh water supply is related to the ecosystem's biotic components and therefore the total quantity of water provided by PAs is often used as an estimate for the socio-economic value of this service. It is useful to realise that, given the above, these estimates might somewhat overestimate the biotic role of nature (habitat structure and species composition) in providing the service.

6.2 Socio-economic importance of benefits

As outlined in Chapter 3, the benefits provided by PAs can be broadly divided into two different types of value: economic and welfare (health and social). This section provides an overview and a range of concrete examples (Table 6.2) of different socio-economic values provided by provisioning services and related goods. Please note also that benefits related to the employment linked with provisioning services and related goods are covered in more detail in Chapter 10.

Economic value

Provisioning services translate directly into good(s) valued by people. In addition, many of these goods are traded on markets determining their (current) economic value. Consequently, it is often feasible to estimate the economic value of PAs' provisioning services and related goods. These services and goods support (local) economies by creating opportunities for different sectors including, for example, fisheries, farming, timber harvesting and extraction of building materials. They also help to create cost-effective solutions for water and energy supply. For example, the Grampians National Park in Australia contributes to the regional economy by supporting honey production for a market value of A$156,000 (US$120,000) per year (WCPA, 1998). Yearly local incomes generated from commercial fishing to local economies in a Southern European Marine Protect Area were estimated to amount to €720,000 (Roncin *et al.*, 2008). More examples of economic value are provided in Table 6.2 below.

Provisioning services from PAs do not always have to be traded in markets to be economically valuable. PAs are an important source of livelihood for local communities, functioning as a source of food, timber, fibres, medicinal plants, building materials and fuel. For example, in Bhitarkanika Conservation Area the resources extracted from the forest contribute more than 14.5 per cent of the total income of the households on average, and 30 per cent for the poor people living nearest to the forest (Hussain and Badola, 2010). Allpahuayo-Mishana National Reserve in Peru supports agriculture, forest products extraction, hunting and fishing by locals, for an overall economic value estimated to be US$105,320 per year (Pyhälä *et al.*, 2006). Finally, water estimated to be worth NZ$12 million (US$9.7 million) per year is delivered by the Te Papanui Conservation Park in New Zealand for irrigating the surrounding rural areas (New Zealand Department of Conservation, 2006).

TABLE 6.2 Examples of economic and welfare values of PAs related to provisioning services and related goods

Protected area	*Benefits (ecosystem services and related goods)*	*Estimated socio-economic value of benefits*	*Method*	*Source*
New Zealand, Te Papanui Conservation Park (grassland)	Water supply	The Te Papanui Conservation Park in New Zealand provides the Otago region with water that would cost NZ$136 million (US$110 million) to service from elsewhere. The 22,000 ha tussock supplies water flows valued at NZ$31 million (US$25 million) for hydroelectricity, NZ$93 million (US$75 million) for urban water supply and NZ$12 million (US$9.7 million) for irrigating.	Replacement cost method	New Zealand Department of Conservation, 2006
US, Eastern Sierra watershed (several)	Water supply	The Eastern Sierra watershed, comprising 891,000 ha, supplies the city of Los Angeles with water. The U.S. Forest Service, U.S. Bureau of Land Management and the Los Angeles Department of Water and Power (LADWP) own 98 per cent of this land and many of California's national forests were created specifically to safeguard and preserve water supplies. LA County's 9.8 million residents live in a semi arid climate and depend on unreliable water sources and a 2 million increase in population is expected by 2030. Due to increases in demand at an unsustainable rate, Los Angeles was rated the city with the greatest chance of running out of water in a 2010 environmental report by 24/7 Wall Street and sustainability group, Ceres. The Safe, Clean and Reliable Drinking Water Supply Act of 2012, on the California ballot in November, would borrow US$11.4 billion in bonds to upgrade California's water system including water storage, water recycling facilities, levees and *ecosystem restoration* if approved by voters.	Replacement cost method	Dudley and Stolton, 2003 Town Hall Los Angeles, 2012

Protected area	*Benefits (ecosystem services and related goods)*	*Estimated socio-economic value of benefits*	*Method*	*Source*
Kenya, Aberdare National Park (forest)	Food Medicine Timber Fuel	In Aberdare National Park (Kenya) the willingness to pay method was used to estimate the value of products collected by locals around the park area. Fuel wood, grazing, charcoal, building materials, honey, medicines, wild foods, hunting, and timber were valued at 18,408 KSH/year or US$306 per household. Methodologically, the general unwillingness to reveal values of products collected was overcome by showing pictures of different types of provisioning services, instead of direct questioning.	Willingness to pay	Emerton, 1996
India, Bhitarkanika Conservation Area (forest, coastal, marine)	Fish Honey Fuel wood Timber Thatching materials	The value of forestry and fishery products harvested in the Bhitarkanika Conservation Area is US$107.3 per household/year. Local consumptive use of mangrove forest products was investigated using structured questionnaires and prices were extrapolated from local markets. Inshore and offshore fishery amounts to 98 kg per household/year of catch equivalent to US$68.6 per household/year. Honey is extracted for 525 kg per household/year, providing US$3.6 per household/year. Fuel wood extracted amounts to 312 kg per household/year, equivalent to US$12.5 per household/year. Timber extraction is 370 kg per household/year, corresponding to US$20.1 per household/year. Thatching materials are collected for 49 kg per household/year, with an economic value of US$2.5 per household/year. The resources extracted from the forest contribute more than 14.5 per cent to the total income of the households on average, and 30 per cent for the poor people living nearest to the forest.	Market price method, information obtained via local surveys	Hussain and Badola, 2010

Continued

Protected area	*Benefits (ecosystem services and related goods)*	*Estimated socio-economic value of benefits*	*Method*	*Source*
Tanzania, Gonabis Game Reserve (several)	Wild meat	In the villages surrounding Gonabis Game Reserve wild products originating from the reserve are sold at a lower price than their actual economic value. Consequently, their use value among locals was estimated through willingness to pay, rather than market price value. As the local communities are comprised mainly of subsistence farmers, the number of working days was considered as an appropriate payment mechanism instead of direct cash payments. The mean value of willingness to pay per household was 14 labour days for wild meat. Given the wage rate in the region (TSh1500/day or US$0.9/day) it is estimated that each household is willing to pay TSh21,000 (US$1.3) in return for such products.	Willingness to pay	Baldus, 2005
Costa Rica, Ostional Wildlife Refuge (marine, coastal)	Egg collection	Ostional Wildlife Refuge in Costa Rica is one of the world's most important nesting sites for Olive Ridley Sea Turtles (*Lepidochelys olivacea*). 235 people are involved in legal egg collection. The eggs are later sold to consumers at markets with around 66 intermediaries selling eggs throughout Costa Rica. In 2001, 4,137,000 Olive Ridley eggs were collected for sale at Ostional. Considering an average price of US$0.12-US$0.37 per egg, gross revenue benefitting villagers and intermediaries is estimated at US$1,011,615 per year. This market also contributed to equitability in terms of creating employment opportunities for both genders.	Market price method	Troëng and Drews 2004

Protected area	*Benefits (ecosystem services and related goods)*	*Estimated socio-economic value of benefits*	*Method*	*Source*
Australia, Grampians National Park (several)	Honey Water supply	The Grampians National Park in Australia contributes to the regional economy with provisioning services such as honey and water production. The Park has 68 sites for beehives, producing honey with a market value of A$156,000 (US$120,000) per year. *Water catchment areas within* the *Grampians National Park* are the *source* of *water supply* for *over 45 towns* and *7,000 rural properties* in *Victoria*. Water data estimates correspond to likely water prices inferred from known water prices.	Market price method	WCPA, 1998 Grampians National Park, undated
Peru, Allpahuayo-Mishana National Reserve (forest, cultivated)	Agriculture Timber Non-timber forest products Game Fish	The economic importance of provisioning services and related goods the Allpahuayo-Mishana National Reserve in Peru was assessed by carrying out a household survey (n = 94) in the six local communities living within the reserve. The survey was designed to reveal how the resources originating from the reserve were used, including extraction for both subsistence and market purposes. While most of the resources originating from the reserve are still used for subsistence (i.e. non-market), the reserve is also a source of income with revenues from agriculture (US$37,380/year), timber (US$55,030/year), non-timber forest products (NTFPs) (US$11,060/year), hunting (US$9,070/year) and fishing (US$3,840/year). The overall economic market-based value of the provisioning services and related goods was estimated to be US$105,320/year. With the estimated mean per capita income being less than one dollar a day (US$261 per year), these values present an important economic buffer for local communities.	Market price method	Pyhälä *et al.*, 2006

Continued

Protected area	*Benefits (ecosystem services and related goods)*	*Estimated socio-economic value of benefits*	*Method*	*Source*
Netherlands, Hoge Veluwe Forest (forest)	Timber Game	In Hoge Veluwe Forest (Netherlands) 2600 m3 of saw wood, 1200 m3 of medium quality wood and 8000 m3 of low quality wood are harvested every year. Considering a price of €90/m3, €40/m3 and €9/m3 for high, medium and low quality wood, the total value for timber is €354,000/year. Operative costs for timber harvesting are borne by the logging company, while the park receives a net benefit for selling the standing stocks. In addition, the park is obliged to hand over the hunted meat, which is sold by the park with zero hunting costs. Meat production and the recreational value from hunting in the park are therefore distinct and potential double counting between these services is avoided. A total of 300 animals are shot each year, yielding around 10,000 kg of gross animal weight, of which around half is sold as meat. The net revenue for the park related to the sale of hunted animals is around €50,000/year.	Market price method	Hein, 2011
Cambodia, Ream National Park (forest, marine, freshwater, cultivated)	Firewood Medicinal plants Fencing materials Fruits Fish Crops Farming animals	A household survey examined local communities' dependency on forest, marine/freshwater and agricultural/farming products in Ream National Park (Cambodia). Some 44,788 m3 of firewood and 47,240 kg of medicinal plants is used yearly by local people, who also collect fencing materials, berries, mangoes, rambutans and other wild fruits. Given the market prices of these forest products, the economic value is estimated at US$170,615/year. Fish quantities range from a few kg/month for subsistence fishermen to several tons/month for big operators. An estimated 3,174 tons/year of catch is harvested in the park for a market value estimated at US $2.0 million/year. Finally, 28 per cent of families rely on farming. Using average market prices, crops in the park and farming animals are valued, respectively, at US$ 316,594 and US$182,317 per year. All market values are adjusted for collection costs.	Market price method	De Lopez, 2003

Protected area	*Benefits (ecosystem services and related goods)*	*Estimated socio-economic value of benefits*	*Method*	*Source*
Southern Europe, Twelve marine Protected Areas (marine, coastal)	Fish	Twelve marine Protected areas (MPAs) in southern Europe were examined to determine the magnitude of benefits provided to local communities. Information concerning the number of commercial fishing boats inside the MPAs comes from marine authorities. Annual turnover and manpower (e.g. a boat with a crew of three persons, including skipper, deriving half of its annual landings from the fishing zone of the MPA is taken to account for 1.5 jobs) within the fishing zone is derived from fishers' answers to a field survey. Yearly local incomes generated by the use of MPA ecosystem services amount to approximately €720,000 per MPA in the commercial fishing industry. Commercial fishing generates approximately 54 jobs per MPA on average. See Chapter 10 for a broader discussion on the impacts of PAs on employment.	Market price method, based on existing statistics and information collected via surveys	Roncin *et al.*, 2008

Welfare value

Human psycho-physical well-being (health value) is greatly dependent on living conditions and surrounding environment. Provisioning services fulfil the most basic human needs, including food, water and sanitation, and energy and therefore their availability greatly affects individual and societal quality of life. The benefits to human health can rarely be captured in monetary terms, but their importance, directly related to the number of beneficiaries and the magnitude of improved living conditions, is consciously perceived and formally recognised by individuals and society (UN, 1974; MDG, 2000).

The role of PAs in supplying drinking water highlights their importance in maintaining human health. A third of the world's largest cities obtain a significant proportion of their drinking water directly from PAs (Dudley and Stolton, 2003). However, many still face a crisis of water shortage with significant deficiencies in sanitation and an estimated 780 million people still lacking safe drinking water in 2010 (WHO and UNICEF 2012). This leads to a burden in terms of short- and long-term health impacts including infant mortality, with effects on the ability to work, industrial productivity and on already overstretched health services. It affects especially the poorest members of society, unable to afford sterilised or bottled water. Similar problems affect the rural poor as well, but in a rapidly urbanising world the scale of the problem facing cities is particularly acute and likely to increase in the future (Dudley and Stolton, 2003).

Provisioning services and related goods from PAs can translate into concrete improvements in communities' welfare and sustainable development, such as direct or indirect employment, creation of infrastructures, support of urban, industrial or rural development and social participation (see also Chapters 8 and 10). In Southern Europe, for example, commercial fishing generates on average approximately 54 jobs (fisheries and scuba-diving tourism) per marine protected area (Roncin *et al.*, 2008). Provisioning services may also be related to social and cultural aspects of human society and therefore closely linked to cultural welfare (Chapter 8).

Similarly to the health-related values above, social values are hard to define in monetary terms and other means are needed to capture their socio-economic importance. An autochthonous wine variety and feral cattle breeds are farmed within Doñana protected area (Spain) (Martín-López *et al.*, 2011 in Annex 2). Despite their modest economic value, maintenance of these land races is kept alive as they have a cultural and historical meaning. These kind of social values are not definable in terms of individual needs, rather they relate to a broader dimension – in the above mentioned case being social and cultural identity – which is explained by the emergent properties of human society as a system.

Finally, it is to be noted that the economic and welfare values of provisioning services are often interlinked. For example, the Makira protected area in Madagascar comprises a high number of medicinal plants that provide economic revenue for the locals and have a potential to be developed into biomedicines with an estimated to be in the order of US$ billions value for pharmaceutical companies (Golden *et al.*, 2012 in Annex 2). In addition to these economic benefits, the extraction and/or commercialisation of these medicinal plants contributed to improved health of individuals, resulting in significant social benefits for the global community.

6.3 Estimating the value of benefits

This section summarises the most commonly used methods for estimating the socio-economic value of provisioning services and related goods, supported by a number of concrete examples in Table 6.2 above and Boxes 6.1-6.3 below. More information, examples and step-by-step guidance on the application of key methods is provided in Annex 2. An introduction to the different valuation methods is available in Chapter 3.

Monetary: market values

Provisioning services result in tangible goods that are frequently traded in established markets. The market price method uses the observed and/or estimated price of goods bought and sold in commercial markets as an indicator of their economic value (see Boxes 6.1-6.2 and Table 6.2 for examples from Thailand, Peru, Cambodia and the Netherlands). While market prices offer a relative reliable and feasible way for demonstrating the socio-economic value of provisioning services a number of considerations should be kept in mind when using the method.

First, it is important to realise that market prices do not by default reflect how sustainably the resource has been extracted and/or cultivated. Therefore, when using market prices to indicate the socio-economic value of PAs great care should be taken to ensure that the extraction/production of these resources is carried out in a sustainable manner, for example within a framework of certification schemes (see Box 6.2).

Market prices might be affected by subsidies and taxes. For example, local governments tend to keep water prices low in order to ensure its affordability to users. Consequently, the observed price of water does not necessarily reflect its true availability and value. Therefore, market prices should preferably be adjusted for any distortions before being used as indicators for the value of ecosystem services and related goods (see Box A2.5 in Annex 2 for an example from Madagascar).

As explained in Chapter 3, the market prices of goods often also integrate a range of human inputs. Therefore, in principle the share of human inputs, such as the costs of extraction and production, should be deducted from the market price in order for it to more accurately reflect the value of the provisioning service itself (the so-called production function method; see Annex 2 for an example from Spain).

Sometimes the current selling price of goods on the market might be lower than their potential value (see Table 6.2 for example from Tanzania). Therefore, using contingent valuation methods to assess consumers' willingness to pay for goods can be used to complement the observed market prices. For example, creating a value range based on observed and estimated prices can provide a more comprehensive picture of the value of provisioning services and related goods.

BOX 6.1

Benefits from provisioning services provided by Skadarsko jezero National Park (Montenegro) based on market valuation

Skadarsko jezero National Park (Montenegro), comprised of Lake Skadar and surrounding areas, supports several local livelihoods and industries, such as inland water fishery (commercial and sports fishing), honey and wine production, and sand and gravel extraction.

Method: Benefits related to the provisioning services and goods were estimated by using the market price method (quantity of resource x their market price). It is to be noted that this study focused only on estimating gross values (i.e. human inputs and/or costs of production were not deducted from market prices). Also, no dedicated assessment was made related to the sustainability of the current level of resource use.

Results: There are 300 bleak (*Alburnus spp.*) fishermen in the PA, each fishing for 95 days/year and landing an average catch of 16 kg of bleak per day trip (total catch of 456 tons/year). In addition, 100 carp fishermen operate for 190 days/year, landing an average catch of 5 kg of carp per day trip (total catch of 95 tons/year). At a market price of €3/kg for bleak and €5/kg for carp this leads to a total income for fishermen of around €1.8 million/year.

Some of the bleak and carp catch are further used to produce value added products, namely canned smoked fish (around 140-200 tons of bleak and 70 tons of carp per year). This translates into around 0.82 million cans of bleak and 0.14 million cans of carp produced per year at a retail price of €1 and €4 per can, respectively. This leads to the gross market value of canned smoked fish production of €1.4 million a year. In addition to commercial fishing, about 150 honey producers maintain 7,500 hives in the PA, with a market value of €0.83 million a year. Finally, land rental from a wine producer generate €61,878 in revenues for the PA and yield grapes and wine for €350,000 for the producer.

Source: UNDP and GEF (2011)

Monetary: non-market values

Contingent valuations using revealed and stated preference methods (Chapter 3) are the most commonly used methods to estimate the value of provisioning services and related goods that are not (yet) traded on markets. For example, the willingness to pay method is frequently used to establish the monetary value of products used for subsistence (see Box 6.2 and Table 6.2 for examples from Kenya and Annex 2). Willingness to pay surveys can also be used to explore potential markets for novel goods originating from PAs, such as certified meat, raw material and timber (e.g. Bensel *et al.*, 2008). Similarly, revealed preferences can form the basis for estimating the monetary value of provisioning services. For example, the time spent collecting fuel wood and other resources from a PA can form a basis for estimating the monetary value of these resources (hours spent x average salary per hour in the area).

The costs of losing, replacing or finding an alternative for provisioning services and related goods can be used to create a monetary estimate for their value. Such estimates include, for example, costs of replacing food, fuel or water originating from a PA by goods sourced elsewhere, costs of establishing an alternative resource such as replacing herbaceous biomass for grazers with synthetic food, or costs to local businesses and industries that are dependent on PA resources (e.g. water) (see Box 6.2 and Table 6.2 for examples from Paraguay and New Zealand). The (estimated) loss of income forgone due to the loss of a provisioning service and related goods, for example, due to excessive hunting of game species and/or overfishing, could also be used as an indication for their value.

BOX 6.2

Economic value of bush meat in Mbaracayú Biosphere Reserve (Paraguay), based on replacement costs and ecological/spatial modelling

The economic benefits of bush meat (and other ecosystem services) were assessed and mapped in Mbaracayú Biosphere Reserve (Paraguay) (2,920 km^2). Hunting for wild meat is the most important economic activity for the local populations in the area. The assessment was a part of a larger study that aimed to determine those areas of Mbaracayú Biosphere Reserve where the benefits from restricting access would outweigh the opportunity costs of restricted access to resources.

Method: To estimate the value of bush meat, the amount/availability of game was assessed based on a spatial model estimating the occurrence of 12 game species within the PA (probability of occurrence of species across habitats x typical size of the group species moves in x average body size of an individual x estimated species-specific fraction of biomass that can be sustainably harvested). Summed up over all species, the modelling resulted in the estimated amount of bush meat per area (kg/ha) that could be sustainably extracted from the PA. Lacking information on market price of bush meat, its economic value was estimated based on the estimated costs of replacement, i.e. by using the local price of store-bought beef (US$1.44/kg as of May 2005). The estimated economic value of the potential sustainable supply of bush meat within the Mbaracayú Biosphere Reserve was then estimated as an average US$/ha.

Results: The economic value of the potential/maximum sustainable supply of bush meat was estimated to be on average US$15.59/ha (292,000ha x US$15.59= US$4.5 million (total)*). However, it is to be noted that the actual level of game extraction is likely to be substantially lower than the sustainable maximum level of extraction estimated by the study.

*Own calculation in the context of this guidance

Source: Naidoo and Ricketts (2006)

Non-monetary values

Information on the non-monetary value of provisioning services and related goods is required to complement and/or form a basis for the monetary estimates. For example, information on the biophysical availability of provisioning services, such as trends in and projection of a PA's capacity to maintain healthy populations of economically important species, helps to determine whether the estimated economic values are based on sustainable use of services and related goods (Box 6.2). Biophysical information, including existing hydrological studies, is required to determine on a case-by-case basis the role of PAs in maintaining water supply and balance in the area (Box 6.3).

Qualitative and quantitative information, including, for example, the number of households using water purified by PAs and the amount of non-timber resources originating from PAs, can also be a valuable way to reinforce and substantiate monetary assessments. For instance, specifying the number of local communities depending on PAs for their livelihoods helps to underline the broader welfare value of these areas (see Table 6.2 for examples from US Eastern Sierra and Australia).

It is important to realise that while some non-monetary information, such as statistics on the volume of production and sales (Box 6.1 and 6.3), might be relatively easily available, obtaining some biophysical information requires investment in field surveys, modelling and elaborate data analysis (Box 6.2). This might have implications for the decisions on the depth and spread of a socio-economic assessment.

Finally, it is difficult – and often not even advised – to estimate the importance of provisioning services in the context of broader welfare (as highlighted in Section 6.2 above) in only monetary terms. Therefore, qualitative and quantitative methods, including stakeholder surveys, interviews and discussions, are required to capture these values (Box A2.3 in Annex 2).

BOX 6.3

Demonstrating the importance of three *Montenegrin* National Parks in maintaining water supply by using non-monetary information

Biogradska gora, Durmitor and Skadarsko jezero National Parks in Montenegro provide, among several other ecosystem services, important sources of water supplies to residents and tourists within the parks and in the surrounding regions.

Method: The following examples from National Parks in Montenegro highlight three non-monetary methods to assess the magnitude of water supply service provided by protected areas to surrounding regions: 1) amount of water tapped/delivered by the park (m3), 2) number of people benefiting from the service and 3) seasonal water deficit in the surrounding region (m3).

Results: Biogradska gora National Park and the contiguous Bjelasica Monument of Nature has four springs which are used by commercial water bottling companies, which use about 15,000 m3 of water a year. In addition, there is considered to be the potential to tap an additional 18,700 m3 of water for bottling. Water sources in Durmitor National Park supply about 2,100 households, 1,600 of which are in the town of Žabljak where 135-230 thousand m3 of water is consumed per year. Moreover, almost 55,000 visitor days are spent each year by tourists staying inside the park, also benefitting from the park's water sources. Finally, Skadarsko jezero National Park and its Lake Skadar provides wells and reservoirs for the estimated 1,500 people or 600 households that live inside the park as well as for the tourists that spend approximately 42,000 visitor days per year in the park area. The region around the park is suffering from increasing problems with water supply, affecting a resident population of around 170,000 people and a significant number of tourists (up to 7.6 million visitor days per year). It has been estimated that the annual summer water deficit is of 24.5 million m3, projected to rise to 27.5 million m3 in 2020 and 38 million m3 in 2033. A project is currently undergoing to supply water from Skadar Lake also to the broader coastal area in order to meet the water deficit in the region.

Source: UNDP and GEF (2011)

References

Baldus, D. (ed.) (2005) *What the Kidunda Dam will Destroy: Ecological and Socio-economic Value of Gonabis, Selous Game Reserve, Tanzania*, Tanzania Wildlife Discussion Paper No. 45, http://www.wildlife-baldus.com/download/nr_45.pdf, accessed 9 June 2013

Bensel, T., Newsom, D. and Bahn, V. (2008) *'Are there economic benefits from Forest Stewardship Council (FSC) Certification?'* An Analysis of Pennsylvania State Forest Timber Sales, Rainforest Alliance. Working Paper, http://www.rainforest-alliance.org/resources/documents/economic_benefits.pdf, accessed 12 November 2012

De Groot, R. S., Fisher, B., Christie, M., Aronson, J., Braat, L., Gowdy, J., Haines-Young, R., Maltby, E., Neuville, A., Polasky, S., Portela, R. and Ring, I. (2010) 'Integrating the Ecological and Economic Dimensions in Biodiversity and Ecosystem Services Valuation', in Kumar, P. (ed.) *The Economics of Ecosystems and Biodiversity (TEEB) Ecological and Economic Foundations*, Earthscan, London and Washington, DC

De Lopez, T. T. (2003) 'Economics and stakeholders of Ream National Park, Cambodia', *Ecological Economics* 46: 269-282

Dudley, N. and Stolton, S. (2003) *Running Pure: The importance of forest protected areas to drinking water*, World Bank/WWF Alliance for Forest Conservation and Sustainable Use

EASAC – European Academies Science Advisory Council (2009) *Ecosystem Services and Biodiversity in Europe*, EASAC policy report 09, The Royal Society, London

Elmqvist, T., Maltby, E., Barker, T., Mortimer, M., Perrings, C., Aronson, J., de Groot, R., Fitter, A., Mace, G., Norberg, J., Sousa Pinto, I. and Ring, I. (2010) 'Biodiversity, Ecosystems and Ecosystem Services', in P. Kumar (ed.) *The Economics of Ecosystems and Biodiversity (TEEB) Ecological and Economic Foundations*, Earthscan, London and Washington, DC

Emerton, L. (1996) *Participatory Environmental Valuation: Subsistence Forest Use Around the Aberdares, Kenya*. African Wildlife Foundation, Nairobi, Kenya

Grampians National Park – water catchment protection (undated), WannonWater and GWM water, http://www.gwmwater.org.au/information/publications/brochures/doc_download/1752-grampians-national-park-water-catchment-protection, accessed 19 November 2012

Hein, L. (2011) 'Economic benefits generated by Protected Areas: the case of the Hoge Veluwe forest, the Netherlands', *Ecology and Society* 16(2): 13

Hussain, S. and Badola, R. (2010) 'Valuing mangrove benefits: contribution of mangrove forests to local livelihoods in Bhitarkanika Conservation Area, East Coast of India', *Wetlands Ecology and Management*, 18: 321–331

Kettunen, M., Vihervaara, P., Kinnunen, S., D'Amato, D., Badura, T., Argimon, M. and ten Brink, P. (2013) *Socio-economic importance of ecosystem services in the Nordic Countries – Synthesis in the context of The Economics of Ecosystems and Biodiversity (TEEB)*, Nordic Council of Ministers, Copenhagen

MA – Millennium Ecosystem Assessment (2005) *Ecosystems and Human Well-being: Biodiversity Synthesis*, World Resources Institute, Washington, DC, http://www.maweb.org/documents/document.354.aspx.pdf, accessed 1 October 2012

Naidoo, R. and Ricketts, T. H. (2006) 'Mapping the economic costs and benefits of conservation', *PLoS Biology* 4(11): e360

New Zealand Department of Conservation (2006) *The Value of Conservation: What does conservation contribute to the economy?* http://www.doc.govt.nz/upload/documents/conservation/value-of-conservation.pdf, accessed 17 November 2012

Pyhälä, A., Brown, K. and Adger, W. N. (2006) 'Implications of Livelihood Dependence on Non-Timber Products in Peruvian Amazonia', *Ecosystems* 9: 1328–1341

Roncin, N., Alban, F., Charbonnel, E., Crec'hriou, R., de la Cruz Modino, R., Culioli, J-M., Dimech, M., Goñi, R., Guala, I., Higgins, R., Lavisse, E., LeDireach, L., Luna, B., Marcos, C., Maynou, F., Pascual, J., Person, J., Smith, P., Stobart, B., Szelianszky, E., Valle, C., Vaselli, S. and Boncoeur, J. (2008) 'Uses of ecosystem services provided by MPAs: How much do they impact the local economy? A southern Europe perspective', *Journal for Nature Conservation* 16: 256–270

Town Hall Los Angeles (2012) *How Can We Secure Future Water Sources?* A Panel Discussion, Los Angeles, http://ens.lacity.org/councilcmte/lariver/events/lariverevents265275942_02242012.pdf, accessed 14 November 2012

Troëng, S. and Drews, C. (2004) *Money Talks: Economic Aspects of Marine Turtle Use and Conservation*, WWF-International, Gland, Switzerland, http://awsassets.panda.org/downloads/moneytalks.pdf, accessed 23 November 2012

UK NEA – National Ecosystem Assessment (2011) *The UK National Ecosystem Assessment: Synthesis of the Key Findings*, UNEP-WCMC, Cambridge, http://uknea.unep-wcmc.org, accessed 18 June 2012

UN (1974) *Universal Declaration on the Eradication of Hunger and Malnutrition*, http://www.ohchr.org/EN/ProfessionalInterest/Pages/EradicationOfHungerAndMalnutrition.aspx, accessed 9 June 2013

UN MDG (2012) UN Millennium Development Goals, http://www.ohchr.org/EN/ProfessionalInterest/Pages/EradicationOfHungerAndMalnutrition.aspx, accessed 12 November 2012

UNDP and GEF (2011) *The Economic Value of Protected Areas in Montenegro*, Report by the UNDP and GEF project Catalyzing financial sustainability of Protected Areas in Montenegro (GED/UNDP PIMS4279), http://www.undp.org.me/ee/projects/GEF/Catalyzing%20financial%20sustainability%20of%20the%20protected%20areas%20system%20in%20Montenegro%20(PAF)/index.html, accessed 9 June 2013

WCPA – Task Force on Economic Benefits of Protected Areas of the World Commission on Protected Areas (WCPA) of IUCN, in collaboration with the Economics Service Unit of IUCN (1998) *Economic Values of Protected Areas: Guidelines for Protected Area Managers*, IUCN, Gland, Switzerland and Cambridge

WHO and UNICEF (2012) *Progress on drinking water and sanitation*, http://www.who.int/water_sanitation_health/publications/2012/jmp_report/en/index.html, accessed 13 January 2013

ANNEX 2

Step-by-step guidance: valuation of biodiversity resources

Kettunen, M. and D'Amato, D.

1 Introduction

This step-by-step guidance supports Chapter 6, providing detailed information on how to apply a number of key methods to assess the socio-economic value of provisioning services and related goods. It covers the following methods: estimates based on market values (revenue and employment) (Section 2 below), estimates based on non-market values (subsistence and household use) (Section 3 below) and cost-based valuation methods (Section 4 below). It also outlines a checklist for things to be kept in mind in the context of valuation, including how to ensure synergies and/or prevent conflicts between socio-economic valuation and PAs' conservation goals.

2 Market values: goods and products with existing/ potential markets

The monetary value of goods and products with existing and/or potential markets can be estimated as follows:

Step 1: Calculate the total gross market value of good. Estimate the total value by multiplying the market price by quantity of good produced or sold (e.g. \$/kg x kg/year = total gross value of good/year). Note that using the quantity of production as a basis for calculation might lead to overestimating the actual market value of a good. This is because not all of the produced good might be sold at the market. Note also that existing market prices should be adjusted for any distortions (subsidies) that artificially lower the price of goods (see Box A2.4 below).

Step 2: If possible, calculate the total net market value of good (total gross market value of good/year – costs of production/year = total net value of good). This so-called production function approach determines the share of the market value that can be attributed to the actual ecosystem service and related good (i.e. deducting the value of human inputs).

Step 3: By using information on status and trends: 1) reflect whether the level of extraction and/or cultivation of the good is sustainable and 2) estimate whether the importance and/or socio-economic value of the good is likely to increase in future (due to increasing scarcity, etc.).

Information required: to carry out the valuation information is required on 1) the quantity of good produced and/or sold, 2) the market price of the good, 3) if possible, the share of human input in market price (costs of extraction and production) and 4) the status of and trends in the PA's ability to supply the good.

The steps above can also be used to estimate the value of potential market goods and products from a PA, including certified products. In this case the potential market volume and price for goods and products can be determined by 1) using information on the market volume and price for similar goods from elsewhere (value transfer, see Chapter 3) or 2) if resources are available, carrying out stakeholder surveys to establish people's willingness to buy (volume) and pay (price) for the good. Note that the latter provides a more accurate and case-specific estimate.

Finally, the market-based value of goods can be complemented by also estimating the (scale of) employment associated with production and sales of the good, including the number of businesses and/or jobs supported by the provisioning of good (see Chapter 10).

Concrete examples: Boxes A2.1, A2.2, A2.4 and A2.5

BOX A2.1

Valuation of provisioning services and related goods in the Doñana PA and surrounding region (Spain)

The Doñana social–ecological system (Spain) provides numerous and consistent benefits to local communities, supporting many human activities in surrounding lands such as agriculture and farming of autochthonous varieties and breeds.

Method: Ecosystem services of Doñana PA were assessed through market-based approach. Values were derived from published and unpublished data (official statistical data and annual reports of the activities of the Doñana PA during the years 2001–2008) and through semi-structured interviews to 24 stakeholders (snowball sampling) canvasing information about 1) profession, residence time, and knowledge of the area, 2) provisioning service exploitation and net profit received/market value, and 3) perceived trend of provisioning service.

Results: The strawberry cultivation area in the PA was approximately 4,970 ha in 2007, representing an average of 60 per cent of the total Spanish production. Between the years 2001 and 2007, the average production was 198,647 tons. The average income received by farmers, without subsidies, in 2008 was €1.06/kg of strawberries (around €210.56 million total/year). The average production costs were estimated at €24,500/ha per year (around €121 million total/year). Based on these figures the net value of the strawberry industry was around €89.6 million in 2008.

During the years 2001–2008, the average area under rice cultivation within the PA was 26,000 ha yielding to an average yield of 208,133 tons/year with a gross benefit of €49.3 million/year (without subsidies). The average cost for the rice fields was approximately €1,496/ha (€38.8 million total/year). Therefore, the average net value of rice production between the years 2001 and 2008 was approximately €10.5 million/year.

In terms of gross value estimates, the reported annual catch between the years 2000 and 2008 in the fish market of Bonanza (in Sanlúcar de Barrameda) was 2,956 tons, with an annual value €14.5 million/year (gross). Of this, 1,032 tons and €8.2 million were related to species that use the Doñana PA's Guadalquivir Estuary as a nursery ground. In addition, harvesting of the edible clam *Donax trunculus* is operated by an average of 78 fishermen per day and the maximum authorised captured is 25kg/day/person over 205 days per year. The harvest reaches a maximum of approximately 350 tons/year, equivalent to €1.5 million/year (gross). Since 2004, on average 1,770 m3 of timber per year was collected inside the PA, yielding a total income of €64,479/year (gross). In addition, 1 million and 0.4 million kg of pine cones were harvested in 2004 and 2008, respectively, for a value of €134,398 and €27,600 per year (gross). Finally, the average honey production for 2001-2008 was estimated to be 79.4 tons per year for a value of €121,256/year (gross).

Cattle farming and viniculture have represented an important income source within the PA in the past. However, the cultural values associated with these activities have long exceeded their economic importance. The livestock in the Doñana PA are mainly comprised of three autochthonous breeds: between 2004 and 2008, 1,523 Retuertas horses, 932 Mostrenca cows, and 579 Andalusian churra sheep were farmed. Based on prices established in the Andalusian statistics, their average economic value was approximately €2.3 million per year (gross). Similarly, between the years 2001 and 2008 the average production of the autochthonous grape variety *Vitis vinifera* cv. *zalema* was 33,957 tons per year, yielding an estimated net average value of €247.4. Despite their low profitability, these vineyards continue to be maintained for their cultural values not reflected by market prices.

Source: calculations based on Martín-López *et al.* (2011)

BOX A2.2

Potential economic value of biomedicines originating from Makira PA (Madagascar)

Makira PA in Madagascar (over 370,000 ha) is estimated to comprise approximately 0.8 per cent of global botanical diversity and, as such, is a potential source of novel pharmaceutical drugs.

Method: The potential value of future biomedicines originating from Makira PA was estimated based on the general likelihood for discovering new biomedicines. In biomedicine testing, each plant species undergoes a screening procedure that tests three different parts of a plant, using two different extraction procedures per part and 500 screens per sample. Considering the number of plant species in the area (around 300) it was estimated that around 918,000 tests would be required. Of these tests, approximately one in 50,000 to one in a million tests may produce an effective and viable commercial pharmaceutical drug. Based on this success rate Makira PA could be a source for 1–18 potential novel drugs. Using the average sales value of novel US Food and Drug Administration approved pharmaceuticals, each new botanically based pharmaceutical is worth $316 million/year in revenue in the US (gross, i.e. without subtracting the costs of research and development).

Results: The potential for developing novel biomedicines from the Makira watershed's unique flora ranges in untapped benefit value from US$0.3–5.7 billion for American pharmaceutical companies, non-inclusive of the broader welfare values related associated with providing novel medicines and improved healthcare to society (see Chapter 6).

Source: Golden *et al.* (2012)

3 How to estimate non-market monetary values: subsistence/household use

The monetary value of goods and products with no existing markets (i.e. resources used for subsistence) can be estimated as follows:

Step 1: Estimate the volume of good extracted, collected, used or produced for subsistence. This can be done based on 1) existing statistics, studies and information from the area and/or 2) stakeholder surveys.

Step 2: Estimate the monetary value of good extracted, collected, used or produced for subsistence. The most commonly used approach makes use of the value of the same or a similar good sold on the market (see Box A2.3).

Step 3: Calculate the gross (or if possible net) total market value of good. This can be done by multiplying the estimated market price by the quantity of good used for subsistence (e.g. \$/kg x kg/year = total gross value of good/year). In addition, if the relative importance of different subsistence goods from PA is known then the estimated total monetary value of one good can also provide a basis for approximating the monetary value of other non-market goods (see Box A2.3).

Information required: to carry out the valuation information is required on 1) the quantity of good extracted, collected, used or produced for subsistence, 2) an indication of the monetary value of the good, including the market price of a similar good, if possible adjusted for possible distortions and reducing the costs of extraction and production, or other proxies and 3) the status of and trends in the PA's ability to supply the good.

Finally, monetary estimates for the value of goods used for subsistence can be complemented by estimating the welfare dependency on the good(s) in question. This can be done by assessing the number and/or share of communities, households and people depending on the provisioning of good(s) (see Chapter 10).

Concrete examples: Box A2.3

BOX A2.3

Estimating the value of protected wetlands to well-being and poverty alleviation in Stoeng Treng Ramsar site, Cambodia

A study to assess the contribution of protected wetlands to local communities' well-being and poverty alleviation was carried out in 2004–2005 in Stoeng Treng Ramsar site, Cambodia. Fisheries form an integrated component of livelihood in the area. Consequently, the study focused on assessing the socio-economic importance of fisheries in maintaining the well-being of the village and its households. In addition, using the socio-economic value of fisheries as a benchmark the study also estimated the value of other wetland resources in the area.

Method: Benefits associated with the PA were determined based on a focus group discussion consisting of the representatives of different wealth groups within the village (three men and three women). The benefits were identified with the help of a resource flow diagram, demonstrating the flow of different benefits from the conserved wetland to the beneficiaries. The identified benefits were then rated according to their importance on a scale of 1–5. Finally, the estimated average monetary values of the fisheries per different wealth groups (consumption and trade) (below) were used as a basis to translate these quantitative rankings into non-market monetary values of other wetland sources.

A village-level survey was carried out in a case study village (Veun Sean village) to gain information about the value of the fisheries to households. Instead of simply estimating the total fish catch in the village and multiplying it by average market prices two distinct direct uses of the resource were considered separately: fisheries for household consumption and as a source of income. This separation enabled a more detailed consideration of the subsistence value of fisheries to individual households (non-market value). It also allowed comparisons between households in terms of the two different uses of the resource (subsistence vs. trade). The following steps were followed to calculate the value of the fisheries resource to households: 1) calculating annual household fish consumption (kg) based on survey responses on the frequency and quantity of fish consumed per household, crosschecking the results with secondary information about fish consumption in Cambodian fishing households, 2) using acquired information on the share of fish sold vs. share of fish consumed to calculate annual fish sales (kg) per individual household and 3) using the market price of KHR3,000/kg (around US$0.75) to calculate the value of fish sales (market) and consumption (non-market) for each household.

Results: The results of the assessment (quantitative and monetary) are presented in tables below. These results indicate that the wetlands play a very critical role to the livelihoods of the rural community. These estimates reveal that in Veun Sean village, the fisheries resource is more valuable to poorer households than wealthier households, partly because the larger household size of poor households means that they consume more fish per household, and partly because a greater proportion of poorer household's fish catch is sold for income.

TABLE A2.1 Identified benefits from Stoeng Treng wetland and their quantified importance on a scale of 1–5

Identified benefits	*Quantified importance*
Resource: fishing	5
Water: washing	5
Water: cooking and drinking	5 for each benefit
Water: transportation	4
Fibre: construction (sand and rock)	3
Fibre: fuel wood	3
Biodiversity resources: aquatic animals	2
Biodiversity resources: water birds	2
Biodiversity resources: reptiles	2
Biodiversity resources: traditional medicines	2
Water: irrigation	2
Resource: floodplain rice	1
Recreation: swimming	1
Other: dolphins	1

TABLE A2.2 Estimated monetary value of fisheries (market and non-market)

Value of fisheries	*Average household*	*Middle or rich household*	*Poor or very poor household*
Annual fish consumption for subsistence	180 kg	150 kg	210 kg
Annual fish sale	420 kg	190 kg	690 kg
Value of fish consumed for subsistence (non-market)	KHR500,000/ US$125	KHR500,000/ US$125	KHR600,000/ US$150
Income from fish sale (market)	KHR1,200,000/ US$300	KHR600,000/ US$150	KHR2,000,000/ US$500
Total monetary value of fish (market and non-market)	KHR1,700,000/ US$425	KHR1,100,000/ US$275	KHR2,600,000/ US$650
Fish sale (market value) as a percentage of total fish value	70%	56%	77%

TABLE A2.3 Estimated monetary (non-market) values of other benefits, calculated based on their relative importance as proportion of the fisheries monetary value

Identified benefits	*Quantified importance*	*Estimated monetary value*
Fishing	5	KHR1,700,000/US$425
Water: washing, cooking and drinking	5	KHR1,700,000/US$425/benefit
Water: transportation	4	KHR1,360,000/US$340
Fibre: construction (sand and rock), fuel wood	3	KHR1,020,000/US$255/benefit
Biodiversity resources: aquatic animals, water birds, reptiles, traditional medicines	2	KHR680,000/US$170/benefit
Water: irrigation	2	KHR680,000/US$170
Other: floodplain rice, dolphins and recreation	1	KHR680,000/US$170/benefit
Total		KHR12,900,000/US$3,229, Approximately US$3,000/household/ year

Source: based on Chong (2005)

4 Cost-based valuation: cost of loss and/or replacement of resource

The cost-based (monetary or quantitative) value of goods and products (market or non-market) can be estimated as follows:

Step 1: Estimate the volume of good used, required, produced or sold, possibly needed to be replaced. This can be done based on 1) existing statistics, studies and information (market and non-market goods) and/or 2) stakeholder surveys (non-market goods).

Step 2: Estimate the monetary value of good were it to be replaced. The most commonly used estimate includes the current value of good or, in the case of non-market goods, the value of similar goods sold on the market.

Step 3: Calculate the gross (or if possible net) total market value of good. This can be done by multiplying the estimated market price by the quantity of good possibly needed to be replaced (e.g. \$/kg x kg/year = total gross value of good/year needed to be replaced) (see Box A2.4). Alternatively, quantitative estimates (percentage/share) on the dependency of sectors, businesses, communities and/or households on the provisioning of good can be used on their own or as a basis for estimating the monetary value of goods to the sector (total economic value of sector x estimated share of this value dependant on a good as an input, see Box A2.5).

Information required: To carry out the valuation information is required on 1) the quantity of good (used, required, produced or sold) possibly needed to be replaced, 2) estimate on costs of replacing/losing the good (e.g. based on economic value (total) and dependency of (share of total economic value) sectors, businesses, communities or households using the good) and 3) the status of and trends in the PA's ability to supply the good. The application of a cost-based approach requires data on the likely alternative. In several cases, this might require additional research and/or collecting some previously unavailable information via surveys or questionnaires.

Concrete examples: Boxes A2.4 and A2.5

BOX A2.4

Estimated value of ethnomedicines at Makira PA (Madagascar) based on costs of replacement

Makira PA in Madagascar (over 370,000 ha), comprising approximately 0.8 per cent of global botanical diversity, has significant socio-economic value in terms of botanical ethnomedicines to local people. Approximately 241 locally recognised species are used as ethnomedicines, including 113 agricultural or weed species.

Method: The value of botanical ethnomedicines in Makira PA was calculated by assessing the costs associated with substituting the resource (replacement costs and choice modelling). Each ethnomedicinal treatment was equated to the monetary value of a comparable pharmaceutical treatment. The comparable treatments were identified based on the actual preferences by local communities within the PA (rather than selecting known or assumed medicinal equivalent treatments). The frequency of ethnomedicine use and personal preferences were estimated by randomly sampling 634 households in 24 villages. The monetary non-market value of ethnomedicines was estimated based on the frequency of ethnomedicine use and their estimated market price (i.e. the price of comparable treatment). Two different market prices (the subsidised price of pharmaceutical treatments in Madagascar and the online price of equivalent pharmaceuticals in America) were used to create a range value that aims to correct the final estimate for existing subsidies.

Results: The value of botanical ethnomedicines to local communities was assessed as US$5.40–7.90/year/individual when estimated based on the price of highly subsidised Malagasy pharmaceuticals and US$100.60–287.40/year/individual when estimated based on the price of American pharmaceuticals. Using local pharmaceuticals as substitutes, the value per household was US$30.24–44.30/year, equivalent to 43 to 63 per cent of median annual household income, demonstrating their local importance. Using the value of American pharmaceuticals, the amount is equivalent to 22 to 63 per cent of the median annual health care expenditures for American adults under 45 in 2006.

Source: Golden *et al.* (2012)

BOX A2.5

Economic valuation of benefits provided by provisioning services and related goods from Leuser Ecosystem (Sumatra) based on the estimated costs of loss

The Leuser Ecosystem in Northern Sumatra is officially protected by its status as an Indonesian national park. Nevertheless, deforestation has already caused a decline in ecological functions and services and it is affecting numerous economic activities in and around the PA. The economic value of several ecosystem services, including a number of provisioning services, was assessed by estimating the costs related to the loss of benefits over the period 2000–2030.

The current levels of extraction were assumed to be sustainable; however, no dedicated assessment was carried out to confirm this.

Methods: In general, the costs of loss due to deforestation were estimated based on the foreseen possible future management regimes: continued deforestation (i.e. loss of benefits) vs. selective use and/or conservation (i.e. maintenance of benefits). In the case of deforestation, the availability of provisioning services and related goods provided by the PA were assumed to continue to decline whereas the market price of these goods was suspected to increase (due to scarcity of resource). The decline in supply and increase of prices were determined for each identified provisioning service, based on expert opinion. Both of these parameters were estimated as percentage changes from the current levels. A discount rate of 4 per cent was used to account for the change in value over the 30 year period (see Chapter 12 and Annex 6).

The estimated value of fresh water supplied by the park was based on the reduction of the amount of water provided per year and the (minimum) price of water to households and industries. In the case of deforestation, it was assumed that 1) the dependency of surrounding areas on water originating from Leuser PA declines from 74 per cent in 2000 to 12 per cent in 2030 and that at the same time 2) the price of water increases by 0.3 per cent/year from the current levels, due to increased costs of retrieving water.

As for water supply to hydroelectricity, increased erosion of the waterways has forced the operators to remove excessive sediments from their turbines. This has led to frequent interruption of the power supply, higher operational costs and damage to the blades of the turbines. The current level of power generation dependent on the PA was estimated as 22 per cent of the total power generation in the area. Due to deforestation scenario, this share was assumed to decline to 16 per cent. Furthermore, the costs of electricity generation were assumed to increase by 2 per cent/year.

The estimated value of fisheries was based on information on the fish catch and its market price. Different fishery sectors (maritime fishery, brackish water fishery, brackish water aquaculture and freshwater aquaculture) were treated separately in the calculations as the dependency of these sectors on Leuser PA varies. The average share of the sector's dependency on the PA has been estimated to be 2 per cent for the maritime fishery, 9 per cent for brackish water fishery and 100 per cent for brackish and freshwater aquaculture. In case of deforestation, the support of Leuser PA to fisheries was estimated to decline 1 per/year (all sectors) while the market prices of fish were to increase 0.5 per cent/year.

The value of non-timber forest products – cotton tree, rattan, resin, palm sugar, cinnamon, nutmeg, honey and vanilla – was estimated based on local statistics, surveys and expert assessments on the level of extraction and market prices for different goods. In the case of deforestation it is assumed that an over-exploitation of resources will occur. As a result, a short-term increase in harvested NTFP will be observed in the first decade after which this sector collapses.

Net value for timber was derived by applying the market price to the estimated quantities that could be sustainably harvested from forested areas, deducting costs of extraction (US$17 per m3). Market prices were established based on local field surveys, expert assessments and local statistics, corrected for market and policy failures as timber prices in Indonesia are far below international standards due to subsidies, non-tariff barriers and illegal timber trade.

Results: The economic values of provisioning services and related goods from Leuser PA are estimated to decrease (2000–2030) due to deforestation as follows:

- water supply (consumption): from US$2,419–2,005 million per year (conservation – selective use) to US$699 million per year (deforestation)
- water supply (hydroelectricity): from US$898–696 million per year to US$252 million per year
- fisheries: from US$674–659 million per year to US$557 million per year
- non-timber forest products: from US$1,222 (selective use) to US$235 (deforestation)

Contrary to the above, the value of timber is foreseen to decrease due to selective use (from US$1,184 to US$825). This is because of the trade-off between timber extraction and the maintenance of the rest of the provisioning (and other) services. Please see Chapter 12 for further detail of the net value of Leuser PA.

Source: van Beukering *et al.* (2003)

5 Key issues to be kept in mind

When using existing market prices care should be taken to 1) adjust them for any distortions (subsidies) that artificially lower the price of goods, this way leading to the depression of its estimated value and 2) deduct the share of human inputs from the market price (costs of extraction and production) in order not to overestimate the value of provisioning service itself.

Information on status and trends of provisioning services and related goods is required to reflect whether the level of extraction and/or cultivation of the good is sustainable and also estimate whether the importance and/or socio-economic value of the good is likely to increase in future (due to increasing scarcity, etc.).

Estimating changes in value over time, to reflect the depletion of services and related goods, might be desirable. This requires the use of a discount rate to account for the change in value over the period of several years (see Chapter 12 and Annex 6). For such calculations it is advised to seek support from an economist.

References

Chong, J. (2005) *Valuing the Role of Wetlands in Livelihoods: Constraints and Opportunities for Community Fisheries and Wetland Management in Stoeng Treng Ramsar Site, Cambodia*, IUCN Water, Nature and Economics Technical Paper No. 3, IUCN — The World Conservation Union, Ecosystems and Livelihoods Group Asia, Colombo

Golden, C. D., Rasolofoniaina, B. J. R., Anjaranirina, E. J. G., Nicolas, L. and Ravaoliny, L. (2012) 'Rainforest Pharmacopeia in Madagascar Provides High Value for Current Local and Prospective Global Uses', *PLoS ONE* 7: 7

Martín-López, B., García-Llorente, M., Palomo, I. and Montes, C. (2011) 'The conservation against development paradigm in Protected Areas: Valuation of ecosystem services in the Doñana social-ecological system (south-western Spain)', *Ecological Economics* 70: 1481–1491

van Beukering, P. J. H., Cesar , H. S. J. and Janssen, M. A. (2003) 'Economic valuation of the Leuser National Park on Sumatra, Indonesia', *Ecological Economics* 44: 43–62

7

REGULATING SERVICES AND RELATED GOODS

Badura, T. and Kettunen, M.

7.1 Overview of benefits

PAs provide several benefits which build directly on their capacity to perform a number of ecological functions and biophysical processes. These benefits are commonly called regulating services and they include, for example, PAs' ability to stabilise the climate, provide protection against natural hazards (flooding, droughts and avalanches, etc.) and regulate the quality of air and water by removing pollutants and excess nutrients (see Table 7.1). PAs can also support the pollination of crops and/or contribute to the control the outbreaks of diseases and pests.

There is a significant interest in exploring the socio-economic value of regulating services and related goods provided by PAs, ranging from direct benefits of water purification to companies and consumers to a fast growth in interest in exploring local, national and global benefits related to climate regulation (linked *inter alia* to UN REDD+ discussions, see Section 15.4). PAs' role in offering coastal protection and mitigating impacts of floods and storms at risk areas is also increasingly studied.

It is important to recognise that regulating services underpin and/or are interlinked with the availability of provisioning and cultural services and related goods. For instance, in many places wild pollination and natural pests control supported by PAs can increase agricultural production. Together with the protection of crop wild relatives (Chapter 6) these beneficial processes underpin the availability of biodiversity resources and help to increase food security. Moreover, ecosystems' role in regulating local and global water cycles has an impact on a range of industries such as energy sector (via cooling or hydropower) or beverage sector. Ecosystems capture, store and filter water and thereby underpin the provision of drinking water (Chapter 6), making water provisioning and water-related regulating services closely interlinked. Similarly, regulating services help to contribute to

TABLE 7.1 Regulating services and related goods

Ecosystem service	*Examples of related goods*
Climate regulation – through maintaining carbon cycles and regulating a number of climate parameters such as humidity, local precipitation and temperature	Carbon storage and sequestration, stable climate (regional and local)
Natural hazard regulation – through mitigating extreme events/providing natural barriers	Flood control, storm protection, wildfire control, prevention of avalanches, prevention of rock falls, avoiding coastal erosion
Purification and detoxification of air, soils and water – through retention of toxins and nutrients Nitrogen (N) retention capacity	Drinking water for human consumption, water for industrial use and to support the production of goods (food, fibre, energy), reduced air pollution levels and related reduced risk of negative impacts on human health
Water/water flow regulation – through regulating surface water runoff, (sub) surface water flow and soil infiltration capacity	Drainage and stabilisation of water flow (non-flood related), drought mitigation, availability of water for irrigation, aquifer recharge
Erosion and soil fertility regulation – through maintaining soil intactness, nutrient content and fauna	Maintenance or increase of agricultural/forestry production
Pollination – through maintaining wild pollinators	Maintenance/increase of agricultural production by supporting the production of food and fibre, aesthetic benefits and inspiration (wildflowers and other ecosystem components forming aesthetic beauty, etc.)
Pest and disease regulation – through maintaining natural enemies of pests	Disease and pest control with beneficial impacts on both agriculture and also human health
Noise regulation – through vegetative cover acting as natural barrier for noise	Noise control with impacts on human health (mental and physical)

Source: own presentation, based on MA (2005a), de Groot *et al.* (2010), UK NEA (2011), Kettunen *et al.* (2013)

aesthetic beauty and tourism and recreation opportunities (Chapter 8), for instance through pollination of wildflowers or via air and noise regulation improving environmental quality. Consequently, the economic value of regulating services and related goods needs to be assessed with due reflection and understanding of these interlinkages, particularly to avoid the problem of double counting in estimating the total economic value of PA benefits (see Chapters 3 and 12).

The understanding of regulating services has been increasing significantly since the publication of the Millennium Ecosystem Assessment (MA, 2005a). The complexity of biophysical interconnections and linkages between natural, social and economic systems still makes the assessment of their socio-economic value challenging. It is, however, possible to develop estimates that capture the level of importance and/or order of magnitude of these benefits, offering a new evidence base for decision-making processes related to the establishment and management of PAs (see Section 7.4 below).

7.2 Socio-economic importance of benefits

The existing evidence indicates that the benefits associated with PAs' regulatory functions are significant, being of fundamental importance to both local and global well-being (environmental security, climate regulation, food security, poverty reduction, opportunities for development, etc.). These benefits can have considerable socio-economic value to society and the economy, for example, through providing clean water to rural and urban populations, helping to avoid losses and costs related to natural hazards, increasing agricultural revenue through maintaining effective pollination and pest control, and decreasing the number of costs of respirational illnesses by helping to regulate air pollution. Some of these benefits are amenable to monetary valuation while several benefits are better captured in a quantitative and qualitative manner (see Section 7.4).

It is important to note that the socio-economic value of benefits associated with regulating services varies greatly, depending on the socio-economic and ecological contexts and their interconnections. For instance, each PA has its unique biophysical characteristics which together with the interconnections to social and economic systems determine the availability, supply and magnitude of regulating services and related goods. Similarly, PAs differ in terms of stakeholders benefiting from and valuing the services (population size, income level, cultural preferences, etc.). Consequently, the magnitude of benefits is very context specific and great care needs to be taken when using the estimated socio-economic value of benefits on one site in the context of another site (see value transfer in Chapter 3).

Economic values

PAs play a key role in sequestering and storing CO_2 and by doing so help mitigate the adverse effects of climate change. Many ecosystems, including sites located on forested lands, wetlands, agricultural lands, and marine and coastal areas, stock vast amounts of carbon and offer significant opportunities for further carbon sequestration. PAs have been estimated to contain about 15 per cent of global terrestrial carbon stock, with an estimated economic value of the magnitude of trillions US$ (Campbell *et al.*, 2008). It has also been estimated that the economic value of carbon stored by the habitats of the European Natura 2000 Network of PAs is between €607 and €1,130 billion in 2010, depending on the carbon prices used (Ding *et al.* in ten Brink *et al.*, 2011) (see Table 7.2 and Annex 4 for further examples).

Ecosystems maintained and protected by PAs are often of paramount importance for the local and global water cycles, helping to regulate the quality and quantity of water (Russi *et al.*, 2012). Wetlands, peat bogs and sandy ridges play an important role in capturing and storing freshwater while the processes within these ecosystems also help to filter out and decompose organic waste, and assimilate and detoxify hazardous compounds from the water (MA, 2005a). In this way, natural ecosystems in many PAs ensure stable supply of drinking water to cities and communities, (Kettunen *et al.*, 2011; ten Brink *et al.*, 2011). For example, about 80 per cent of the water for the city of Quito in Ecuador comes from three PAs in the vicinity of the city (Dudley *et al.*, 2010, see also Box 15.1 in Chapter 15). PAs also play a crucial role in regulating the quantity and quality of water consumed by world's capital cities such as New York, Bogota, Vienna, Oslo and Berlin (Table 7.2). The role of PAs in maintaining water quality is mainly related to the ability of soil and vegetation to purify water by removing nutrients from polluted water. Excessive nutrient load in the water (nitrogen, phosphorous, etc.) reduces the quality of water for drinking, recreation, and agricultural and industrial purposes and hence their removal can provide significant economic benefits. For example, the estimated value of waste purification and nutrient retention services provided by the Nakivubo Swamp in Uganda was valued between US$1 million and US$1.75 million a year (Table 7.2 and Box A3.2 in Annex 3).

PAs are recognised to maintain healthy, intact and robust ecosystems, which help to reduce the probability and severity of extreme events and moderate their impacts (Kettunen *et al.*, 2011; MA, 2005a; Mulongoy and Gidda, 2008). A significant number of human settlements are located in regions that are exposed to natural hazards (MA, 2005a). In addition, climate change is expected to increase the frequency of extreme events (e.g. IPCC, 2012). Consequently, PAs' role in protecting against natural hazards and supporting ecosystem-based adaptation to climate change is likely to increase in the future (Kettunen *et al.*, 2011). For example, a significant portion of Swiss forest is protected for its protective functions against avalanches, rock falls and landslides, with an estimated value between US$2 and US$3.5 billion per year (ISDR, 2004; Dudley *et al.*, 2010). Similarly, many coastal ecosystems are known to act as defence against storm and tidal surges while at the same time helping to avoid coastal erosion. For example, existing evidence indicates that mangrove forests act as natural buffers against typhoons and hurricanes (Table 7.2).

Furthermore, healthy ecosystems within PAs also help to protect communities and infrastructure against floods, thereby providing an important benefit to approximately 2 billion people living in high flood risk zones (MA, 2005b). For example, inland waters, such as lakes and wetlands, are traditionally considered to be very important for the temporal regulation of water flow, mainly by accumulating water during wet periods (reducing peak flow). This often represents a benefit of considerable economic value. For instance, the protection of Whangamarino Wetland in New Zealand (Ramsar site) has been estimated to have saved NZD millions in disaster prevention. Similarly, the benefits of Muthurajawela-Negombo Wetland Sanctuary in Sri Lanka for flood prevention have been estimated to be around US$5.4 million per year (Table 7.2 and Box 7.2).

PAs can also help to regulate the occurrence and intensity of wild fires. Ecosystems' ability to mitigate wild fires can be defined as the capacity of ecosystems to maintain natural fire frequency and intensity (MA, 2005a). One of the main ecosystem factors affecting fire resistance is the amount of vegetation (i.e. fuel) in the system. This is in turn linked to climate conditions, land cover and land use. The economic damages of wild fires are significant: the forest fires that took place in Greece in summer 2007 caused damages worth €5 billion, 110 villages were damaged or destroyed and livestock, cultivated land and forests were burned down. The events lead to a particularly high death toll (60-68 people) (Bassi *et al.*, 2008). A recent report highlighted the role of invasive grass species (*Bromus tectorum*) in the spread and initiation of wildfires in the US (Balch *et al.*, 2013), contributing to the significant damages and financial losses during the past years. While no documented examples yet exist on PAs role in mitigating wildfires, the presence of native fire resistant species within PAs combined with appropriate management practices (e.g. removal of excess biomass by extensive grazing) can regulate the risk and/or intensity of wildfires, helping to avoid damages to property and people as well as avoid release of CO_2 (see climate regulation above).

The role of PAs in regulating erosion is increasingly recognised. Healthy and/or undisturbed vegetative cover of protected ecosystems binds soil particles effectively and protects it against the effects of wind and rainfall, helping to retain top soil, prevent erosion and mitigate landslides. This regulative function can have significant economic value, for example by helping to avoid the negative impacts of erosion on agriculture or on water quality. While no examples specific to PAs are available, it has been estimated that the value of erosion control in in Europe was €44/ha, at 2008 prices (Ruijgrok, 2006 in Braat *et al.*, 2008).

Wild pollination represents an essential ecosystem service for human well-being, being a key ecological process on which natural and agricultural systems depend (e.g. MA, 2005a; Balmford *et al.*, 2008). It was estimated that insect pollinators were directly responsible for 9.5 per cent (around €153 billion) of the total value of the world's agricultural food production in 2005 (Gallai *et al.*, 2009). Insect pollination has also been estimated to increase the yields of 75 per cent of globally important crops and being responsible for an estimated 35 per cent of world crop production (Klein *et al.*, 2007). The type and quality of habitats surrounding crop fields have been found to correlate with pollen deposition provided by bees (Kremen *et al.*, 2004). Hence, many of the PAs located in close proximity to agricultural fields are likely to play a significant role in maintaining pollination, contributing to its economic value. However, studies specific to PAs are still lacking. It is also important to note that, beyond its impact on agriculture, pollination's contribution to human well-being has a broader scope. Even though currently unknown, pollination value and services to wildflowers and for recreational and other cultural services is expected to be significant (UK NEA, 2011).

Welfare benefits

The benefits provided by regulating services have a close association with human health and broader welfare. First, PAs' ability to sustain air and water quality (above) reduces the risk of respiratory diseases and illnesses caused by contaminated water. The availability of clean air and water has a direct influence on human health, particularly in the developing countries.

Second, well-functioning ecosystems within PAs can also help to control the range and abundance of species that are hazardous to human health. For example, a number of species (birds and insects, etc.) are known to be vectors of human diseases (malaria, dengue fever, Lyme disease, etc.). Under natural conditions ecosystem functioning, including predation and competition on resources, tends to keep the populations of these vector species under control. However, in a changed situation the populations of these harmful species might increase exponentially causing an epidemic of the disease they carry. For instance, a study from the Peruvian Amazon found that the biting rate of malaria vector *Anopheles darling* in deforested areas was 278 times higher than in the areas that were heavily forested (Vittor *et al.*, 2009).

Finally, well-functioning ecosystems are also often better 'equipped' against the invasion of alien species (IAS) with harmful health impacts, such as exotic pathogens, disease vectors and allergenic species. This is because, in comparison to disturbed areas, natural ecosystems tend to have a higher capacity to sustain their typical functioning under changed conditions. For example, several IAS are known to cause allergies and skin damage, including giant hogweed (*Heracleum mantegazzianum*), common ragweed (*Ambrosia artemisiifolia*) and silver wattle (*Acacia dealbata*) (Kettunen *et al.*, 2008).

The evidence based on PAs' contribution to human health is growing however more work is still needed to translate these benefits into quantitative and monetary values (Stolton and Dudley, 2010). Such benefits are important to be explored, especially in the light of increasing urbanisation of the global population.

7.3 Role of different stakeholders in maintaining and using benefits

Regulating services and related goods benefit a range of stakeholders, mostly at local or regional level but also on national and global scales. Benefits are often enjoyed outside PAs, though the majority of them occur in rather close proximity of a site. Benefits such as water purification and wastewater treatment, natural hazards mitigation, pollination, and air quality and noise regulation benefit mostly communities in the vicinity of PAs (or in case of water regulation downstream of PAs). Many of these benefits have also broader impacts. For instance, pollination and the regulation of pests provided by networks of naturally functioning ecosystems within PAs help to maintain national food security. Similarly, water purification services underpin water security as well as food and energy production. Finally on a global scale, carbon sequestration and storage by PAs provides ecosystem-based adaptation and mitigation strategies to climate change.

TABLE 7.2 Examples of economic and welfare values of PAs related to regulating services and related goods

Protected area	*Benefits (ecosystem services and related goods)*	*Estimated socio-economic value of benefits*	*Method*	*Source*
Natura 2000 network of PAs, EU (several ecosystems)	Carbon storage	Natura 2000, the European network of protected areas, currently stores around 9.6 billion tons of carbon, equivalent to 35 billion tons of CO_2, which is estimated to be worth between €607 billion and €1,130 billion (stock value in 2010), depending on the price attached to a ton of carbon (see Annex 4). These estimates should be seen as a first indicative assessment of the value of the stock of carbon in the Natura 2000 network, and as a conservative estimate given that part of the network (marine and some terrestrial habitats) could not be fully covered in the analysis.	Value transfer/ extrapolation based on the price of carbon (see Annex 4)	Ding *et al.*. in ten Brink *et al.*, 2011
USA, Catskill State Park (several ecosystems)	Water purification and supply	The Catskill/Delaware watershed provides 90 per cent of the 1.4 billion gallons of drinking water consumed by New York City. The ecosystems within the watershed capture, store and naturally purify the water and hence represent one of the few sources of natural, unfiltered water in the country. In 1997 the City launched a long-range watershed protection programme aimed at improving the quality of its water that was being threatened by waste treatment plants, and agricultural and urban inputs, increasing the risk of contamination above acceptable quality standards. The strategy's start-up cost was estimated at US$1 to US$1.5 billion over ten years, as opposed to US$6–8 billion, plus an annual US$300–500 million operating costs, that would have been required for a treatment plant.	Replacement cost method	Wunder and Wertz-Kanounnikoff, 2009 Dudley and Stolton, 2003 Mehaffey *et al.*, 1999

Protected area	*Benefits (ecosystem services and related goods)*	*Estimated socio-economic value of benefits*	*Method*	*Source*
Chingaza National Park, Bogota, Colombia (several ecosystems)	Water purification and supply	The city of Bogota, Colombia, gets most of the drinking water for its 8 million citizens from the watershed within the Chingaza National Park. The wetland ecosystem within the area and its vegetation absorbs, filters and releases the clean water at a reliable rate of 28 m3 per second with little seasonal variation and minimal need for treatment. The delivered raw water is so clean that the only treatment given is chlorine for disinfection.	Quantitative assessment	Postel and Thompson, 2005
Nakivubo Swamp, Uganda (wetland)	Wastewater treatment	The economic value of the wastewater treatment services provided by the Nakivubo Swamp in Uganda was estimated to be between US$1 million a year (using replacement cost methods) and US$1.75 million a year (using mitigation expenditures to offset the effects of wetland loss). The data for the assessments was already available from a previous feasibility study. Based on the economic argument and recognising the role of the swamp in Kampala's water and sanitation infrastructure, Nakivubo was designed as part of the city's greenbelt zone (UNDP-UNEP 2008).	Replacement costs Mitigation expenditures	Emerton, 2003b UNDP-UNEP Poverty-Environment Facility, 2008
Mangroves forests, North Vietnam	Natural hazards protection	Mangroves act as natural protection against the impacts of tidal and storm surges. This was illustrated after the devastating typhoon Wukong in Vietnam in 2000: areas where mangrove forests acted as natural buffer zone remained relatively unharmed whereas neighbouring areas without this protection suffered significant losses of life and property. The plantation of the mangrove forests in the northern parts of the country was the result of the efforts of the Vietnam National Chapter of the Red Cross since 1994. Note: While highly relevant in the context of this guidance this example is not specifically related to PAs.	Qualitative assessment	Brown *et al.*, 2006

Continued

Protected area	*Benefits (ecosystem services and related goods)*	*Estimated socio-economic value of benefits*	*Method*	*Source*
Forests and mangroves cover, Haiti	Natural hazards protection	In Haiti, Hurricane Jeanne and related floods killed approximately 5,400 people as a consequence of the destruction of mangroves and the loss of soil-stabilising vegetation, causing landslides that led to most casualties. The link is being made between the level of environmental degradation and potential impacts of natural hazards, particularly when looking at significantly lesser casualties/impacts in neighbouring Dominican Republic.	Qualitative assessment	Peduzzi, 2005
Protection forests, Switzerland	Natural hazards protection	Around 17 per cent of Swiss forests are protected and managed mainly for their protective function against avalanches, rock falls and landslide protection. The benefits provided by these forests were valued at around US$2 and US$3.5 billion per year.	Avoided damage	ISDR, 2004 Dudley *et al.*, 2010
Hikkaduwa Marine Park Sri Lanka (mangroves)	Natural hazards protection, storm damage	Following the 2004 tsunami studies in Hikkaduwa marine park, Sri Lanka, noted that damage reached only 50 m inland and waves were only 2–3 m high. At nearby Peraliya, where reefs have been extensively affected by coral mining, the waves were 10 m high, and damage and flooding occurred up to 1.5 km inland.	Quantitative assessment	Dudley *et al.*, 2010
Whangamarino wetland, New Zealand	Natural hazards protection, floods	Ramsar's Whangamarino Wetland in New Zealand is the second largest in the country. The study by the Department of Conservation looked at the values provided by the wetland, including the flood control benefits. Looking at the past flood events in the area and the foreseen costs to the public sector to provide flood mitigation schemes of similar magnitude, the Whangamarino Wetland was estimated to have saved the regional council costs in the magnitude of several $US millions.	Avoided damage method	Department of Conservation, 2007 Dudley *et al.*, 2010

Protected area	*Benefits (ecosystem services and related goods)*	*Estimated socio-economic value of benefits*	*Method*	*Source*
Muthurajawela-Negombo wetland, Sri Lanka	Natural hazards protection, floods	Muthurajawela-Negombo wetland in Sri Lanka is one of the priority wetlands in the country, declared as a Wetland Sanctuary. An assessment of the economic benefits of wetland conservation and economic costs of wetland degradation and loss in Muthurajawela has been carried out. Among other benefits, the wetland provides flood attenuation services to the city of Colombo valued at around $US5,4 million per year, translating into $US1,700 per hectare per year. See Box 7.1 for further detail.	Replacement cost method	Emerton and Kekulandala, 2003 Emerton, 2003a
San Isidro del General, Costa Rica (forest)	Pollination	Forest-based pollination in Costa Rica increased the yields of coffee by 20 per cent within 1 km from the forest. It was estimated that the value of pollination services from nearby patches of forests to a local coffee plantation was around €50,000 per year. Note: While highly relevant in the context of this guidance this example is not specifically related to PAs.	Production function approach	Ricketts *et al.*, 2004

PA managers and/or other land users (farmers, foresters, landowners, local communities, etc.) play a key role in maintaining regulating services (protection, sustainable management and/or restoration of ecosystems' natural functioning, etc.). For example, management practices influencing the vegetation cover in PAs have a direct impact on climate, air and noise regulation, as well as on erosion control and natural hazards mitigation. The maintenance of healthy and diverse vegetation cover supports ecosystems' ability to remove pollutants, pests and assists disease regulation or pollination services. Maintaining riparian vegetation and avoiding deforestation has a direct beneficial influence on water purification and wastewater treatment.

As highlighted in Chapter 2, there can be conflicts between optimising the provisioning of different benefits. These conflicts should be particularly considered in the case of regulating services. For instance, maximising timber provision or agricultural production might adversely affect some of the regulating services, such as water purification or natural hazards mitigation.

7.4 Estimating the value of benefits

The following section will outline the main tools used in estimating the socio-economic value of regulating services and related goods, supported by a range of concrete examples. Detailed step-by-step guidance and further case studies on water and climate related regulating services are provided in Annexes 3 and 4. An introduction to the different valuation methods is available in Chapter 3.

A range of tools exists to estimate the socio-economic value of regulating services and related goods. A combination of methods, including both monetary and non-monetary valuation, is needed to provide a comprehensive picture of regulating services provided by PAs. The choice of methods needs to be considered with respect to time, expertise and resources available. A number of basic valuation methods (see below and Annexes 3 and 4) can help to provide a useful first-cut assessment of the socio-economic value of regulating services. However, more comprehensive assessments require dedicated information and expertise, both in terms of biogeographical information and economic valuation of benefits. Such assessments also entail the need for cooperation with relevant researchers and professionally trained experts (e.g. expert in ecological-economic modelling and GIS methods, see for example UK NEA, 2011 and Natural Capital Project, 2012).

Monetary: market values

The production function approach isolates the effects of ecosystem services as inputs to the production process. It is a potential tool for estimating the value of such regulating services that support the production of goods sold on established markets. For instance, isolating the effects of pest control or pollination as an input to agriculture can indicate the value of these ecological processes for food production (see Box 7.1). The production function approach can also be applied in the context of natural hazards mitigation (see below and Annex 3). Constructing production function requires detailed understanding of the ecological processes of PAs and the contribution of a particular service into the production of a given good. As such, it has particular demands in terms of ecological and economic data.

BOX 7.1

Estimating the monetary value of Anolis lizard as a pest control agent by using production function approach

The Greater and Lesser Antilles, which are home to the Anolis lizard, rely heavily on the export of primary commodities like sugar cane, banana and cocoa for their economic well-being. The Anolis lizard preys on pests that adversely affect the production of these commodities, this way providing a valuable regulating service to the area and decreasing the costs of cultivation by reducing the demand for pesticides. A study by Arain and Fisher (1995) provides an interesting example on how the production function might be applied to estimate the economic value of regulating services. However, it is to be noted that the results of the study are limited by the lack of available data.

Method: The output from agricultural sector was estimated as a function of a number of key inputs: labour, capital and fertilisers. The value of the Anolis lizard was determined as the value of the modelled change in output caused by the change in the lizard population (value of lost output = decrease in output x unit price of output). This approach assumes that no substitute for the lizard is available at a cost less than the value of lost output. Please see the full article by Narain and Fisher (1995) for more detailed technical information on the approach.

Results: Assuming a decline of 50 per cent in the lizard population, the estimated value of lost output would be US$67 million. In the case of a loss of all output – following a hypothetical catastrophic decline in the lizard population – the estimated value would be US$455 million.

Source: Narain and Fisher (1995)

Monetary: non-market values

Most of the benefits provided by regulating services and related goods embody indirect use values not associated with existing markets (see Chapter 3). However, cost-based approaches and stated preference methods can be used to demonstrate their economic value.

Cost-based approaches

Cost-based approaches are often used to assess the economic value of ecosystems to mitigate natural hazards, control erosion, maintain pollination and regulate water quality. In essence, these approaches estimate 1) costs of providing benefits by different, usually artificial, means such as substituting regulating services via water treatment plants or manual pollination (substitute or replacement costs); 2) costs of restoring lost benefits (restoration costs); or 3) avoided costs of damage that would have occurred in the absence of regulating ecosystems (avoided costs). Cost-based valuation methods are based on the assumption that the costs incurred to avoid potential damages and/or to replace benefits provide a fair reflection of the socio-economic value of benefits currently provided by ecosystems. It is to be noted that the estimated value of regulating services depends significantly on the alternative solution chosen to replace the service. For example, in the case of water purification a range of alternative pre-treatment solutions, with a range of different costs, could be used to achieve the same end quality.

Cost-based approaches are a relatively easy means to obtain a first-cut estimate of the socio-economic value of regulating services and related goods. However, given the range of assumptions affecting the valuation the results should be considered indicative only. Boxes 7.2 and 7.3 and Annexes 3 and 4 provide more detailed information on how cost-based approaches can be applied.

Finally, the production function approach (above) can be used to estimate the damages expected to be avoided by conserving ecosystems' regulatory functions (e.g. Barbier, 2007; Bateman *et al.*, 2011). While this so-called 'expected damage function approach' provides a more nuanced way for estimating the economic value of regulating services it also requires detailed understanding of a PA's ecological characteristics and their role in preventing damage (see Annex 3 for a detailed example).

BOX 7.2

Assessment of the economic value of Muthurajawela Wetland Sanctuary based on the replacement costs of flood attenuation

An economic assessment of the benefits provided by the Muthurajawela Wetland Sanctuary (Ramsar site), Sri Lanka, has been undertaken by the IUCN. In this context, benefits related to flood attenuation were valued using the replacement cost approach. Flood mitigation is of significant importance in the area because the continued degradation of Muthurajawela has led to an increase in the frequency of floods, with around 1,000 people being affected by flooding during every rainfall period.

Method: An artificial drainage system supported by a pumping station would need to be constructed to replicate the flood attenuation benefits currently provided by the wetland. To estimate the costs of such engineering works, the study looked at the occurred construction costs of a drainage system in a nearby area. These measures, including widening and deepening of water-courses and retaining and pumping out water, had resulted in costs of Rs2.11 million (around US$233,000) per hectare. In addition to these one-off costs, the maintenance of flood protection infrastructure resulted in reoccurring costs of around 3 to 4 per cent of the investment costs per year. Extrapolating these costs to the Muthurajawela sanctuary provided an indication of the economic value of flood mitigation benefits provided by the wetland.

Results: The economic value of flood attenuation benefits was estimated to be around US$5.4 million per year (or US$1,700 per hectare per year). Given the overall estimated benefits of the wetland at around US$8 million, flood attenuation represents the most economically valuable benefit provided by the marsh.

Source: Emerton (2003a), Emerton and Kekulandala (2003)

BOX 7.3

Using replacement cost approach to estimate the value of pollination to the fruit industry in Western Cape, South Africa

Insect pollination plays an important role in the agricultural production of crops used for human consumption. However, studies aiming to estimate the economic value of insect pollination are scarce. Moreover, they seldom distinguish between the value of managed and wild pollination, the latter providing clear evidence in support of conservation initiatives and sustainable land management. The study in Western Cape, SA used a replacement approach to value both managed and wild pollination services for the deciduous fruit industry within the region.

Method: Pollination by mechanical means was considered to be the only viable alternative to insect pollination, including hand pollination or pollen dusting via aircraft/helicopters supported by hand-operated pollen blowers. The study estimated the value of pollination (wild insects and managed honeybees) based on the costs of pollen dusting and three hand-pollination methods. A set of assumptions was adopted related to the relative efficacy of these methods and their costs (e.g. hand pollination providing equal outputs to insect pollination with pollen dusting resulting in significantly lower yields).

Results: The study estimated the value of the managed honeybee pollination to be between US$28–US$122.8 million and for wild pollination to be between US$49.1–US$310.9 million.

Source: Allsopp *et al.* (2008)

Revealed and stated preference methods

Hedonic pricing, a commonly used revealed preference method, establishes a relationship between property prices and surrounding environment by assuming that the price of a property reflects the environmental attributes of the area. Hedonic pricing can be used to assess the contribution of regulating services into these prices (Chapter 3). For instance, the proximity of well-functioning ecosystems maintained by PAs might have a positive impact on the property prices whereas increased vulnerability to environmental hazards and/or high noise or pollution levels might decrease the prices. It was estimated that the floods of the Meuse River in the Netherlands in 1993 and 1995 resulted in a permanent devaluation of the houses affected by flooding (around a 9 per cent decrease in property prices) (Daniel *et al.*, 2009). While no concrete examples related to PAs yet exist, revealed preference method could provide a suitable method for estimating the monetary value of PAs in mitigating natural hazards.

Most of the regulating services and related goods do not have established markets and therefore stated preference methods often provide the only suitable approach for estimating their monetary value (Box 7.4). The outcomes of stated preference studies depend heavily on respondents' subjective perceptions of value and therefore the less people are familiar with the benefits – and their impacts on well-being – the less accurate the results are. This might limit the ability of these methods to fully reflect the value of regulating services and related goods. Careful study design, planned and conducted by trained experts, is therefore a prerequisite for robust and reliable results.

BOX 7.4

Using contingent valuation to estimate the shoreline protection function of mangroves in Benut, Malaysia

A contingent valuation of mangrove ecosystems in Benut, Malaysia, was carried out to raise awareness of their socio-economic importance and benefits (53,000 hectares, 48 villages, 1,640 inhabitants). Among other things, the study focused on estimating the value of mangroves as a means to protect the shoreline, safeguarding the agricultural land situated behind the mangroves and preventing shoreline erosion and possible saltwater intrusion.

Method: A willingness-to-pay survey was distributed to 243 households. Maps, written descriptions and graphics were used to communicate the benefits of mangroves to local communities. Based on the information provided, the respondents were then asked to state their willingness-to-pay to a fund supporting mangrove protection.

Results: The study estimated the value of the shoreline protection of mangroves to be US$845 per hectare per year. In general, the shoreline protection was estimated to be the most valuable benefit provided by coastal mangroves, making up more than half of the total estimated benefits in the study area (US$1,375/ha/year). The second most significant benefits were capture fisheries, estimated at US$526 per hectare per year. The study argues that given the high economic values associated with mangroves in the area – and indeed its exceptional biodiversity value – Benut should be designated into a PA in the future.

Source: Bann (1999), Kumar *et al.* (2011)

Non-monetary values

Due to the range of identified context specificities and challenges (see Section 7.1 above), non-monetary valuation has an important role to play when assessing the socio-economic value of regulating services and related goods. Moreover, these valuation approaches are useful in complementing – or even substituting – monetary valuation.

The quantification of PAs' ecological properties and their ability to provide regulating services (current or potential) forms the basis for monetary valuation (see above and Box 7.5). In addition, quantitative evidence can be directly used to highlight the socio-economic importance of PAs' regulative functions. The ability to attenuate floods, for instance, can be assessed by measuring the amount of water that the ecosystem within PAs can store. Similar biophysical assessments can be carried out for most of the regulating services and related goods (water quality, removal of nutrients, rate of air pollution removal, etc.), however they require detailed understanding of the ecological properties of the ecosystems, often supported by field studies. Employing geographic information system (GIS) tools can help to highlight the spatial dimension of regulating services and related goods (Box 7.6). GIS tools combined with biophysical assessments serve as a crucial input into some of the more advanced monetary valuation approaches (Box 7.5).

Finally, qualitative valuation methods, such as participatory approaches supporting the recognition and valuation of benefits, are important tools in assessing the socio-economic dimensions of regulating services and related goods. In many cases, such assessment might sufficiently highlight the regulatory benefits obtained from PAs, helping to identify socio-economic values which are not easily amenable to monetary valuation (health benefits, etc.) or providing a stimulus for further studies in the area (e.g. Espinosa, 2005). For instance, recognising the ecological connection between a PA and surrounding agricultural areas can serve as a first indication of a site's role in supporting the pollination of crops in the area.

BOX 7.5

Valuation of wetland flood mitigation based on biophysical indicators and geospatial mapping in Momoge National Nature Reserve, China

Located in the alluvial floodplain of the Nen River in North-eastern China, the Momoge National Reserve is comprised of inland saline wetlands typical to the region. During the last century large areas of the Nen River floodplain were drained and converted into farmlands, making the area one of the most important sources for producing grain in China. However, agricultural production in the area is frequently threatened by floods from the Nen River, with the floods in 1998 causing over $US3.75 billion direct negative impacts to the economy. A valuation of the flood mitigation benefits of the Momoge National Reserve was carried out, including an ecological assessment of the site as well as monetary estimation of the benefits.

Method: An existing ecological assessment of the site's ability to mitigate floods formed the basis for monetary valuation. This included: 1) dividing the area into three distinct areas according to ecological characteristics (soil, vegetation and hydrological characteristics), 2) obtaining a total of 245 soil samples from the area, 3) determining the key soil characteristics based on the samples (saturation water content, natural water content, bulk weight) and 4) carrying out statistical analysis of the data collected. Employing spatial analysis, the volume of flood mitigation (maximum level of water saturation) for each of the three areas was calculated. Building on this information, the benefits of natural flood mitigation were calculated by using the replacement cost approach. The costs were estimated as the average costs of investment in man-made reservoir construction in China per cubic metre of water throughout the period 1988–2004. The economic value of flood mitigation by the Nen River was then estimated by multiplying by the amount of water (m^3) stored in the wetlands by the average costs of flood mitigation infrastructure per m^3.

Results: The study estimated the flood mitigation benefits in terms of the overall quantity of water stored in the wetland (i.e. flood mitigation capacity) – to be 10.3 billion m³. When the replacement cost method was applied ($US0.08/m³ representing the unit cost of replacement), the value of the flood mitigation in Momoge National Reserve was estimated to be $US57/km² year.

Source: Ming *et al.* (2007)

BOX 7.6

Mapping the pollination potential of ecosystems

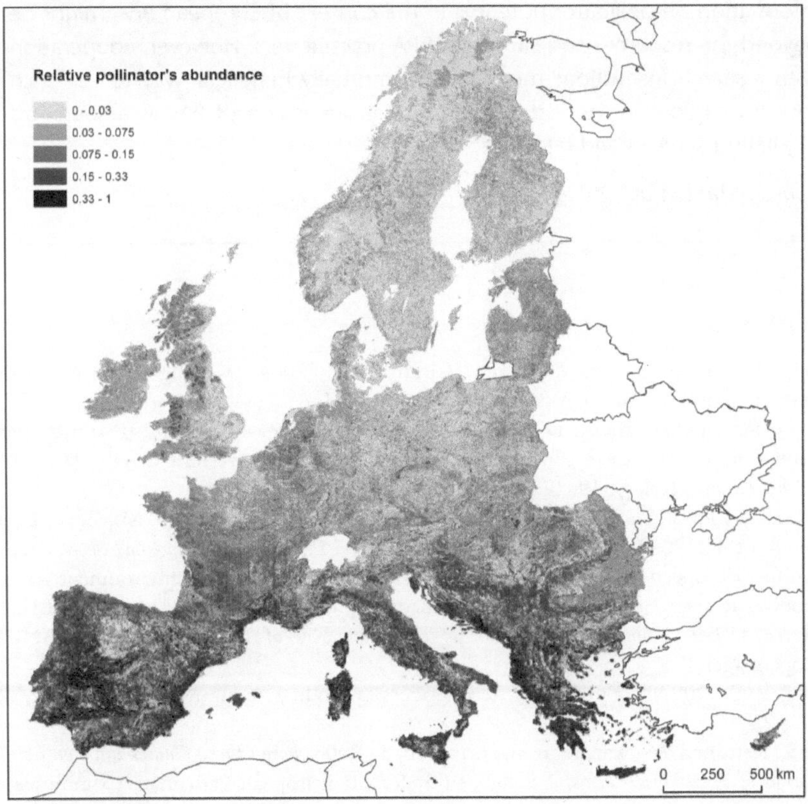

FIGURE 7.1 Relative pollinator abundance across Europe

Work in the area of mapping the biophysical availability of ecosystem services has been carried out within the PRESS project (PEER Research on EcoSystem Services – Phase 2). Alongside case studies looking at the pollination ecosystem service across Europe, a map for the pollination potential of Europe has been developed (Maes *et al.*, 2012; see Figure 7.1 above). The map was developed based on the following steps. Spatial data on land cover and land use were transformed into indicators for nesting suitability and floral resource availability. These indicators were combined with climate data in order to simulate the pollinator activity and map the relative pollination abundance at a landscape scale. The mapping approach used was based on the InVEST model of the Natural Capital Project (see Chapter 5). The report notes that around 23.6 per cent of the total production of crops, depending on pollination, could be assigned to insect pollination at aggregated EU level; corresponding to a production deficit if no pollination services were offered by insects.

Using elaborate mapping approaches, including attempting to use the information on pollinator potential in the context of European PAs, might be beyond the resources and capacity of PA practitioners. However, cooperation with research institutions might offer a mutually beneficial way to use such tools to explore in more detail the spatial dimension of PAs as a source of regulating services and related goods.

Source: Maes *et al.* (2012)

References

Allsopp, M. H., de Lange, W. J. and Veldtman, R. (2008) 'Valuing Insect Pollination Services with Cost of Replacement', *PLoS ONE*, 3(9)

Balch, J. K., Bradley, B. A., D'Antonio, C. M. and Gómez-Dans, J. (2013) 'Introduced annual grass increases regional fire activity across the arid western USA (1980–2009', *Global Change Biology*, 19: 173–183

Balmford, A., Rodrigues, A. S. L., Walpole, M., ten Brink, P., Kettunen, M., Braat, L. and de Groot, R. (2008) *The Economics of Biodiversity and Ecosystems: Scoping the Science*, a report for the European Commission, Cambridge, http://ec.europa.eu/environment/nature/biodiversity/economics/pdf/scoping_science_report.pdf, accessed 17 December 2012

Bann, C. (1999) *Contingent Valuation of the Mangroves of Benut*, report to DANCD, Copenhagen

Barbier, E. B. (2007) 'Valuing ecosystem services as productive inputs', *Economic Policy*, 22: 177–229

Bassi, S., Kettunen, M., Kampa, E. and Cavalieri, S. (2008) *Forest Fires: Causes and Contributing Factors to Forest Fire Events in Europe*, study for the European Parliament Committee on Environment, Public Health and Food Safety, http://edz.bib.uni-mannheim.de/daten/edz-ma/ep/08/EST19989.pdf, accessed 23 December 2012

Bateman, I., Mace, G., Fezzi, C., Atkinson, G. and Turner, K. (2011) 'Economic Analysis for Ecosystem Service Assessments', *Environmental & Resource Economics*, 48(2): 177-218

Braat, L. and ten Brink, P. (eds) with J. Bakkes, K. Bolt, I. Braeuer, B. ten Brink, A. Chiabai, H. Ding, H. Gerdes, M. Jeuken, M. Kettunen, U. Kirchholtes, C. Klok, A. Markandya, P. Nunes, M. van Oorschot, N. Peralta-Bezerra, M. Rayment, C. Travisi, M. Walpole. (2008) *The Cost of Policy Inaction. The case of not meeting the 2010 biodiversity target*, project for the European Commission, http://ec.europa.eu/environment/nature/biodiversity/economics/teeb_en.htm, accessed 10 December 2012

Brown, O., Crawford, A. and Hammill, A. (2006) *Natural Disasters and Resource Rights: Building Resilience, Rebuilding Lives*, International Institute for Sustainable Development, Manitoba, Canada

Campbell, A., Miles, L., Lysenko, I., Hughes, A. and Gibbs, H. (2008) *Carbon Storage in Protected Areas: Technical Report*, UNEP-WCMC, http://www.unep-wcmc.org/medialibrary/2010/09/24/d8a43698/Carbon_storage_PAs.pdf, accessed 12 December 2012

Daniel, V. E., Florax, R. J. G. and Rietveld, P. (2009) 'Floods and Residential Property Values: A Hedonic Price Analysis for the Netherlands', *Built Environment*, 35: 563–576

Department of Conservation (2007) *The economic values of Whangamarino Wetland*, http://www.doc.govt.nz/publications/conservation/benefits-of-conservation/the-economic-values-of-whangamarino-wetland, accessed 12 December 2012

Ding, H., Nunes, P. and Markandya, A. (2011) 'Climate regulation: carbon storage' in P. ten Brink *et al. Estimating the Overall Economic Value of the Benefits provided by the Natura 2000 Network*, final report to the European Commission, Institute for European Environmental Policy/GHK/Ecologic, Brussels, http://www.ieep.eu/publications/2012/06/estimating-the-overall-economic-value-of-the-benefits-provided-by-the-natura-2000-network, accessed 12 December 2012

Dudley, N. and Stolton, S. (2003) *Running Pure: The importance of forest protected areas to drinking water*, World Bank/WWF Alliance for forest conservation and sustainable use, WWF, Gland, Switzerland

Dudley, N., Stolton, S., Belokurov, A., Krueger, L., Lopoukhine, N., MacKinnon, K., Sandwith, T. and Sekhran, N. (eds) (2010) *Natural Solutions: Protected areas helping people cope with climate change*, IUCN-WCPA, TNC, UNDP, WCS, The World Bank and WWF, Gland, Switzerland, Washington, DC and New York

Emerton, L. (2003a) 'Muthurajawela Marsh, Sri Lanka: Safeguarding wetland protected areas in cities', in *Integrating Wetland Economic Values into River Basin Management: Case Studies in Wetland Valuation #1*, Water and Nature Initiative and Ecosystems and Livelihoods Group, IUCN Asia, Colombo, http://cmsdata.iucn.org/downloads/casestudy01muthurajawela.pdf, accessed 23 December 2012

Emerton, L. (2003b) 'Nakivubo Swamp, Uganda: managing natural wetlands for their ecosystem services', Integrating Wetland Economic Values into River Basin Management: Case Studies in Wetland Valuation #7, Water and Nature Initiative and Ecosystems and Livelihoods Group IUCN Asia, Colombo, http://cmsdata.iucn.org/downloads/casestudy07nakivubo.pdf, accessed 23 December 2012

Emerton, L. and Kekulandala, L. D. C. B. (2003) *Assessment of the Economic Value of Muthurajawela Wetland*, IUCN — The World Conservation Union, Sri Lanka Country Office and Regional Environmental Economics Programme Asia, Colombo

Espinosa, C. (2005) *Payment for Water-Based Environmental Services: Ecuador's Experiences, Lessons Learned and Ways Forward*, IUCN Water, Nature and Economics Technical Paper No. 2, IUCN — The World Conservation Union, Ecosystems and Livelihoods Group Asia, Colombo, http://data.iucn.org/dbtw-wpd/edocs/2005-104.pdf, accessed 23 December 2012

Gallai, N., Salles, J. M., Settele, J. and Vaissiere, B. E. (2009) 'Economic valuation of the vulnerability of world agriculture confronted with pollinator decline', *Ecological Economics*, 68(3): 810–821

IPCC – Intergovernmental Panel on Climate Change (2012) 'Summary for Policymakers,' in C. B. Field *et al.* (eds) *Managing the Risks of Extreme Events and Disasters to Advance*

Climate Change Adaptation, A Special Report of Working Groups I and II of the Intergovernmental Panel on Climate Change, Cambridge University Press, Cambridge and New York

ISDR – International Strategy for Disaster Reduction (2004) *Living with Risk: A global review of disaster reduction initiatives*, UN/ISDR, Geneva, Switzerland, www.preventionweb.net/files/657_lwr1.pdf, accessed 23 December 2012

Kettunen, M., Genovesi, P., Gollasch, S., Pagad, S., Starfinger, U., ten Brink, P. and Shine, C. (2008) *Technical support to EU strategy on invasive species (IAS) – Assessment of the impacts of IAS in Europe and the EU*, final report to the European Commission, Institute for European Environmental Policy (IEEP), Brussels, Belgium, http://ec.europa.eu/environment/nature/invasivealien/docs/Kettunen2009_IAS_Task%201.pdf, accessed 23 December 2012

Kettunen, M., Berghöfer, A., Brunner, A., Conner, N., Dudley, N., Ervin, J., Gidda, S. B., Mulongoy, K. J., Pabon, L. and Vakrou, A. (2011) 'Recognising the value of protected areas', in P. ten Brink (ed.) *The Economics of Ecosystems and Biodiversity in National and International Policy Making*, Earthscan, London and Washington, DC

Kettunen, M., Vihervaara, P., Kinnunen, S., D'Amato, D., Badura, T., Argimon, M. and ten Brink, P. (2013) *Socio-economic importance of ecosystem services in the Nordic Countries – Synthesis in the context of The Economics of Ecosystems and Biodiversity (TEEB)*, Nordic Council of Ministers, Copenhagen

Klein, A., Vaissière, B. E., Cane, J. H., Steffan-Dewenter, I., Cunningham, S. A., Kremen, C. and Tscharntke, T. (2007) 'Importance of pollinators in changing landscapes for world crops', *Proceedings of the Royal Society of London B*, 274: 303-313

Kremen, C., Williams, N. M., Bugg, R. L., Fay, J. P. and Thorp, R. W. (2004) 'The area requirements of an ecosystem service: crop pollination by native bee communities in California', *Ecology Letters*. 7: 1109–1119

Kumar, P., Verma, M., Wood, M. and Negandhi, D. (2011) *Guidance manual for the valuation of regulating services*, Ecosystem Services Economics (ESE) Working Paper Series, Nairobi, UNEP

MA – Millennium Ecosystem Assessment (2005a) *Ecosystems and Human Well-being: Synthesis*, Island Press, Washington, DC

MA – Millennium Ecosystem Assessment (2005b) *Ecosystems and Human Well-Being: Wetlands and Water Synthesis*, World Resources Institute, Washington, DC

Maes, J., Hauck, J., Paracchini, M. L., Ratamäki, O., Termansen, M., Perez-Soba, M., Kopperoinen, L., Rankinen, K., Schägner, J. P., Henrys, P., Cisowska, I., Zandersen, M., Jax, K., La Notte, A., Leikola, N., Pouta, E., Smart, S., Hasler, B., Lankia, T., Andersen, H. E., Lavalle, C., Vermaas, T., Alemu, M. H., Scholefield, P., Batista, F., Pywell, R., Hutchins, M., Blemmer, M., Fonnesbech-Wulff, A., Vanbergen, A. J., Münier, B., Baranzelli, C., Roy, D., Thieu, V., Zulian, G., Kuussaari, M., Thodsen, H., Alanen, E. L., Egoh, B., Sørensen, P. B., Braat, L. and Bidoglio, G. (2012) *A spatial assessment of ecosystem services in Europe: methods, case studies and policy analysis – phase 2. Synthesis report.* PEER Report No 4. Ispra: Partnership for European Environmental Research, http://www.peer.eu/fileadmin/user_upload/publications/PEER_report_4_phase_2.pdf, accessed 10 January 2013

Ming, J., Xian-guo, L., Lin-shu, X., Li-juan, C. and Shouzheng, T. (2007) 'Flood mitigation benefit of wetland soil — A case study in Momoge National Nature Reserve in China', *Ecological Economics*. 61: 217–223

Mulongoy, K. J. and Gidda, S. B. (2008) *The Value of Nature: Ecological, Economic, Cultural and Social Benefits of Protected Areas*, Secretariat of the Convention on Biological Diversity, Montreal

Narain, U. and Fisher, A. (1995) Modelling the value of biodiversity using a production function approach: the case of the Anolis lizard in the lesser and greater Antilles', in Perrings, C. A. (ed) *Biodiversity Conservation*, Kluwer Academic Publisher. http://books.google.be/books?id=5Ee0PTsH5B4C&printsec=frontcover&hl=it&source=gbs_ge_summary_r&cad=0#v=onepage&q=anolis&f=false, accessed 23 December 2012

Natural Capital Project (2012) InVEST tool, http://www.naturalcapitalproject.org, accessed 1 December 2012

Peduzzi, P. (2005) 'Tropical cyclones: Paying a high price for environmental destruction', *Environment and Poverty Times*, 3:3, www.grida.no/publications/et/ep3/page/2587.aspx, accessed 23 December 2012

Postel, S. L. and Thompson, B. H. (2005) 'Watershed protection: Capturing the benefits of nature's water supply services', *Natural Resources Forum*, 29: 98–108

Ruijgrok, E. M. C. (ed.) (2006) *Kentallen Waardering Natuur, Water, Bodem en Landschap. Hulpmiddel bij MKBAs*, http://www.mkbainderegio.nl/docs/Kentallen_waardering_natuur_water_bodem_en_landschap.pdf, accessed 23 December 2012

Ricketts, T. H., Daily, G. C., Ehrlich, P. R. and Michener, C. D. (2004) 'Economic value of tropical forest to coffee production', *Proceedings of the National Academy of Sciences*, 101: 12579–12582

Russi, D., ten Brink, P., Farmer, A., Badura, T., Coates, D., Förster, J., Kumar, R. and Davidson, N. (2012) *The Economics of Ecosystems and Biodiversity for Water and Wetlands.* Final Consultation Draft, http://www.ramsar.org/pdf/TEEB/TEEB_Water-Wetlands_Final-Consultation-Draft.pdf, accessed 23 December 2012

Stolton, S. and Dudley, N. (2010) *Vital Sites: The contribution of protected areas to human health*, WWF International, Gland, Switzerland

ten Brink, P., Badura, T., Bassi, S., Daly, E., Dickie, I., Ding, H., Gantioler, S., Gerdes, H., Kettunen, M., Lago, M., Lang, S., Markandya, A., Nunes, P. A. L. D., Pieterse, M., Rayment, M. and Tinch, R. (2011) *Estimating the Overall Economic Value of the Benefits provided by the Natura 2000 Network*, final Report to the European Commission, Institute for European Environmental Policy/GHK/Ecologic, Brussels, http://www.ieep.eu/publications/2012/06/estimating-the-overall-economic-value-of-the-benefits-provided-by-the-natura-2000-network, accessed 12 December 2012

UK NEA – National Ecosystem Assessment (2011) *The UK National Ecosystem Assessment: Synthesis of the Key Findings*, UNEP-WCMC, Cambridge, http://uknea.unep-wcmc.org, accessed 18 June 2012

UNDP–UNEP Poverty-Environment Facility (2008) *Making the Economic Case: A Primer for Mainstreaming Environment in National Development Planning*, http://www.undp.org.gy/web/documents/bk/Making-the-economic-case-primer.pdf, accessed 23 December 2012

Vittor, A. Y., Pan, W., Gilman, R. H., Tielsch, J., Glass, G., Shields, T., Sánchez-Lozano, W., Pinedo, V. V., Salas-Cobos, E., Flores, S. and Patz, A. J. (2009) 'Linking deforestation to malaria in the Amazon: Characterization of the breeding habitat of the principal malaria vector, *Anopheles darling*', *American Journal of Tropical Medicine and Hygiene*, 81: 5-12

Wunder, S. and Wertz-Kanounnikoff, S. (2009) 'Payments for ecosystem services: A new way of conserving biodiversity in forests', *Journal for Sustainable Forestry*, 28: 576–596

ANNEX 3

Step-by-step guidance: water-related ecosystem services

Badura, T. and Kettunen, M.

1 Introduction

This step-by-step guidance supports Chapter 7, providing a more detailed description of the key valuation methods available for assessing the socio-economic value of water-related regulating services. It outlines the use of the production function approach (Section 2) and gives step-wise guidance on how to apply replacement and avoided cost methods (Section 3). A number of concrete case studies are provided to illustrate the use of these methods in practice including storm protection (Box A3.1), water purification, nutrient retention and waste water treatment (Box A3.2), and flood mitigation (Box A3.3).

It has to be noted that valuing regulating services and related goods is considered to require resources and dedicated expertise (see Chapter 7). A solid understanding of both economic analyses and the ecological characteristics of a PA is often required. Nonetheless, the approaches outlined below (the cost-based approaches in particular) can also be carried out based on a more limited expertise, providing a useful first-cut estimate and possibly supporting and/or initiating further and more detailed analyses.

2 Production function approach

Barbier (2007) outlined a variation of the production function approach – so-called expected damage function – to value some of the ecosystem services which protect people and infrastructure against potential damage caused by natural hazards (flood and storm surge protection, prevention against erosion, etc.). In essence, the approach estimates the value as the reduction in the expected damage due to the presence of the ecosystem's regulating services. This is done through using econometric analysis to analyse the relationship between the ecosystem (e.g. mangrove forest cover, see

Box A3.1), expected disaster occurrence and their impacts. Crucially, the expected disaster occurrence and their impacts can be estimated from the past disaster events and their economic repercussions. For calculations of potential future costs, the use of a discount rate is necessary – preferably employing a set of discounts rates (for further guidance see Chapter 12 and Annex 6).

While providing an economically justified and robust approach for estimating the value of ecosystems' protective features, the production function method requires significant inputs with regards to both expertise and data and, consequently, requires the assistance of a trained expert. A detailed account of the methodology for using this approach is provided in Barbier (2007).

Concrete example: Box A3.1

BOX A3.1

Using expected damage function approach to estimate the value of Thailand's mangroves in protecting coastal areas against storm surges

Mangroves are known to act as natural buffer zones helping to decrease the damaging impacts of storm surges on coastal areas. In turn, the loss of coastal mangroves cover is likely to lead to an increase in the impacts of natural disasters on coastal communities. A study by Barbier (2007) in Thailand illustrated how the production function approach can be used to determine the role of coastal mangroves in mitigating damage to coastal areas. While focused on a broader coastal zone, the study is considered to provide a good example of the use of the method applicable also in the context of PAs.

Method: The economic value of mangroves' protective function was estimated by using the so-called 'expected damage function' approach. This method establishes a statistical relationship between mangrove forest cover and the occurrence and impacts of storm surges in Thailand. The study employed a so-called *count data model* to look at how a change in mangrove area influences the expected incidence of economically damaging natural disasters. The relationship between mangrove cover and level of damage is established based on a set of econometric/statistical methods, using data on population density, mangrove cover in the 21 coastal provinces of Thailand, and number of coastal disasters occurring in Thailand from the International Disaster Database.

Results: The results indicate that loss of mangrove area in Thailand increases the expected number of economically damaging natural disasters in coastal provinces. In particular, the estimations suggest that one km^2 loss of mangrove area results in an increase of expected storm damages of about US$585,000 per km^2. Moreover, looking at FAO and Thailand (Royal Thailand Forestry departments) estimates of the average annual rates of deforestation, it was possible to compute the welfare loses in storm protection in Thailand for the period 1996-2004. Taking the FAO deforestation rate estimate of 18 km^2 per year, the annual welfare loss was estimated to be around US$3.4 million and the net present value (see Annex 6) of this loss over the entire period ranged between US$16.1-US$19.5 million, depending on the choice of discount rate. Similarly, taking the Thailand annual deforestation estimate of 3.4 km^2, the estimated annual welfare loss is at US$0.65 million and the loss over the whole period ranges between US$3.1 to US$3.7 million.

Source: Barbier (2007)

3 Cost-based approaches

Cost-based approaches can offer a potentially useful tool to provide an indication of the socio-economic value of regulating services and related goods (see Chapter 7 and Section 7.4.2 for a general description of the approach). This is particularly the case when the data, recourse and/or expertise available do not allow for other, more detailed types of assessments. However, the result of cost-based approaches should be considered as indicative only as they might overstate the value of the benefits (see Barbier, 2007 for further details). The procedure for replacement and avoided costs valuation approaches is summarised in the following text.

Replacement cost method

Three key conditions form the basis for using the replacement costs method in the context of ecosystem services: 1) ensuring that the alternative, used to estimate the costs of replacement, results in similar benefits as the ecosystem service, 2) opting for the least-costly suitable alternative, to avoid deliberately inflating the costs and 3) ensuring that there is likely public demand for the suitable alternative (Barbier, 2007).

Step 1: Carry out an ecological assessment of the current biophysical availability of regulating services and related benefits. This is necessary to be able to estimate the costs of providing such benefits via alternative means. If relevant, this exercise might also assess the level of the benefit provided if the site would be better managed to increase the level of benefits. Examples of the information needed include the volume of water purified, amount of nutrients retained or level of flood attenuation supported by a site (Box 7.5 in Chapter 7). In some cases this type of assessment might be already available (see Box A3.2), whereas in others cases, proceeding with the valuation might require a dedicated biophysical assessment and/or study to be carried out. If resources are limited it might be possible to use data obtained from a similar site. However, estimates based on value transfer should be treated as indicative only and great care needs to be taken to ensure that the biophysical characteristics of the two sites are as comparable as possible.

Step 2: Identify the least-costly alternative to provide the benefits in question and estimate associated costs (constructing a water treatment plant, building and maintaining walls or levees against impacts of floods, etc.). The total costs of an alternative usually comprise costs of construction (one-off costs) and annual maintenance (reoccurring costs). As above, such cost assessments might be available (Box A3.2) or it might be also possible to use costs of construction and maintenance from a similar case elsewhere (Box 7.2 in Chapter 7).

Step 3: Finally, compare the estimated replacement costs with the costs of maintaining the PA (or a certain habitat within the PA) in a state it can deliver the regulative benefits (Box A3.2). The latter costs refer *de facto* to the estimated costs of PA management outlined in Chapter 11.

Concrete example: Box A3.2

Avoided costs

Step 1: Firstly, carry out an ecological assessment to identify and assess the benefit provided (current and potential level of flood protection provided by an ecosystem, etc.) – see Step 1 under the replacement cost method above.

Step 2: Identify and estimate the foreseen avoided costs, such as estimated property damage costs from flooding. This step generally builds on the ecological assessment and understanding of the site's ability to provide particular benefits. For instance, this step might require modelling of flood risks and likely impacts on the affected population centres or looking at past instances of such events and estimating their probable future impacts. If natural protective functions in the area are already scarce, it is also possible to use existing expenditures on mitigating the potential damages as an indication of costs (value of private insurance taken against floods or investment and maintenance costs of man-made levees, etc.). Summing up the existing expenditures across the population affected and by the risk of flooding can provide an indicative estimation of the economic value of the benefits of existing and/or increased flood protection. However, it is important to note that the two above cost estimates will result in different types of final results.

Step 3: Finally, compare the estimated (expected) avoided costs with the costs of maintaining the PA (or a certain habitat within the PA) in a state it can deliver the regulative benefits (Box A3.2). The latter costs refer *de facto* to the estimated costs of PA management outlined in Chapter 11.

Concrete example: Box A3.3

BOX A3.2

Using replacement and avoided costs to estimate the benefits provided by Nakivubo wetland in Uganda in treating wastewater and retaining nutrients

The Nakivubo wetland, lying at the southeast of Kampala District in Uganda, provides a range of important ecosystem services and related goods. Besides supporting a range of small-scale income generating activities for adjacent slum dwellers, the wetland performs an important socio-economic role in treating and purifying domestic and industrial waste and effluents, hence maintaining the quality of urban water supplies.

Method: The waste treatment and water purification benefits of the swamp were assessed using cost-based approaches. In particular, the study looked at the costs of replacing these services by artificial means (replacement costs approach) and the costs of measures required to be adopted to mitigate the effects of their loss (avoided costs approach). The data on such costs was already available from previous studies.

The quantity of wastewater loads entering the Nakivubo swamp was assessed. Based on a previous study on water quality monitoring, these were estimated to be equivalent to the raw sewage produced by around 465,000 people, 40 per cent of the total population of Kampala. Of these, around one fourth of the nutrients entering the wetland were estimated to be from industrial waste. Consequently, two valuation approaches were used to value the benefits of water purification and nutrient retention of the site.

Replacement costs: Two types of costs were considered to be required to replace the waste treatment and water purification benefits of the Nakivubo wetland. These were 1) construction of proper sewage and sanitation facilities in low-costs settlements around the wetland and 2) connecting the Nakivubo Channel to a sewage treatment plant and its extension to cope with the additional wastewater load. Based on the costs of an elevated pit latrine, shared between five households (around USh625,000 per household, around US$235), the average annual costs for the whole wetland-adjacent population were estimated at USh98 million a year (around US$36,800). Given that at the time of analysis the sewage treatment facility served around a fourth of the potentially required capacity, the costs of its extension were estimated to be around USh1.5 billion (around US$564,000). Consequently, resulting total estimated benefits using the replacement costs approach totalled around USh1.6 billion (around US$600,000).

Mitigating expenditures: Offsetting the loss of the waste treatment provided by the wetland would require moving the inflow for Kampala's water supply to an alternative location situated further away from the outlet of wastewaters. The existing estimates for the construction of a new water treatment plant translate into an average annual cost of USh2.65 billion (US$996,000).

In addition, the study assessed the costs of maximising the benefits of water treatment and purification at the scale required for an increasing population. This comprised of spreading the waste over a greater area, hence increasing the time for wastewater to run through the wetland. Costs associated with such works were estimated at USh350,000 (around US$130) and these should be subtracted from the estimated benefits.

Results: Summarising the above (i.e. estimated benefits using both techniques minus the costs of maximising the benefits provided by the wetland to the required level), the study estimated the value of water treatment and purification services of Nakivubo wetland at USh1,250–USh2,300 million per annum (around US$470,000–US$860,000).

As regards the practical implications of the study, it is important to consider the various trade-offs between the services provided by the site as well as certain limitations to the site's utilisation. Maximising the wastewater treatment of the swamp can threaten crop cultivation. In addition, there might be thresholds as to how much waste the wetland can assimilate which, when exceeded, might threaten ecological integrity of the site and have significant negative impacts on both local and/or regional biodiversity and the site's ability to provide ecosystem services.

Source: Emerton *et al.* (1998)

BOX A3.3

Assessing the economic value of flood control in That Luang Marsh in Vietnam by using avoided damage costs

That Luang Marsh is the largest of the urban wetlands in the Vientiane Prefecture, being around 20 km^2 in size and located on the eastern edge of the capital city of Lao PDR. Part of the wetland has been converted to rice cultivation and the remaining area is covered with permanent and seasonal aquaculture ponds, shrub and grassland and peat land. The marsh is judged to provide a range of ecosystem services and related goods, including provision of economically valuable goods (fisheries, farming and natural resource collection activities, etc.), maintenance of water quality and supply (sanitation and wastewater purification) and flood protection. A study conducted by IUCN, WWF, the Lao National Mekong Committee, and the Science Technology Environmental Office of Vientiane Municipality, aimed at assessing the values of the ecosystem services of the marsh, including its role in controlling floods.

Method: The economic value of flood control was valued using the avoided damage costs approach. The areas benefiting from flood protection and potential damage associated with flooding were identified based on existing information. Costs of potential damage were estimated for different units on the basis of land use and inundation potential (factory, house, crops, etc.). The future damage was estimated on the basis of land use projections and the unit value of damage. Finally, considering the hydrology of the area, the study also took into account that not even with an adequate system of drainage a complete reduction of inundation (and hence potential damages) cannot be fully avoided. As such, it considered different rates of inundation reduction given different scenarios.

Results: Based on the above, the study estimated the annual value of damages avoided in the That Luang Marsh 'regulating basin' by 2020 to be around US$2.8 million.

Source: Gerrard (2004)

4 Other possible valuation methods

A range of other methods can also be used to estimate the value of regulating services, including water-related services and goods. Some of these are described in other Annexes; see Annex 5 for the use of revealed and stated preference methods and Annex 4 for the estimation of carbon sequestration and storage values. Furthermore, as mentioned in Chapter 7, integrated modelling tools, use of GIS methods or non-monetary tools can also be applied.

Finally, cooperation with academia and/or non-profit sector might provide a good solution for addressing the level of complexity and overcoming the requirements for resources and expertise to estimate the socio-economic value of regulating services.

5 Key issues to be kept in mind

It is important to keep in mind that PAs' ability to regulate water underpins and/or is also closely linked with the availability of provisioning and cultural services and related goods. Consequently, PAs' economic value needs to be assessed with due reflection and understanding of these interlinkages, when estimating the economic value of multiple benefits (see Chapters 3, 7 and 12).

References

Barbier, E. B. (2007) 'Valuing ecosystem services as productive inputs', *Economic Policy*, 22: 177–229

Emerton, L., Iyango, L., Luwum, P. and Malinga, A. (1998) *The Present Economic Value of Nakivubo Urban Wetland, Uganda*, IUCN — The World Conservation Union, Eastern Africa Regional Office, Nairobi and National Wetlands Programme, Wetlands Inspectorate Division, Ministry of Water, Land and Environment, Kampala, http://data.iucn.org/dbtw-wpd/edocs/1999-047.pdf, accessed 26 December 2012

Gerrard, P. (2004) *Integrating Wetland Ecosystem Values into Urban Planning: The Case of That Luang Marsh, Vientiane, Lao PDR*, IUCN – The World Conservation Union Asia Regional Environmental Economics Programme and WWF Lao Country Office, Vientiane, http://www.mekongwetlands.org/Common/download/WANI_economics_ThatLuang%20Marsh.pdf, accessed 26 December 2012

ANNEX 4

Step-by-step guidance: climate change mitigation (carbon storage and sequestration)

Bassi, S. and ten Brink, P.

1 Introduction

Terrestrial ecosystems can regulate the Earth's climate by adding and removing carbon C (or its compound carbon dioxide, CO_2), from the atmosphere and storing it in biomass (wood, etc.) and soil. Many systems also continue to accumulate carbon in plants and soil over time, thereby sequestering additional carbon each year (Tallis *et al.*, 2011). Marine ecosystems – although not discussed in this guidance – also affect climate regulation via their effect on biogeochemical cycling and carbon sequestration (MA, 2005).

The first part of this guidance provides simplified step-wise guidance on how to assess the economic value of the carbon *stored* in PAs, providing a snapshot of PAs' role in mitigating climate change in the present time.

Ecosystems within PAs, if well managed and/or intact, will also sequester additional quantities of carbon over time through plant growth and ageing. This carbon will be gradually added to the existing carbon storage, combining both carbon in the living biomass and 'dead' carbon (dead wood, litter, soil). It is useful to assess both the level and value of carbon stored and sequestered by PAs over time.

The second part of this guidance provides a simplified site-level approach for estimating the economic value of carbon *sequestration*, aimed at supporting the identification of changes in the carbon balance with respect to physical changes in the size of a PA. These estimates can measure the carbon-related benefits of keeping PAs intact or focus on assessing the costs associated with the loss or mismanagement of PAs. For example, if a PA is under a threat from deforestation, one could calculate the potential carbon losses from losing parts of forested areas over time. Similarly, if a protected wetland is under threat from degradation, one could usefully assess the possible losses of carbon capture and storage due to degradation. In addition, assessing the value of the potential carbon sequestration associated with a possible increase in PA size can be useful to substantiate the benefits of the area increase and justify investment.

Some data on carbon stored and sequestered (per hectare) by different types of vegetation within different climatic and geographic conditions is already available. This data can be used in the context of the approaches outlined in this guidance.

In general, the characteristics of different plant species are a key determinant of how much carbon is taken up from the atmosphere and stored in the vegetation. An important feature is, for example, plant growth rate, which governs carbon inputs. Climatic factors also play an important role. Climates with relatively high temperatures, sufficient rainfall and long day lengths are most favourable for rapid plant growth, hence leading to the highest carbon sequestration rates (Alexandrov *et al.*, 1999, 2002). Equatorial regions where both rainfall and temperatures are high have the highest biological potential for carbon sequestration. Therefore, when choosing carbon storage and sequestration data, it is important that these are consistent with the habitat types and climate conditions which are present in the PA.

Furthermore, storage and sequestration rates tend to change over time, for example with vegetation age and with changes in climate. More sophisticated assessments of carbon sequestration should take into account such dynamic changes, integrating foreseen environmental and ecosystems changes in PAs into the future projection (see Chapter 13). However, such assessments require a significant amount of additional site-specific data and assumptions (climate conditions, composition of the forest, age of plants on the site and decomposition processes, etc.) and the complexity of these estimations is beyond the scope of this guidance.

2 Information required (carbon storage and price)

Carbon storage (quantity)

Values for ecosystems' capacity to store carbon can be obtained from existing literature and databases. Some examples of data sources include:

- A first approximation of per hectare values of carbon stocks for broad biome categories, based on WBGU (1998); these are shown in Table A4.1
- Estimates of minimum, medium and maximum carbon densities in 32 major world ecosystems, based on Olson *et al.* (1983 and 1985), as in CDIAC (undated). These data can be used to increase the sensitivity of an assessment by creating a range estimate (high – low). Alternatively, it is possible to simply rely on a central value (the 'revised medium carbon density' by Olson *et al.*) to produce a single average value.
- Country-specific data on carbon stocks in forest areas can be found in online databases, such as Mongabay database (2012), Food and Agriculture Organization (FAO) databases or in country/region specific reports and statistics (e.g. FAO, 2010).

TABLE A4.1 Estimates of global carbon stocks in vegetation and soils to 1 metre depth

Biome	*Area*	*Total carbon stocks (billion tons C)*			*Carbon stock per hectare (tons C/ha)*		
	Billion ha	*Vegetation*	*Soil*	*Total*	*Vegetation*	*Soil*	*Total*
Tropical forests	1.76	212	216	428	120	123	243
Temperate forests	1.04	59	100	159	57	96	153
Boreal forests	1.37	88	471	559	64	344	408
Tropical savannahs	2.25	66	264	330	29	117	147
Temperate grasslands	1.25	9	295	304	7	236	243
Deserts & semi-deserts	4.55	8	191	199	2	42	44
Tundra	0.95	6	121	127	6	127	134
Wetlands	0.35	15	225	240	43	643	686
Croplands	1.6	3	128	131	2	80	82
Total	15.12	466	2,011	2,477	31	133	164

Source: based on WBGU (1998) as in IPCC (2001)

Carbon value (monetary)

A number of different estimates of the economic value of carbon exist. These can be used to estimate the value of both carbon storage and sequestration. Some of these estimates try to capture the marginal cost of the damage caused by carbon emissions to society (social cost of carbon) whereas others reflect the marginal costs of reducing emissions in a specific country. Alternatively, if carbon taxes exist in a given country, the tax rate per ton of carbon could be used as a proxy for carbon value. In European countries, the current market price of carbon in the EU Emissions Trading Scheme (ETS) can also be used as a reference.

In general, the estimated social costs of carbon provide the most appropriate basis for assessing the benefits of carbon storage/sequestration, which reflects the avoided marginal damage from climate change to society. The marginal cost of reducing emissions instead can be very country specific and usually only reflects the cost of technologies and other measures available to mitigate emissions. This cost is therefore often smaller than the costs of marginal damage, leading to a lower estimated socio-economic value of maintaining carbon storage and/or sequestration capacity. Finally, the price of carbon (based on ETS) is simply a result of the interaction between supply and demand on (imperfect) markets whereas the level of carbon taxes depends largely on the results of political negotiations. Consequently, it does not directly and/or fully capture the benefits associated with maintaining carbon storage and sequestration and may therefore not be the most suitable approach for assessments.

Some examples of existing key sources and carbon values include (see also Table A4.2):

- The World Bank (2002) Adjusted Net Saving index relies on a conservative figure of US$20 per ton of carbon emitted (equivalent to US$5.45 per ton of CO_2) (in 2005 prices), based on Fankhauser (1994)
- The Stern Review (2007) estimates a social cost of carbon of around US$91.75 per ton of carbon (US$25 per ton CO_2) (in 2005 prices)
- Nordhaus (2008) proposes a price of US$11.43 per ton of carbon (US$42 per ton of CO_2) (in 2005 prices)
- EU ETS – the price of carbon in the ETS market in December 2012 was around €6 per ton of CO_2 (Point Carbon, 2012).

Several other estimates are also available. For example, early calculations by the IPCC (1996) estimate that the average social cost of carbon ranges between US$5 and US$125 per ton of carbon (in 1990 prices). After surveying the literature, Clarkson and Deyes (2002) propose a central value of US$105 per ton of carbon (in 2000 prices), with upper and lower values of US$90 and US$210. Tol (2005) calculates a mean value of US$93 per ton of carbon, based on over 100 estimates of the social cost of carbon in 2003. Other peer-reviewed studies generally report lower values, the mean being US$43.

Note that in some cases carbon prices can refer to CO_2 rather than C. CO_2 values can be converted into C by simply multiplying them by 3.67. For example, a carbon price of US$5.45 per ton of CO_2 is equivalent to US$20 per ton of C.

In practice, carbon values vary over time (generally increasing) and it is therefore prudent to estimate the socio-economic value of carbon by using a range of values, both for the current time and future periods (see Box A4.1).

Finally, it is to be noted that the estimates of carbon value need to be continuously revised in the light of improved understanding of the likely impacts of climate change and required policy action. The existing estimates have focused on meeting a 2 degree average global temperature rise. However, it now seems that a 4 or even 5 degree average rise in global temperature might be more likely (World Bank, 2012). Consequently, the estimated marginal costs of damage will be much higher, requiring higher carbon prices to be used.

BOX A4.1

Review of carbon values over time

A study by Bassi *et al.* (2011) carried out a literature review in order to determine a suitable monetary value for carbon stored in forests for the period of 2010-2020. Table A4.2 shows the range of carbon values used by Bassi *et al.* (2011), whereas Table A4.3 summarises the ranges from international studies reviewed by the study. While this study did not focus on PAs in particular, it is a useful example of making practical choices as regards to the selection of carbon values.

TABLE A4.2 Carbon value used in Bassi *et al.* (2011) (€/t)

GHG	*Range*	*2010*	*2020*
Carbon dioxide (CO_2) or CO_2 equivalent	Low	€17.2	€39
	High	€32	€56

TABLE A4.3 Range of values for CO_2 from international studies (a selection)

	Date	*2009 GBP*	*€/ton CO_2*
New Carbon Finance	May–09	36.8	46.1
DB Research	May–09	34.9	43.8
Barclays Capital	May–09	28.8	36.1
Société Générale	May–09	27.3	34.2
European Commission	IA 2008 (for 2020)	31.1	39.0
DECC (UK)	Latest: core	25.1	31.5
	Low	14.2	17.8
	High	31.3	39.3
French government	Value for 2010		32
	Value for 2020		56

Source: Bassi *et al.* (2011) based on EC (2008), DECC (2009) and Centre d'analyse stratégique (2009)

3 Calculating the value of carbon storage

The quantity and value of carbon stored in a given area can be estimated following a few simple steps outlined below. Box A4.2 provides a simple example of a theoretical application of this method. Box A4.3 shows a real example, based on the carbon storage value of a national park in Turkey, which follows a similar approach.

Step 1: Identify the size of the area under observation – typically in hectares or km^2.

Step 2: Identify the storage capacity of the ecosystem(s) under analysis, typically in terms of tons of carbon (C) or carbon dioxide (CO_2) per hectare or per km^2. This depends on a range of variables, such as the type of habitat under analysis (forest, peat land, soil, or marine), and the type, woodiness and age of vegetation cover.

While carbon storage data can be collected from existing sources (Section 2 above), care should be taken to choose data from areas with similar characteristics (biome type, latitude, altitude, aridity/humidity, plant growth rate and woodiness). For example, mid-latitude forests sequester more carbon annually than high latitude forests while middle-aged forests sequester more annually than old forests. In general, older forests store more carbon than younger forests.

Step 3: Choose a carbon value. The range of available estimates for the cost of carbon is very broad, and hence it is important to clearly state when (year of reference) and where (publication/authors) the values are taken from. The value of carbon can be based on relevant studies, literature and/or existing carbon taxes and market prices (Section 2 above). Each of these proxies for carbon value build on a series of assumptions and it is crucial to reflect them correctly when interpreting the results. It is also advisable to estimate a value range for carbon storage, rather than providing a single estimate.

Step 4: Calculate the amount of carbon stored (storage capacity x total area)

Step 5: Express the amount stored in monetary terms using carbon values estimates (total amount of carbon stored x carbon value)

Finally, it is to be noted that there are some general methodological concerns about how to calculate the value of total current carbon stocks. Current carbon stocks are 'static' (total amount of carbon currently stored), whereas the monetary values of carbon – based on the available literature – are typically based on marginal changes in carbon stock/sequestration over time (marginal costs per ton of carbon 'lost' from the storage). The marginal costs of damage increase when carbon stock decreases (i.e. emissions increase). In other words, not every ton of carbon stored has the same value, and therefore using existing marginal cost estimates leads to an underestimate of the total value of carbon. However, while this is an important methodological issue at global level, it is not considered to lead to significant errors on the scale of PAs as individual PAs are typically too small to have a tangible effect on the marginal damage costs and hence cost of carbon.

BOX A4.2

Example of estimating the economic value of carbon storage

The below calculations illustrate how to assess the value of the carbon stored in a temperate forest of 100 hectares (vegetation and soil).

Forest carbon (vegetation)

- *Step 1:* Identify the area size: for example 100 ha
- *Step 2:* Choose an estimate of carbon storage suitable to temperate forests, for example based on WBGU (1998) 57 tons C/ha (vegetation only)
- *Step 3:* choose a carbon value, for example use a range based on Nordhaus (2008) and Stern (2007) US$11.43/tC and US$91.75/tC
- *Steps 4 and 5:* Calculate C stock (range) as low C value = 100ha x 57tC/ha x US$11.43/tC = US$65,151 and high C value = 100ha x 57tC/ha x US$91.75/tC = US$522,975

Soil carbon

- *Step 1:* Identify the area size (100 ha)
- *Step 2:* Choose an estimate of carbon storage suitable to temperate forests, for example based on WBGU (1998), 96 tons C/ ha (soil only)
- *Step 3:* choose a carbon value, for example use a range based on Nordhaus (2008) and Stern (2007) US$11.43/tC and US$91.75/tC
- *Steps 4 and 5:* Calculate C stock (range) as low C value = 100ha x 96tC/ha x US$11.43/tC = US$109,728 and high C value = 100ha x 96tC/ha x US$91.75/tC = US$880,800

Based on the above the total value of carbon (vegetation and soil) (low) equals US $65,151 + US $109,728 = US$174,879. Total cost of carbon (high) = US $522,975 + US $880,800 = US$1,403,775

BOX A4.3

Estimating the economic value of carbon storage for the Küre Mountains National Park in Turkey

A study by Bann (2010) assessed, among other benefits, the economic value of carbon storage in the Küre Mountains National Park, Turkey. The park covers an area of 37,000 ha, 90 per cent of which is forested. It can be classified as a temperate mountain system with temperatures >10°C 4–8 months of the year and altitudes approximately >800 m.

Methodology: At the time of the analysis there were no site-specific estimates for the above or below ground biomass carbon content for the Küre Mountains National Park. Consequently, default values were drawn from the estimates provided by the Intergovernmental Panel on Climate Change (IPPC 2006), based on regional-specific climate and vegetation data. The IPCC default carbon fraction value for above ground biomass (ton C per ton of dry matter (d.m.) for temperate and boreal areas was 0.47 tC/d.m. The value for typical above-ground biomass of temperate mountain areas was 130 d.m./ha. The value of carbon was conservatively estimated based on the price of carbon credits in the voluntary offset market which were around 4.50 TL (US$3) per ton of carbon dioxide equivalent (CO_2e) when the study was conducted.

Results: The amount of carbon stored in above-ground biomass was estimated to be 61.1 tons/C/ha (0.47 tC/d.m. x 130 d.m.). In terms of CO_2, this was equivalent to 224 t/CO_2/ha (61.1t/C/ha x 3.67). The approximate carbon value for the national park was therefore estimated to be 1,008 TL/ha (US$672) (224 t/$CO_2$/ha x 4.50 TL/$CO_2$e). Assuming a forest area of 33,363 ha (90 per cent of the park), the total carbon storage value for the park was 33,529,104 TL (US$22,352,736) (1,008 TL/ha x 33,363 ha).

Source: Bann (2010)

4 Calculating the value of carbon sequestration

The step-by-step approach presented below helps to assess the amount of carbon sequestered by a PA when the area size changes. A simple application of the methodology is provided in Box A4.4. Box A4.5 shows a real example applying slightly more sophisticated methodology to identify sequestration benefits from Mexican forests. While not directly related to PAs, this example is considered a valuable illustration of how to estimate the economic value of carbon sequestration in practice.

The approach below has been chosen in order to keep the guidance as simple and pragmatic as possible. However, it is important to understand that focusing on the change in PA size only means that sequestration capacity of the area that remains unchanged will not feature in the calculations. This is clearly an oversimplification as growing forests or even many mature forests continue to sequester carbon. Calculations that take into account changes in sequestration within a location make use of marginal sequestration rates over time (or marginal loss rates in the context of degradation of an ecosystem). Such complex calculations fall outside the scope of this guidance.

Step 1: Choose the time period the analysis should cover, for example – 10, 20, 30 years or more.

Step 2: Identify the rate of yearly change, i.e. how much will the PA increase or decrease every year. This can be based on site-specific data, such as existing information on the annual deforestation rate (percentage) of a given site, or it can be chosen arbitrarily, for example to assess the effect on carbon if 1 per cent of PA was lost/increased every year.

Step 3: Identify the storage capacity of the ecosystem under analysis. For simplicity, the same values can be used for per hectare storage capacity, as in Section 2 above. This implies assuming that the hectares of ecosystems lost/added to the PA will have the same ability to store carbon as the initial plot of land. Note that this can be a strong oversimplification as the lost/additional area could have a different storage capacity. For example, when the area increase consists mainly of young forest it can store substantially less carbon than old forest.

Step 4: Choose a carbon value as in Section 2 above.

Step 5: Calculate the amount of carbon sequestered (or sequestration capacity lost) at the end of the given period. This is either the amount of carbon lost if the PA size has decreased, or the additional amount of carbon stored if the area has increased. The following formula can be used:

$$[\mathrm{PA}_0(1+r)^n - PA_0] \times S$$

Where PA_0 is the initial PA size, r is the rate of change, S the storage capacity per hectare, and n the number of years under observation.

Step 6: Calculate the value of the carbon additionally sequestered (or sequestration capacity lost) in the given period. This will require using a discount rate that will reduce the value of carbon sequestration over time.

The total value of the carbon additionally sequestered or lost every year will be equal to the sum of:

- The additional/lost carbon in the first year: $[\mathrm{PA}_0(1+r) - PA_0] \times S \times P \times \frac{1}{1+d}$

Where PA_0 is the initial PA size, r is the rate of change, S the storage capacity per hectare, P the carbon price and d the discount rate

- The additional/lost carbon in the second year: $[\mathrm{PA}_0(1+r)^2 - PA_0] \times S \times P \times \frac{1}{(1+d)^2}$
- The additional/lost carbon in the third year: $[\mathrm{PA}_0(1+r)^2 - PA_0] \times S \times P \times \frac{1}{(1+d)^3}$
- And so on until the last year under analysis

The overall formula can be written as:

$$\sum_{n=0}^{y} [\mathrm{PA}_0(1+r)^2 - PA_0\,(1+r)^{n-1}] \times S \times P \times \frac{1}{(1+d)^n}$$

Where $\sum$ is the summation of all the elements above, and n is the number of years up to the total length of the period under observation. Note: in case of area decrease, the rate of change will be negative (-r).

A range of discount rates can be adopted (see Chapter 12 and Annex 6). The final choice will depend on the country and landscape being evaluated and the preferences of the evaluator. For example, 7 per cent per year is one of the market discount rates recommended by the U.S. government for cost-benefit evaluation of environmental projects. However, ethical arguments have been made for using a lower discount rate when modelling climate change related dynamics (see for example Stern, 2012). If the rate is set equal to 0 per cent then monetary values are not discounted (Tallis, 2011). As a rule of thumb, it may be preferable to use a range of values, for example 7, 3 and 0 per cent.

BOX A4.4

Example of estimating the economic value of carbon sequestration

PA size increase (temperate forest)

The following steps outline the calculations required to estimate the value of the carbon sequestrated in a temperate forest of 100 hectares, with a planned increase in forest area by 2 per cent every year for 10 years. The calculations can be done relatively easily within an Excel spread sheet, allowing for integration of changing carbon values over time.

Step 1: Choose the time period under observation: 10 years

Step 2: Identify the rate of change: 2 per cent

Step 3: Choose an estimate of carbon storage suitable to temperate forests, for example based on WBGU (1998) 57 tons C/ ha (vegetation only)

Step 4: Choose the carbon value: for illustrative purposes the value by Fankhauser (1994) of US$20/tC

Step 5: Additional carbon sequestered after 10 years: this is the height of the last dark grey rectangle in Figure A4.1 below.

$$[100(1+0.02)10 - 100] \times 57 = (121.9 - 100) \times 57 = 1{,}248.3$$

Step 6: Overall (discounted) value of C sequestered. In this example, a discount rate (d) of 3 per cent used.

$$\sum_{n=0}^{y} [100\,(1+0.02)^{n} - 100\,(1+0.02)^{n-1}] \times 57 \times 20 \times \frac{1}{(1+0.03)^{n}} = \$21{,}193$$

Note this is equivalent to:

$$[100\,(1+0.02)-100\times 57\times 20\times\frac{1}{(1+0.03)}+[100\,(1+0.02)^2-100(1+0.02)]\times 57\times 20$$

$$\times\frac{1}{(1+0.03)^2}+[100\,(1+0.02)^3-100\,(1+0.02)^2]\times 57\times 20\times\frac{1}{(1+0.03)^3}+\textit{etc.}$$

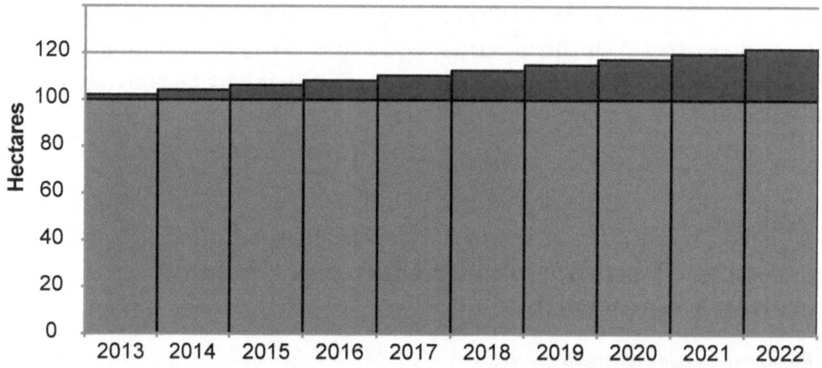

FIGURE A4.1 Example of increase in area size at a rate of 2 per cent over 10 years

Source: own calculations

The following steps outline the calculations required to estimate the value of the carbon lost in a temperate forest of 100 hectares that is at risk of deforestation at a rate of 2 per cent each year, for 10 years.

Step 1: Choose the time period under observation: 10 years

Step 2: Identify the rate of change: –2 per cent

Step 3: Choose an estimate of carbon storage suitable to temperate forests, for example, based on WBGU (1998) 57 tons C/ha (vegetation only)

Step 4: Choose the carbon value: assume Fankhauser (1994) value of US$20/tC

Step 5: Carbon lost after 10 years due to deforestation: this is the height of the last light grey rectangle in Figure A4.2 below. Note: since the rate of change is negative (–2 per cent), we can rewrite [(1+(–r)] as: (1–r)

$$[100-100\,(1-0.02)^{10}]\times 57=(100-81.7)\times 57=1{,}042.7$$

Step 6: Overall (discounted) value of C lost. In this example we choose a discount rate (d) of 3 per cent.

$$\sum_{n=0}^{y} [100\,(1+0.02)^{n-1} - 100\,(1-0.02)^{n}] \times 57 \times 20 \times \frac{1}{(1+0.03)^{n}} = \$17{,}876.2$$

Note this is equivalent to:

$$[100-100\,(1-0.02)\times 57\times 20\times \frac{1}{(1+0.03)} + [100\,(1-0.02)-100(1-0.02)^2]\times 57\times 20$$

$$\times\frac{1}{(1+0.03)^2} + [100\,(1-0.02)^2 - 100\,(1+0.02)^3]\times 57\times 20\times \frac{1}{(1+0.03)^3} + \textit{etc.}$$

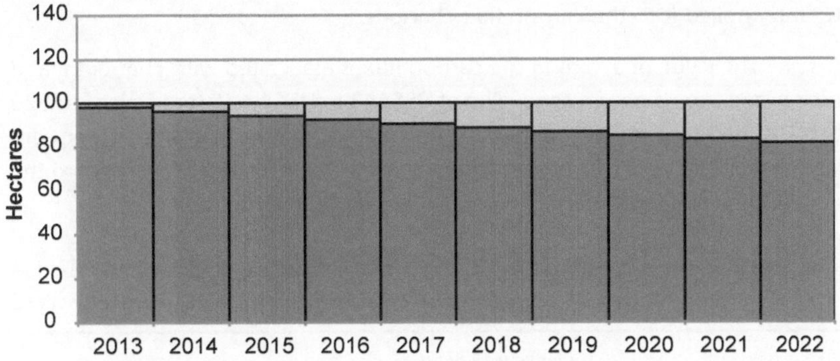

FIGURE A4.2 Example of decrease in area size at a rate of 2 per cent over 10 years

Source: own calculations

BOX A4.5

Estimating the carbon sequestration benefits of Mexican forests

Mexican forests cover an area around 50 million ha. They are an important resource of livelihood for almost one quarter of the Mexican population but are under significant threat of deforestation. A study by Adger *et al.* (1994) demonstrates the economic techniques for estimating the Total Economic Value (TEV) (see Figure 3.2 in Chapter 3) of the Mexican forest estate, including the value of carbon sequestration from avoided land use change.

Method: The carbon sequestration benefits of forest conservation and of plantation forestry were estimated through a two stage process. In the first stage, physical models were used to capture the land use change per different forest type. In the second stage, the monetary value of forest was estimated in terms of avoided damage caused by global warming. It was assumed that in the case of land use change caused by deforestation the new land use continues for an indefinite period. This was because the major changes in carbon sequestration (fluxes) were assumed to occur at deforestation and therefore the subsequent changes were estimated to be insignificant. The estimated changes in carbon sequestration from forest land conversion into pastures and agriculture is shown in Table A4.4 below. In order to estimate the monetary value of avoided damage due to global warming when maintaining forest the paper used Fankhauser's (1994) central estimate for current emissions of US$20 per tC and applied it to the carbon flux changes.

Results: The value of avoiding the carbon fluxes associated with changing land use ranged from a net present value of US$643 to US$3,630 (equivalent to $20 to $100 per ha per year), depending on the forest type and the subsequent land use (Table A4.4). These represent a social value, in that they capture the benefit of avoiding damage by global warming in the global sense.

TABLE A4.4 Estimated carbon fluxes caused by changing land uses in Mexico and estimated value of carbon sequestration if forests were conserved

Forest type	*Converted land use*			
	Permanent pasture		*Agriculture*	
	(tC/ha)	($/ha)	(tC/ha)	($/ha)
Temperate coniferous	168.4	3,436	167.1	3,410
Temperate deciduous	34.0	693	31.5	643
Tropical evergreen	164.8	3,633	163.5	3,337
Tropical deciduous	92.5	1,887	91.25	1,863

Sources: Adger *et al.* (1994), Brown (1992), Houghton *et al.* (1991) and Masera *et al.* (1992)

5 Other possible valuation methods

This guidance outlines a few simple approaches (i.e. assuming no changes in ecosystems characteristics over time) for assessing the socio-economic value of carbon stored in and sequestered by PAs. Similar approaches could also be used to assess more complex scenarios, for example integrating changes within a PA into the calculation (changes in ecosystem coverage and related changes in carbon stored). This, however, may lead to more complex calculations, as different carbon storage values would be needed for different land use types. Additional sophistication could be introduced by using different, tailored values for storage and for annual sequestration. As noted earlier, there is an increasingly good literature and monitoring data on carbon storage and sequestration values which can facilitate this task, but care is needed to ensure that the right values are used for the PA being looked at, given important differences across biomes, climatic conditions and state and maturity of the ecosystems.

It is, of course, possible to adopt more sophisticated modelling that will help take better account of geographic characteristics, for example using Geographic Information System (GIS) maps, and more refined assumptions. Several of these models are available, but they require a deeper understanding of statistical and econometrics methods (see for example Tallis *et al.*, 2011).

6 Key issues to be kept in mind

It is important to ensure that the units used in the above calculations are consistent. For example, if storage capacity is measured as tons of C per hectare, the area size will have to be calculated in hectares and the resulting amount of carbon sequestered will be in tons of C. The carbon value chosen for the monetary assessment will have to be expressed in terms of a given currency per ton of C.

Furthermore, it should be noted that the cost of carbon chosen can lead to very different values, hence potentially overestimating or underestimating the actual value of the service. In some cases, it may be useful to use a range of values (e.g. a lower bound, a higher bound and an average value) to provide a broader picture. The same may be said about discount rates.

It is also important to ensure that one is clear about the distinction between stock of carbon and the annual sequestration of carbon, as well as whether the analysis only covers carbon in living biomass (vegetation) or also carbon in soils and litter ('dead carbon'). It is also fundamentally important to check, when looking at the value of stock of carbon, whether the use of marginal carbon values is acceptable or is an inappropriate approximation, and ensure that results are presented with due caveats and context.

References

Adger, N., K. Brown, R. Cervigni, and D. Moran (1994) *Towards Estimating Total Economic Value of Forests in Mexico*, GEC 94-21, Centre for Social and Economic Research on the Global Environment, University of East Anglia and University College London

Alexandrov, G., A., Oikawa, T. and Yamagata, Y. (2002) 'The scheme for globalization of a process-based model explaining gradations in terrestrial NPP and its application', *Ecological Modelling*, 148(3): 293-306

Alexandrov, G., A., Yamagata, Y. and Oikawa, T. (1999) 'Towards a model for projecting net ecosystem production of the world forests', *Ecological Modelling*, 123:183–191

Bann, C. (2010) *Developing a Business Plan for Küre Mountains National Park and its Buffer Zone*, Project report, http://www.kdmp.gov.tr/BUSINESS_PLAN_FOR_KMNP_Final_Report.pdf, accessed 17 December 2012

Bassi, S., ten Brink, P., Farmer, A., Tucker, G., Gardner, S., Mazza, Van Breusegem, W., Hunt, A., Lago, M., Spurgeon, J., Van Acoleyen, M., Larsen, B. and Doumani, F. (2011) *Benefit Assessment Manual for Policy Makers: Assessment of Social and Economic Benefits of Enhanced Environmental Protection in the ENPI countries*, a guiding document for the project 'Analysis for European Neighbourhood Policy (ENP) Countries and the Russian Federation on social and economic benefits of enhanced environmental protection, IEEP, Brussels, http://www.ieep.eu/assets/945/ENPI_Benefits_Assessment_Manual_for_Policy_Makers.pdf, accessed 17 December 2012

Brown, K. (1992) *Carbon Sequestration and Storage in Tropical Forests*, Global Environmental Change Working Paper 92-24, Centre for Social and Economic Research on the Global Environment, University of East Anglia, University College London

Carbon Dioxide Information Analysis Center (CDIAC) (undated) *Major World Ecosystem Complexes Ranked by Carbon in Live Vegetation*, http://cdiac.ornl.gov/ftp/ndp017/table.html, accessed 17 December 2012

Centre d'analyse stratégique (2009) *Rapports et documents N.16/2009 – La valeur tutélaire du carbone.* Rapport de la commission présidée par Alain Quinet, http://www.ladocumentationfrancaise.fr/var/storage/rapports-publics/094000195/0000.pdf, accessed 17 December 2012

Commission of the European Communities (EC) (2008) *Commission Staff Working Document – Impact Assessment – Document accompanying the Package of Implementation measures for the EU's objectives on climate change and renewable energy for 2020.* SEC(2008) 85/3, http://ec.europa.eu/clima/documentation/docs/sec_2008_85_ia_en.pdf, accessed 17 December 2012

DECC – Department for Energy and Climate Change (2009) *Carbon Valuation in UK Policy Appraisal: A Revised Approach*, https://www.gov.uk/government/uploads/system/uploads/attachment_data/file/41798/1_20090715105804_e____carbonvaluationinukpolicyappraisal.pdf, accessed 17 December 2012

FAO – Food and Agriculture Organisation (2010) *Global Forest Resources Assessment*, http://www.fao.org/forestry/62318/en/aze, accessed 13 January 2013

Fankhauser, S. (1994) 'The economic costs of global warming damage: A survey', *Global Environmental Change*, 4(4): 301-309

Houghton, R. A., Skole, D. L. and Lefkowitz, D. S. (1991) 'Changes in the landscape of Latin America between 1850 and 1985: II Net release of CO_2 to the atmosphere', *Forest Ecology and Management*, 38: 173–199

IPCC – Intergovernmental Panel on Climate Change (2001) 'Climate Change 2001: The Scientific Basis', in Houghton *et al.* (eds), *Contribution of Working Group I to the Third Assessment Report of the Intergovernmental Panel on Climate Change*, Cambridge University Press, Cambridge

IPCC – Intergovernmental Panel on Climate Change (1996) *Climate Change 1995: Economic and Social Dimensions of Climate Change.* Contribution of WGIII to the Second Assessment Report of the Intergovernmental Panel on Climate Change, http://www.ipcc.ch/ipccreports/sar/wg_III/ipcc_sar_wg_III_full_report.pdf, accessed 3 June 2013

Masera Cerutti, O., Ordonez, M. d. J. and Minjarez, R. D. (1992) *Carbon Emissions and Sequestration in Forests: Case Studies from Seven Developing Countries. Volume 4: Mexico*, USEPA and Energy and Environment Division, Lawrence Berkeley Laboratory, Berkeley, CA

Mongabay database (2012) *Online Database on Deforestation Figures for Selected Countries*, http://rainforests.mongabay.com/deforestation, accessed 20 December 2012

Nordhaus, W. D. (2008) *A question of balance*, Yale University Press, New Haven and London

Point Carbon (2012) *News on update carbon prices*, http://www.pointcarbon.com/news/1.2086140, accessed 20 December 2012

Stern, N. (2007) *The Stern Review of the Economics of Climate Change*, Cambridge University Press, electronic version, www.hm-treasury.gov.uk/stern_review_report.htm, accessed 20 December 2012

Stern, N. (2012) *Ethics, equity and the economics of climate change*, Centre for Climate Change Economics and Policy Working Paper No. 97, Grantham Research Institute on Climate Change and the Environment Working Paper No. 84, http://www.cccep.ac.uk/Publications/Working-papers/Papers/90-99/WP97-ethics-equity-economics-of-climate-change.pdf, accessed 20 December 2012

Tallis, H. T., Ricketts, T., Guerry, A. D., Nelson, E., Ennaanay, D., Wolny, S., Olwero, N., Vigerstol, K., Pennington, D., Mendoza, G., Aukema, J., Foster, J., Forrest, J., Cameron, D., Lonsdorf, E., Kennedy, C., Verutes, G., Kim, C. K., Guannel, G., Papenfus, M., Toft, J., Marsik, M., Bernhardt, J., Wood, S. A. and Sharp, R. (2011) *InVEST 2.1 beta User's Guide*, The Natural Capital Project, Stanford, http://stanford.edu/~woodsp/natcap/invest/docs/21/index.html, accessed 20 December 2012

WBGU – Wissenschaftlicher Beirat der Bundesregierung Globale Umweltveränderungen (1998) *Die Anrechnung biologischer Quellen und Senken im Kyoto-Protokoll: Fortschritt oder Rückschlag für den globalen Umweltschutz?* Sondergutachten, WBGU, Bremerhaven, Germany, http://www.wbgu.de/fileadmin/templates/dateien/veroeffentlichungen/sondergutachten/sn1998/wbgu_sn1998.pdf, accessed 20 December 2012

World Bank (2002) *Manual for Calculating Adjusted Net Savings*, World Bank, Washington, DC, http://siteresources.worldbank.org/INTEEI/1105643-1115814965717/20486606/Savingsmanual2002.pdf, accessed 20 December 2012

World Bank (2012) *Turn Down the Heat – Why a 4°C Warmer World Must Be Avoided*, International Bank for Reconstruction and Development / The World Bank, Washington, DC, http://climatechange.worldbank.org/sites/default/files/Turn_Down_the_heat_Why_a_4_degree_centrigrade_warmer_world_must_be_avoided.pdf, accessed 20 December 2012

8

CULTURAL SERVICES AND RELATED GOODS

Gantioler, S. and D'Amato, D.

8.1 Overview of benefits

Cultural services and related goods are mostly related to the pleasure, inspiration, or spiritual or religious enrichment people derive from spending time in nature, particularly PAs (Table 8.1). The drivers behind these (often intangible) benefits are as manifold as the motivations of people visiting PAs. The benefits derived can relate to enjoying nature's amenities and tranquillity, observing different habitats and species, and gaining a deeper understanding of nature's functions. Nature is also often used as a destination for religious pilgrimages or as a 'medium' for building social relations. Some visitors may be driven by educational reasons, others in search of artistic inspiration – or even for technological innovations. In practice, the visits to PAs often combine and interlink several different benefits.

Recreation and tourism are among the key ecosystem services from PAs. These benefits are frequently used interchangeably, with several valuation studies assessing the two together. Arguably the main difference between the two is that tourism generally refers to activities (also recreational) occurring outside people's usual everyday environment, often including travelling (day trips or overnight stays) (UNWTO, 2008). The same recreational opportunities can be available within the usual environment and local communities, though then they are not considered touristic (Bio Intelligence Service, 2011).

PAs can be important destinations within the fast growing sector of nature tourism (Bell *et al.*, 2007), i.e. tourism relying primarily on the natural environment as an attraction to visitors. Tourism related benefits can be based on both use and non-use values (Chapter 3) with the former including fishing and hunting and the latter being focused on bird and nature watching or more sporting activities such as biking, swimming, horse riding, and mountain biking (Bell *et al.*, 2007).

TABLE 8.1 Cultural services and related goods

Ecosystem service	*Examples of related goods*
Opportunities for recreation and tourism	Hiking, camping, nature walks, jogging, winter sports, water sports, angling, wildlife watching, horse riding, hunting, cycling and climbing resulting in increased mental well-being and health, including avoided stress and related mental illnesses, increased physical health, increased productivity, decreased sick days/leave, decreased health costs to employers and society
Aesthetic values	Enjoyment of rural, diversified, unique and colourful landscapes, individual habitats and species, and tranquillity supporting mental well-being (see above)
Inspiration for the arts, science and technology	Writing, painting, designs, documentaries, movies, non-fiction and fiction books; biomimicry for technology, engineering, materials and architecture
Information for education and research	Educational trips by schools and other groups; employee training; research (on and off site) related to ecosystem functions, species and genetic material leading to scientific understanding, publications and patents
Spiritual and religious experience	Natural and built sacred places, philosophy and faith; support to mental well-being (see above)
Cultural identify and heritage	Landscapes and habitats formed by human activities; species of spiritual importance; traditional and indigenous knowledge; traditional crops, cuisine and culture

Source: own presentation, based on MA (2005), de Groot *et al.* (2010), UK NEA (2011), Bio Intelligence Service (2011), Kettunen *et al.* (2013), Gantioler *et al.* (2010)

Visits to PAs can also be motivated by aesthetic pleasure and the desire for inspiration (arts, science and technology). For example, nature can influence architectural, graphic and clothing design or provide a topic and/or backdrop for books, painting, films and TV programmes. Furthermore, PAs can function as a 'living library of science' that conserves knowledge for future generations. They can provide an ideal setting for promoting and increasing the scientific understanding on ecosystem functioning, ecological interactions and provisioning of ecosystem services while also ensuring that traditional knowledge continues to be conveyed to future generations.

In more concrete terms, genetic material from PAs can provide a valuable resource for research supporting, for example, the development of pharmaceutical products (see Chapter 6). Genetic material from PAs can also be used in a number of other contexts such as in biocontrol, industrial biotechnology and plant breeding (ten Brink *et al.*, 2012). Discoveries from PAs can result in new technological developments. Biomimicry, i.e. using nature's solutions as an inspiration to solve human problems, is likely to be a key source for future innovations, offering alternatives to the traditional technological approaches that require significant amounts of energy, raw material, toxic chemicals and heavy machinery while polluting the soil, water and air (Lovins, 2008). Biomimicry is expanding in areas as diverse as architecture, engineering and product development (Biomimicry Institute, 2010a, b; van Nierop *et al.*, 2008; BusinessGreen, 2010).

Protected ecosystems can also be important repositories of cultural knowledge, particularly for indigenous people, and form an important part of cultural identity. The availability of features of historic, cultural or spiritual importance such as monasteries, churches, sacred places and species can also attract additional visitors to a PA.

Factors influencing the availability of benefits

The benefits derived from the cultural services above are influenced by interactions between people and nature, more than for any other ecosystem service. Income and education, socio-economic factors, household characteristics, type of accommodation, and seasonal changes can be drivers to visit a specific site (Bio Intelligence Service, 2011). PAs' potential to provide recreational and cultural benefits also largely depends on site-specific environmental and socio-economic aspects that increase visitors' affinity for the site. Such factors include environmental settings (i.e. the presence of attractive natural and historical features and components), available infrastructure and accessibility, and absence or presence of PA designation (e.g. UK NEA, 2011).

Habitats with flagship species such as big mammals, birds, and/or game species might increase a visitor's affinity to a PA compared to another site with possibly more diverse but less 'visible' or 'attractive' composition of species. Rural customs and local traditions interlinked with site's natural features might offer additional opportunities for visitors. As explained in Chapters 7 and 9, the availability of cultural services and related goods, affecting the attractiveness of PAs to visitors, is underpinned by a range of regulating and supporting services that ensure the stability and quality of natural environment. For example, the maintenance of water quality is a prerequisite for fishing and swimming. The availability of locally and sustainably produced resources can also increase the attractiveness of PAs, contributing to their 'branding' and influencing people's choices to visit a site (see Chapter 6). Consequently, there are significant synergies between cultural benefits and the maintenance and availability of other ecosystem services and related goods (Chapter 2).

Finally, the designation of an area as a PA can significantly influence visitor affinity. Besides increasing a site's general visibility, designation is often associated with a promise of attractive unique and wild features and/or vast natural areas, such as with World Natural Heritage Sites and national parks.

PA managers as well as farmers, forest owners, or local communities help to maintain natural and man-made features in PAs that influence visitors' attraction to a site. This includes the maintenance and restoration of picturesque and diversified nature settings of a certain quality, as well as the development and maintenance of relevant infrastructure that help to regulate accessibility. Beneficiaries of PAs' cultural benefits range from tourists – global and national – to local people using sites for everyday recreation. In addition, stakeholders and businesses providing goods and services to visitors also benefit from PAs' abilities to provide cultural services. Such benefits can be directly linked to a site and/or region (revenue to hotels, restaurants and transport providers) or they can be national and global (outdoor industry offering means and equipment for recreational activities).

Given the trade-offs between different ecosystem services (Chapter 2), the enjoyment of cultural benefits, tourism, and, in particular, recreational activities can have potentially negative impacts on PAs and their ability to conserve biodiversity and maintain other benefits. Uncontrolled visitor flows and intensified land use (infrastructure development, increased road traffic, etc.) can lead to serious disturbances of wild species and also negatively affect the ecosystem services from PAs to people living close to a site. Tourism can, for example, result in the competition and/or overuse of resources (water), facilitate the spread of invasive alien species (IAS), increase waste and pollution and even lead to the 'erosion' of cultural identity. Certain areas may not at all be suitable for recreation or nature tourism due to their fragile ecosystems. For other sites, potential negative impacts need to be factored in when looking at the social, ecological and economic impacts resulting from PAs. Finally, it is important to understand that the role and socio-economic importance of PAs in providing cultural services often needs to be considered in the regional context. The presence of several PAs can attract more visitors to the region but might cause them to 'compete' with each other for visitors (see Chapter 4 and 12 for more detailed advice and considerations).

8.2 Socio-economic importance of benefits

The following section outlines the main socio-economic values provided by cultural services and related goods, including an overview of concrete examples (see Table 8.2). Following Chapter 3, the values have been broadly divided into two categories: economic and welfare (health and social) values. Please note that employment benefits linked to cultural services and related goods are covered in more detail in Chapter 10.

Economic value

The economic (market) value of cultural services and related goods is captured by the flow of income into local communities and/or regions, as well as related profits to the tourism and outdoor industry. This flux of income is related to the expenditure of tourists and visitors to PAs on food and beverages, accommodation, transport services, additional attractions, and shopping for outdoor equipment. The economic impacts can be direct or indirect/secondary, with the former increasing the economic activity within the tourism and recreation sector and the latter affecting the sales and income of other sectors in a region (Bio Intelligence Service,

2011). In addition to spending by PA visitors, the economic effects of the expenditure by employees of tourism and recreational sectors can also contribute to the overall economic value of PAs' cultural services (induced impacts) (see Chapter 11).

For example, the direct and indirect economic income generated through tourism by the Niedersächsisches Wattenmeer National Park (Germany) was estimated at €500 million for 2007, of which almost €60 million was created by visitors with high national park affinity (Mayer *et al.*, 2010, see Box 8.1). Visitor spending for the European network of protected areas, Natura 2000, was estimated at €50 to €90 billion in 2006 for 1.2 to 2.2 billion visitor days per annum, which generated an additional income between €50 and €85 billion (Bio Intelligence Service, 2011). For those with a high affinity to the Natura 2000 network, around €15 billion were spent for 230 to 520 million visitor-days, generating an additional income from €9 to €20 billion.

Second, the economic value of cultural benefits (including market value) can also be estimated by the income directly flowing to PAs via on-site spending by visitors, such as entrance and excursion fees. For example, in 2002 National Parks in South Korea had more than 32 million visitors yearly, resulting in an annual income of roughly US$26 million from entrance fees (Lee and Han, 2002) (Table 8.2). The Lake Nakuru National Park in Kenya hosts around 149,500 international visitors every year. They are charged an entrance fee of US$80 which leads to revenues of roughly US$12million (Ramsar and UNWTO, 2012, see Table 8.2).

Income can also be generated by the purchase of licences and permits (for example, for hunting and fishing) and new markets created if tradable permit schemes are put in place. The Selous Game Reserve is the largest protected area in Tanzania with a long history of trophy hunting. The hunting value from fees amounted to US$3.5 million in 2003 (Bohne, 2008, see Table 8.2).

In-site market opportunities can also result from the set-up of compensation schemes that pay the land managers and/or owners within PAs for services they provide for the maintenance and restoration of the environmental quality of a site. With regard to recreation and tourism, this refers to all site management activities that help to conserve landscapes, habitats, or species that are attractive for residents, visitors and tourists. Payments could come either from public authorities or the tourism and recreational sector.

In the Komodo National Park (Indonesia), for example, a joint venture between an international NGO (the Nature Conservancy – TNC) and a local tourism company (Jaytasha Putrindo Utama) was formed. It was responsible for running a tourism concession, which aimed at conserving the park's biodiversity and generating environmentally sound, socially responsible and economically viable revenues required for the financing of the park (Font *et al.*, 2004).

The economic value of PAs in providing aesthetic pleasure and inspiration (education, research and arts) as well as spiritual and cultural enrichment is more difficult to capture. While these benefits are often the main drivers for recreation and tourism, the number of markets that would capture their direct value is limited. However, the educational value of the Navakavu Locally Managed Marine

Area of Viti Levu Island (Fiji) (O'Garra, 2007, see Table 8.2) has been estimated by looking at the payments provided by researchers and students (food, lodging and fees paid, capital equipment donations, access fees to coastal waters and other direct payments). Other methods, such as increasing rent and purchasing prices for houses next to PAs, are sometimes used to reflect the economic value of amenity benefits provided (see Section 8.3 below).

Welfare value

Although the economic value of recreation and tourism is quite prominent, it captures only a fraction of the total socio-economic value of cultural benefits associated with PAs. Cultural services provide a range of more intangible goods and values derived from recreational and aesthetic pleasures, inspiration and spiritual enrichment.

These less tangible – but equally valuable – benefits include, for example, the role of PAs in improving psychological well-being. Enjoying the amenities and tranquillity offered by a PA can help diminish stress levels, have therapeutic effects on human psyche and promote social integration. The availability of an attractive setting for recreational activities can also help to increase physical activity and as such physical health, enhancing immune and cardiovascular systems and helping to combat some deficiencies of the modern life style such as obesity. In the United Kingdom, for example, the national health agenda has been more closely aligned with the government agencies responsible for environmental protection in the different countries, in order to shift emphasis from treatment and cure to prevention. Consequently, schemes in National Parks and National Nature Reserves have been put in place to promote outdoor health and fitness opportunities (Stolton and Dudley, 2010, see Table 8.2). However, it needs to be noted that a higher number of visits to PAs does not automatically increase physical activity. For example, if no PAs are close by, people tend to travel long distances with the car. Accessibility to PAs is therefore a key factor determining their contribution to people's physical health (Bell *et al.*, 2007).

The socio-economic value of cultural services in supporting human health is difficult to monetise. It is challenging to quantify the impacts of individual visits to a site on a person's health and to distinguish them from other socio-economic factors influencing health. However, information on healthcare costs can provide an indication of the potential magnitude of the benefits. The World Health Organisation (WHO) estimated that physical inactivity is responsible for 3.5 per cent of the disease burden and up to 10 per cent of deaths in Europe (WHO, 2007). Studies also showed that it accounts for 2 to 2.5 per cent of total national healthcare costs in several Western countries (Proper and van Mechelen, 2007). As iterated by the WHO, physical inactivity has also economic consequences due to illness, work disabilities, and premature death. Increased mental health, on the other hand, can have a significant impact on labour productivity (Proper and van Mechelen, 2007).

The above health benefits are all important factors that can determine people's quality of life, particularly in cities. Singapore, for example, has recognised the potential of PAs in enhancing the quality of urban life and is preserving the biological diversity of four nature reserves covering more than 3,000 ha. This can contribute to urban prosperity if it facilitates an equitable access to a city's environmental 'commons' (UN-HABITAT, 2012). It can also attract businesses to locate and invest in an area associated with a high quality of living and a working environment which stimulates regeneration (Mazza *et al.*, 2011).

Furthermore, the money that is spent on research and education in and on PAs (see example from Fiji in Table 8.2) is likely to significantly underestimate PAs' role in increasing the understanding of ecosystems and their interactions within socio-economic systems. Nature has shaped cultural identities and given birth to and/or hosted a range of belief systems. Sacred areas are one of the oldest forms of protected natural areas (Dudley *et al.*, 2006); the effort that is put into conserving those places and the effect this has on indigenous people is demonstrated by the example of Indigenous Protected Areas in Australia. The areas are managed by the indigenous owners under the National Reserve System (NRS) and deliver benefits in health, education, employment, and social cohesion for Indigenous communities. Since 1996/97, 66 per cent of the NRS was contributed through the IPA Programme (see Table 8.2). PAs can also provide testing grounds for sustainable management linked to religious practices. In La Sepultura and El Triunfo Biosphere Reserves (Mexico), indigenous and rural communities work together to sustainably extract non-timber forest products used by both traditional and Christian faith groups with 300 million palms imported into the United States every year (see Table 8.2).

The importance of the welfare value compared to the economic value, the potential risks of focusing only on the latter, and the limits of market methods (see below) are illustrated by an example of the Yosemite National Park in the United States (Sandel, 2012). The park is one of the oldest parks in the world and attracts millions of visitors every year. Its most important campsites can be reserved in advance for US$20 per night. Given the strong demand particularly over summer time, the campsites are fully booked as soon as they become available. Though resale is not allowed, people professionally offer to organise the reservations for US$100 to US$150 per night. As a result, the national park could theoretically have increased the prices, maximising their revenue. However, a huge uproar occurred, highlighting the broader welfare (non-economic) value of the park and emphasising how the wonders of Yosemite belong to all and not just to those who can afford them. Another similar example is trophy hunting of walrus – an Arctic game species with high 'status' value – in Canada under the supervision of Inuit (Sandel, 2012). While the tradable permits and hunting fees profit the community and the hunting is carried out within sustainable limits, it is also important to ask what the moral limits of such markets are and to what extent the sport of trophy hunting and the resulting markets are a reflection of welfare and cultural identity.

TABLE 8.2 Examples of economic and welfare values of PAs related to recreational and cultural benefits

Protected area	*Benefits (ecosystem services and related goods)*	*Estimated socio-economic value of benefits*	*Method*	*Source*
Australia, Carnarvon Gorge National Park (several)	Tourism and recreation	By applying the travel cost method, demand curves for both camping and day visits have been estimated and consumer surplus computed for the Carnarvon Gorge National Park (Australia) during 1993–1994. A questionnaire was posted to selected campers including questions on distance and chief reason for visiting the park. The travel costs were estimated based on the average number of people in each vehicle, distance travelled, and minimum number of nights staying. The estimated (minimum) net present value (NPV) of recreational use of the park was AU$40 million (US$38 million) in 1993–1994 (one-year period).	Travel Cost Method	Beal, 1995
Tanzania, Selous Game Reserve (several)	Trophy hunting	The Selous Game Reserve (SGR) is the largest protected area in Tanzania with a long history of trophy hunting. The direct value from trophy hunting was determined by looking at the fees that every hunting tourist and safari company is obliged to pay by law (whether to obtain a permit or for the trophy and trophy handling, etc.) as a reflection of the willingness-to-pay (WTP) Between 1988 and 2003, when the 'Selous Conservation Programme' (SCP) was running, the hunting value increased around 347.7 per cent, from US$1 million in 1988 to US$3.5 million in 2003.	Market price method	Bohne, 2008

Continued

Protected area	*Benefits (ecosystem services and related goods)*	*Estimated socio-economic value of benefits*	*Method*	*Source*
South Korea, five national parks (several) (several)	Tourism and recreation Preservation value	In 2002 National Parks in South Korea (NPSK) had more than 32 million visitors yearly, but they required annually KRW71 billion (US$59 million), equivalent to KRW3700 (US$3) per visitor, for their maintenance. The admission fee was KRW1000 (US$0.83) per visitor. The value of natural/cultural resources in the national parks however was thought to exceed maintenance costs and to justify an increase in current admission fees in order to maintain the quality of the natural environment. A discrete choices (yes/no) survey was used to estimate willingness to pay for recreation use and preservation values in five national parks, representing different types of recreation resources (mountain/marine-oriented, culture-oriented, etc.).The admission fee was chosen as payment vehicle. Use values were estimated between KRW17,208 (US$14.3) and KRW5,758 (US$4.8) per visitor, while preservation value ranged from KRW16,198 (US$13.5) to KRW12,756 (US$10.6) per person. Note: Since the publication of this study, entrance fees to NPSK have been removed. However, the study still provides a relevant example on the use of choice modelling to estimate the economic value of tourism and recreation.	Choice modelling method	Lee and Han, 2002

Protected area	*Benefits (ecosystem services and related goods)*	*Estimated socio-economic value of benefits*	*Method*	*Source*
Canada, four northern parks, (arctic)	Existence value linked to aesthetic and heritage benefits	The economic value derived from the creation of four hypothetical national parks in northern Canada was quantified using contingent valuation methods. The questionnaire provided respondents with information about the parks to be valued and asked 1) how much they would be willing to pay as a one-time tax to have the new parks established or maintained 2) how much compensation respondents would want if an existing park ceased to exist. Irrespective of the opportunity/likelihood to visit the parks, respondents considered them as places of rare or irreplaceable natural wonder or as a fragile ecosystem over which humans should exercise stewardship. The study found that respondents were willing to pay CA$244 (US$236) per household to establish the four additional northern parks. Generalised across the Canadian population, this yields a value of CA$2.80 billion (US$2.7 billion).	Contingent valuation method	CAFF, 2002

Continued

Protected area	*Benefits (ecosystem services and related goods)*	*Estimated socio-economic value of benefits*	*Method*	*Source*
Australia, National Reserve System and Indigenous Protected Areas (several)	Economic/social participation and local development	Indigenous Protected Areas (IPAs) are located on land held by the traditional Aboriginal owners, who have agreed to manage their country as part of the National Reserve System (NRS) in order to protect its significant natural and cultural values. For the management of IPAs traditional knowledge and practices are combined with modern scientific methods for effective land management. At the same time, IPAs deliver benefits in health, education, employment and social cohesion for Indigenous communities. Australian Indigenous Protected Areas deliver improved social outcomes: 95 per cent of IPA communities report economic participation and development benefits from involvement in the programme; 60 per cent of IPA communities report positive outcomes for early childhood development resulting from IPA activities; 85 per cent of IPA communities report that IPA activities improve early school engagement; 74 per cent of IPA communities say that their IPA management activities make a positive contribution to the reduction of substance abuse. Out of the 20.8 million hectares of land added to the NRS since 1996–97, 13.8 million hectares (66 per cent) was contributed through the IPA Programme. The Australian Government funding to the NRS and IPA programmes amounts to AU$13.6 million (US$14.2 million) during 1996-2006.	Non-monetary quantitative methods such as available statistics; provided funding	Gilligan, 2006

Protected area	*Benefits (ecosystem services and related goods)*	*Estimated socio-economic value of benefits*	*Method*	*Source*
Russia, model areas in Tazovsky and Olyutorsky districts (arctic)	Religious value Spiritual value Cultural identity	A pilot project to map and identify the conservation value of sacred sites of indigenous peoples in two arctic model areas in Tazovsky and Olyutorsky districts, Russia, has been carried out as a collaborative effort between the Russian Association of Indigenous Peoples of the North (RAIPON), the Arctic Council's Conservation of Arctic Flora and Fauna Programme (CAFF) and the Arctic Council Indigenous Peoples Secretariat. A total of 96 indigenous people – elders, reindeer herders and fishermen – were interviewed and 347 sacred sites were documented and mapped. A standardised methodology with photo documentation and audio and video recording has been applied to the description of the sacred sites and a preliminary classification of each model area has been developed on the basis of existing publications, archives, and research. The work has already had a positive impact on legislation work in the model areas and other northern regions in Russia have expressed interest in conducting similar research. While highly relevant in the context of this guidance this example is not specifically related to PAs.	Non-monetary qualitative methods such as documentation and mapping	CAFF, 2002

Continued

Protected area	*Benefits (ecosystem services and related goods)*	*Estimated socio-economic value of benefits*	*Method*	*Source*
Mexico, La Sepultura and El Triunfo Biosphere Reserves (forest)	Non-timber forest products supporting religious value	Protected areas can provide testing grounds for sustainable management linked to religious practices. In Chiapas, Mexico, a GEF project is strengthening management in La Sepultura and El Triunfo Biosphere Reserves. It is working with indigenous and peasant communities to set up extractive reserves for non-timber forest products used by both traditional and Christian faith groups, including orchids, bromeliads, cycads and palms. The project is developing links with the Lutheran Church to develop trade in certified palm products, used in Christian ceremonies during Easter. Some 300 million palms are imported into the United States every year and the project aims to introduce sustainable management principles and nurseries to ensure that trade can continue without depleting the species.	Non-monetary quantitative methods such as available statistics	Stolton *et al.*, 2010
Italy, Val d'Agri – Lagonegrese National Park (mountain)	Tourism and recreation	An open-ended questionnaire with entrance fee as payment method was used to measure willingness to pay of tourists for three different management scenarios of the Val d'Agri – Lagonegrese National Park in South Italy: a status quo scenario and two different scenarios involving improvement of tourism services, through different means. Respectively, 83, 92 and 88 per cent of the respondents were willing to pay €7, €10 and €9 for the first, second and third scenario. Tourism and recreational values for the three scenarios was estimated by multiplying the number of tourists visiting the park yearly (66,000), the percentages of respondents willing to pay entrance fees and relative prices. The first, second and third scenarios have an estimated value of, respectively, €380,000, €600,000 and €520,000/year.	Choice modelling method	FEEM, 2010

Protected area	*Benefits (ecosystem services and related goods)*	*Estimated socio-economic value of benefits*	*Method*	*Source*
England, National parks, National Nature reserves and local/private reserves (several)	Health benefits	One move towards shifting emphasis from treatment and cure to prevention has been to link the national health agenda more closely with the government agencies responsible for environmental protection in the different countries that make up the UK. Several schemes, taking place in a range of areas throughout England, including National Parks, National Nature Reserves and many local and private reserves, are being developed in the UK to promote outdoor health and fitness opportunities. Although the overall effectiveness of individual programmes' contribution to improved health may be hard to assess, the overall role of green spaces in the health of the English is clear.	Non-monetary qualitative method	Dudley and Stolton, 2010
Kenya, Lake Nakuru National Park (inland water)	Tourism and recreation	Lake Nakuru receives around 149,500 international and 95,500 domestic visitors every year, who are charged an entrance fee of US$80 and US$11 respectively. The revenues from entrance fees and concession fees from the lodges contribute to paying the costs of park management. Awareness of the importance of nature is promoted with large-scale environmental education programmes, involving about 100,000 students each year and with inexpensive wildlife viewing tours that are run by the park for residents.	Market price method	Ramsar and UNWTO, 2012

Continued

Protected area	*Benefits (ecosystem services and related goods)*	*Estimated socio-economic value of benefits*	*Method*	*Source*
Fiji, Navakavu Locally Managed Marine Area in Viti Levu Island	Research and educational value	The economic value of research and education activities to locals in the Navakavu Locally Managed Marine Area in Viti Levu Island (Fiji) was estimated by assessing gross revenues from visiting researchers and students (payments received for food, lodging and fees paid; capital equipment donations; access fees to coastal waters; and other direct payments) and costs (mainly food preparation – labour costs are considered to be negligible). Data on number of research/education visits and payments made per visit has been obtained from official data and from researchers who have worked in the field over the past 6 months. The amount of money paid to the community during 2006 was FJ$7,625 (US$4,308). The cost of food production amounts to FJ$815 (US$460). When deducting costs from the gross expenditures, the net benefit from research activities is FJ$6,810 (US$3,847). Assuming that this figure is representative of net benefits for the community from fieldwork research/education activities over the following years, the overall economic value from research and education ranges (depending on the discount rate) between FJ$59,395 (US$33,558) and FJ$40,988 (US$23,158) over the next 10 years, and FJ$91,678 (US$51,798) and FJ$49,436 (US$27,931) over the next 20 years.	Market price method	O'Garra, 2007

Protected area	*Benefits (ecosystem services and related goods)*	*Estimated socio-economic value of benefits*	*Method*	*Source*
Bolivia, Kaa-Iya National Park (forest)	Cultural identity and religious values	Kaa-Iya National Park is Bolivia's largest protected area and it was established and is now managed jointly by indigenous people and the government. The cultural identities and religious values tied to Kaa-Iya National Park, threatened by external interests, have been officially recognised, spatialised in terms of land entitled to the indigenous people and quantified in monetary terms through a compensation scheme negotiated with petroleum companies for gas extraction. The Isozu-Guarani people were able to negotiate with the government their entitlement to Territorio de Comunidad de Origin (TCO), meaning land of indigenous communities, a 1.9 million ha buffer zone adjacent to the park. The park benefits from a US$1 million trust fund established as part of mitigation compensation negotiated for construction of gas pipelines through the park. A Global Environmental Fund (GEF) project supports the park by providing 70 per cent of the annual recurrent management costs, with the remainder coming from the trust fund. The TCO has benefited from a bank allocation of US$3.7 million for implementation of an Indigenous People Development plan (as part of the pipeline mitigation package).	Market price method (compensation)	Stolton *et al.*, 2010

Continued

Protected area	*Benefits (ecosystem services and related goods)*	*Estimated socio-economic value of benefits*	*Method*	*Source*
Forest ecosystems, Japan	Health regulation	Shinrin-yoku is a Japanese practice of taking in the atmosphere and energy of the forest to improve health and reduce stress. A recent study concluded that this practice balanced out acute emotions, particularly amongst those suffering from chronic stress, and that therefore forest landscapes could be considered as therapeutic landscapes. Another study, which focused on diabetic patients, found that the Shinrin-yoku practice of forest-air bathing and walking substantially decreases blood glucose levels, which in the tests dropped on average from 179 to 108. It was suggested that this might be due to changes in hormonal secretions and autonomous nervous function which result from the presence in the forest environment. Shinrin-yoku might therefore represent an important type of aromatherapy which has potential for many health disorders. Note: While highly relevant in the context of this guidance this example is not specifically related to PAs.	Quantitative analysis based on evidence on human health	Morita *et al.*, 2007 Ohtsuka *et al.*, 1998 Dudley *et al.*, 2010

8.3 Estimating the value of benefits

This section summarises the most commonly used methods for estimating the socio-economic value of cultural services and related goods, supported by a number of concrete examples in Table 8.2 above and Boxes 8.1-8.3 below. More information, examples, and step-by-step guidance on the application of key methods is provided in Annex 5. An introduction to the different valuation methods is available in Chapter 3.

Monetary: market values

The market price method is the most often used approach to estimate the economic value of recreation and tourism related to PAs. It usually links data on the number of people visiting an individual PA and/or PAs in a certain region with their average spending during the visits (sales and rent), ideally subtracting the related occurring costs (such as on equipment or food) to derive a net estimate of the socio-economic value. As such, the market price method reflects the profit gained from goods and services bought and sold on commercial markets; such as for accommodation, transport, food, events, special equipment, licences, or excursions. It includes simple approaches such as multiplying entrance fees by the number of visitors. Other methods look into the revenues generated by visitors on the site, such as the direct spending by researchers for food, access fees, and equipment donations (Table 8.1). More complex approaches aim to determine the impact of spending on the regional economy, by applying multipliers to the number of tourists and average spending per visitor (Box 8.1) or using input-output models.

It needs to be considered that prices are not always a good reflection of the actual value people give – or could give – to a service or related good. This may be linked to market imperfections such as distortions caused by taxes and subsidies. Additionally, there might be a difference between price on the market and what consumers would, in theory, be willing to pay (consumer surplus). For the purpose of valuation, prices can usefully be adjusted to take into account these differences and/or complemented by additional methods, such as contingent valuation studies, that help determine consumer surplus (see non-market methods below). Difficulties can also arise in separating activities specifically related to a PA from the general recreational and tourism visits that may occur to a site. One way forward is to determine the affinity of visitors to an area by using comprehensive surveys, as has been done in the example provided in Box 8.1.

BOX 8.1

The economic impact of tourism in six German National Parks

The designation of National Parks in Germany has historically been strongly motivated by the promotion of regional development in often peripheral regions. Though the primary aim of the 14 existing National Parks is to protect endemic species and ecological integrity on large-scale territories of often pristine character, they are also open to scientific research, environmental education, and allow the general public to experience nature.

Method: The study aimed at calculating the economic impact of tourism related to 6 out of the 14 national parks by linking known income multipliers to tourist expenditures, i.e. by looking at the amount of income generated in an economy by an additional unit of tourist expenditure. Income referred to salaries, wages, and income from operations and investment, but not devaluation (aging infrastructure, appliances, etc.) or subsidies. Capital investment in tourist infrastructure was also not considered. Relevant data were mainly obtained by comprehensive on-site surveys at the six national parks including Niedersächsisches Wattenmeer, Bayerischer Wald, Eifel, Müritz, Hainich and Kellerwald-Edersee between 2004 and 2007. This included information on variables such as the total number of visitor days per year (including information on type, age, gender, and accommodation), the mean daily expenditure per visitor by economic sector, the regional average income per capita, and the share of visitors with high national park affinity. Multipliers, i.e. fixed variables which reflect the amount of additional income generated by 1 unit of spending, were provided by a consultancy that collects long-term economic data on tourism and tourism businesses in Germany – including inter-firm linkages. Based on available data, gross and net tourist spending (without VAT) was calculated based on the number of visitors and expenditure in a specific sector. The direct regional income resulted from multiplying net tourist spending with a factor that reflected the proportion of income in a first round of cash flow (direct multiplier). Indirect regional income is calculated by subtracting the direct regional income from net tourist spending and multiplying by a factor that reflects the income generated by a second round of cash flow (indirect multiplier).

Results: The national park Niedersächsisches Wattenmeer was ranked first, generating an estimated income of €500 million created through tourism in the park, of which almost €60 million is created by visitors with high national park affinity. The Bayerischer Wald and Müritz national parks came second, with respectively €13.5 million and €6.9 million for all visitors and €6.5 million and €2.8 million for those with high affinity. All others had a total regional income of less than €5 million each. The regional average income per capita (income equivalent) was highest for the Niedersächsisches Wattenmeer, with roughly €30,000, followed by the Bayerischer Wald with €900. The designation of the national park Niedersächsisches Wattenmeer is responsible only for a minor influx of visitors, as it was already an established tourist destination. The national park Bayerischer Wald is the most successful in attracting visitors and as a result its specific contribution to regional economic development was ranked higher. In this regard, efforts to increase the visibility of a site were deemed decisive, as was the support by local stakeholders – especially tourism businesses. In addition, more mature holiday destinations seemed to attract a higher share of overnight visitors. It was also underlined that, in order to reduce the negative impacts of increased visitor streams, national parks in Germany should aim for a qualitative upgrading rather than a quantitative one; i.e. to increase the actual spending per visitor whether by promoting overnight stays or enhancing the quality of goods and services and, as such, increase prices.

Source: Mayer *et al.* (2010)

Monetary: non-market values

Monetary estimates of cultural, inspirational, and amenity values often require more complex and data intensive valuation methods. This is the case because either no markets exist that directly capture these values in the form of a price that people are willing to pay, or because available markets do not necessarily reflect their true value due to existing distortions (see above and Chapter 3). As such, non-market valuation methods focus on revealing the value of cultural, inspirational, and amenity values by looking into choices made within markets that are closely linked to these benefits (revealed preference), or by creating hypothetical markets (stated preference).

Cost-based approaches are often used to provide a monetary estimate of health benefits linked to cultural goods. Health costs can be calculated by looking into illness rates caused, for example, by stress or physical inactivity, and linking it to the costs of healthcare services such as physicians, hospital visits/stay, and medication. The information can be complemented by considering the amount of productive days lost due to absence from work or lower productivity (Proper and van Mechelen, 2007) (see Box 8.3).

With regard to recreational and cultural benefits, the travel cost method (revealed preference) is often used. It looks at the expenses that people are willing to incur or the costs they are willing to bear to travel to a site. Surveys are carried out to gather relevant information such as on the number of visitors, demographic aspects, purpose of the visit, duration of the journey, time spent at the PA, direct travel expenses, and value placed on time. Additional aspects, such as entrance fees, are also often included in the calculation (see Box 8.2, Table 8.2 and Annexes 1 and 5 for further information).

The hedonic pricing method is often used to reveal the economic values of amenity benefits linked to a PA, assuming that demand for the benefit is reflected in alternative markets. This can refer to demand for wildlife photography, landscape paintings, and, often to the demand for properties situated in the vicinity of PAs. It is assumed that amenity benefits would affect the price of the property, which would be higher than one located elsewhere without the feature. In order to determine whether this is the case, data on property prices and sales in a region are necessary. Also, additional information on the location is needed and included in the analysis, as values can be influenced by different factors. These include structural (age of the house), neighbourhood (school quality, crime or demographic aspects), and environmental attributes (air quality, landscape and amenity). These attributes can be grouped in different categories, representing different variables of a statistical analysis and supported by geospatial data. The statistical analysis determines the level of correlation between the different categories, including the proximity to a protected area. Though only partially related to a PA, a related study in Australia found that by being 1 metre closer to a wetland can increase sales by AU$829 (US$798), and a second wetland will increase the property price by AU$6,081 (US$5,855), leading to a total premium of AU$140 million (US$135 million) in sales due to wetland proximity (Tapsuwan *et al.*, 2007).

Revealed preferences approaches are particularly susceptible to market imperfections as they may distort the estimated monetary value (Pascual *et al.*, 2010). PAs might also negatively affect land use prices, due to the fear of restricted development opportunities going along with increased bureaucratic burdens. In addition, it needs to be noted that, though the value of property might be used as a proxy to reflect the amenity benefits of a site, strongly increasing prices can have social impacts and raises again the question whether it can only be afforded by a few.

The creation of hypothetical markets via stated preference methods can also be used to reflect the economic (non-use) value of cultural services. For example, the economic value derived from the creation of four hypothetical national parks in northern Canada was quantified using contingent valuation methods. The value amounted to CA$244 (US$235) per household or CA$2.80 (US$2.77) billion among the national population. Irrespective of the opportunity/likelihood to visit the parks, respondents considered them as places of rare or irreplaceable natural wonder or as a fragile ecosystem over which humans should exercise stewardship (CAFF, 2002, see Table 8.2). Stated preference methods can also be used to complement existing market values (above) and explore whether there is a difference between current prices (admission fees, etc.) and what visitors would be willing to pay. For example, according to a survey carried out in national parks in South Korea, visitors were willing to pay an admission fee up to 15 times higher than the current fee (Table 8.2).

BOX 8.2

The recreational value of Kakum National Park (Ghana)

The Kakum National Park is one of the most important parks in Ghana that protects a vast rainforest including a wide diversity of species, habitats, and natural processes. The park also hosts the only canopy walkway in Africa, which allows visitors to explore a tropical rainforest canopy from suspension bridges and has a high recreational value with thousands of people visiting the park every year.

Method: The economic value of the park to 'recreationers' was assessed by applying a travel cost method. Key data for the study was obtained through a survey of 224 randomly sampled individuals who visited the site from all over the world; the questionnaires were distributed in both peak and off peak periods. The collected information included: the number of visitors to the site, place of origin, demographic aspects (income, age, education, sex and type of recreation activities preferred), duration of the journey, time spent at the site, direct travel expenses, values placed on time by the respondents, purpose of the visit (other than visiting the site), etc. By considering the number of site visits made by each visitor over a specific period, the total travel cost was estimated as the sum of transport costs and entrance fee.

Results: 126,065 visitors from 15 countries visited the park during the period of data collection, the majority of which were from Europe and Ghana. One result was that older foreigners and younger Ghanaians visit the park more often. Educated people had a stronger appreciation for leisure and the environment, as over 72 per cent of the visitors attained tertiary education. From the statistical analysis, however, demographical aspects were found to be not strictly related with the visit rate; there was a negative relationship between the visit rate and the travel cost and a positive relationship between the visit rate and knowledge of the site. The minimum and maximum travel costs by visitor were GH¢1.70 (US$ 1.21) and GH¢800.20 (US $571.57) respectively, while the mean travel costs were GH¢52.73 (US$31). The annual recreational value of the site was estimated at GH¢67.28 (US$46.40) per person, which translates into an annual total value of GH¢8,481,653.20 (US$5,849,416).

Source: Twerefou and Ababio (2012)

Non-monetary values

Non-monetary values on cultural services and related goods often form an important basis for and also a necessary complement to any monetary estimates. Quantitative information, such as the average number of visitors, is a key variable of any monetary calculation. At the same time it can also help to emphasise the importance of PAs by looking at how these numbers have developed over several years, including what impact the designation of a site as a PA might have had. Information on the frequency of visits, the length of the stay, and the time taken to get to a site can be helpful indications of the recreational and tourism value of PAs. Qualitative information on the demographic aspects of visitors (education, age, nationality) and reasons for visits are needed to determine the affinity of visitors to a site, including which features are considered particularly attractive. Relevant statistical data is usually gathered by organising comprehensive surveys (see Boxes 8.1 and 8.2 and Table 8.2).

Establishing citizens' juries or carrying out structured interviews are also ways forward to get an informed opinion; for example, considering the importance of a protected area in forming the cultural identity of a community or in determining its spiritual value. Religious or cultural identity values can also be assessed at a qualitative level by recognising the importance of PAs to local or indigenous traditions, heritage or practices and at a quantitative level for example by mapping the sacred areas (see examples from Russia and Bolivia in Table 8.2). Reviewing the number of studies that referred to PAs as well as determining the number of TV programmes that have referred to a site can be used as simple methods to underline the inspirational and educational value of an area.

Effects on human health can be assessed by using surveys and statistics to qualify and quantify the estimated increase in physical and mental health and quality of life related to the access to a PA. The possible information to qualify and quantify health impacts includes observations, expert views and research on restorative effects such as decreased levels of stress and mental fatigue, increased levels of physical activity stimulated by the presence of attractive nature in the nearby work and living environment, lower levels of aggression and criminality, lower substance abuse, increased social integration due to natural settings functioning as social meeting places, improved motoric development in children who regularly engage in outdoor activities or increased worker productivity and creativity in offices with views of nature. For example, Box 8.3 provides an example of how mortality records were used to investigate links between mortality, income, and access to green space. Please note that while highly relevant this example is not specifically related to PAs.

BOX 8.3

Effect of exposure to natural environment on health: an observational population study

The costs of avoidable physical and mental ill health for the UK publicly funded National Health Service are considered significant, with estimates ranging from £11-50 billion per year in terms of health costs and productive days lost. Most of these avoidable diseases are a consequence of physical inactivity, with 70 per cent of the UK population having sedentary lives (Stolton and Dudley, 2010). A major research project reviewed the records of more than 360,000 pre-retirement age deaths from all causes in England looking for links between mortality, income, and access to green space.

Method: The study used small geographical units called lower level super output areas (LSOA) as they are usually used in England to provide small area statistics. Five exposure groups to green areas were defined; each group including 20 per cent of the study population. Mortality records matching the LSOAs were gathered from the UK Office for National Statistics, including information on the age at death, sex, cause of death and LSOA of residence. As inequalities are mostly visible in the working age population, populations older than retirement age were excluded. The authors used the English index of multiple deprivations to represent the proportion of low-income families in an area. Stratified statistical models were then applied to explore significant interactions by group of exposure to green space, and adjusted by age group, sex, deprivation in education, skills and training, deprivation in living environment, population density, and urban or rural classification.

Results: The total study population amounted to 40,813,236, including 366,348 deaths. Results showed that people with greater exposure to green space were more likely to be less deprived than those with little exposure to such areas. The inequality in disease mortality related to low income was found to be lower in populations who live in the greenest areas than in those who have less exposure to green space. Physical inactivity and response to stress are factors contributing to circulatory disease. It is assumed that their reduction might have contributed to lower inequalities in income that were recorded in greener areas.

Source: Mitchell and Popham (2008)

References

Beal, D. J. (1995) 'A travel cost analysis of the value of Carnarvon Gorge National Park for recreational use', *Review of Marketing and Agricultural Economics*, 63(2): 292-303

Bell, S., Tyrväinen, L., Sievänen, T., Pröbstl, U. and Simpson, M. (2007) 'Outdoor Recreation and Nature Tourism: A European Perspective', *Living Rev. Landscape Res. 1*, http://www.livingreviews.org/lrlr-2007-2, accessed 12 November 2012

BIO Intelligence Service (2011) '*Estimating the economic value of the benefits provided by tourism/recreation and employment supported by Natura 2000*', final Report prepared for the European Commission, http://ec.europa.eu/environment/nature/natura2000/financing/docs/Estimating_economic_value.pdf, accessed 4 June 2013

Biomimicry Institute (2010a) *Architecture: Learning from Termites How to Create Sustainable Buildings*, www.biomimicryinstitute.org/case-studies/case-studies/architecture.html, accessed 12 December 2012

Biomimicry Institute (2010b) *Energy: Learning from Humpback Whales How to Create Efficient Wind Power*, www.biomimicryinstitute.org/case-studies/case-studies/energy.html, accessed 12 December 2012

Bohne, A. (2008) '*Wildlife value, trophy hunting and rural development – with illustrations from Tanzania*', Conference on International Research on Food Security, Natural Resource Management and Rural Development, http://www.agrar.hu-berlin.de/fakultaet/departments/daoe/ihe/Veroeff/Tropentag_Paper.pdf, accessed 7 December 2012

BusinessGreen (2010) *Biomimicry: Hippy nonsense or corporate game-changer?* www.businessgreen.com/business-green/analysis/2264291/biomimimetics-green-business, accessed 12 December 2012

CAFF (2002) '*Protected Areas of the Arctic: Conserving a Full Range of Values*', Conservation of Arctic Flora and Fauna, Ottawa, http://library.arcticportal.org/1294/, accessed 7 December 2012

De Groot, R., Fisher, B., Christie, M., Aronson, J., Braat, L., Gowdy, J., Haines-Young, R., Maltby, E., Neuville, A., Polasky, S., Portela, R. and Ring, I. (2010) 'Integrating the Ecological and Economic Dimensions in Biodiversity and Ecosystem Services Valuation', in Kumar, P. (ed.) *The Economics of Ecosystems and Biodiversity (TEEB) Ecological and Economic Foundations*, Earthscan, London and Washington, DC

Dudley, N., Higgins-Zogib, L. and Mansourian, S. (2006) '*Beyond Belief: Linking faiths and protected areas to support biodiversity conservation*', a research report by WWF, Equilibrium and the Alliance of Religions and Conservation (ARC), http://assets.panda.org/downloads/beyondbelief.pdf, accessed 4 June 2013

FEEM (2010) *Le potenzialità turistiche del Parco Nazionale dell' Appennino Lucano Val D'Agri-Lagonese*, Rapporto delle attivita' relative al progetto, Fondazione Enrico Mattei, Italy, http://www.feem.it/UserFiles/File/Pub%20Bas/RapportoParcoNazionaleAppenninoLucano.pdf, accessed 7 December 2012

Gantioler, S., Rayment, M., Bassi, S., Kettunen, M., McConville, A., Landgrebe, R., Gerdes, H. and ten Brink P. (2010) '*Costs and Socio-Economic Benefits associated with the Natura 2000 Network*', final report to the European Commission, Institute for European Environmental Policy/GHK/Ecologic, Brussels, http://ec.europa.eu/environment/nature/natura2000/financing/docs/natura2000_costs_benefits.pdf, accessed 8 December 2012

Gilligan, B. (2006) '*The National Reserve System Programme 2006* Evaluation', Department of the Environment and Heritage, Canberra, Australia, http://www.environment.gov.au/parks/publications/nrs/pubs/evaluation-2006.pdf, accessed 7 December 2012

Kettunen, M., Vihervaara, P., Kinnunen, S., D'Amato, D., Badura, T., Argimon, M. and ten Brink, P. (2013) *Socio-economic importance of ecosystem services in the Nordic Countries*

– *Synthesis in the context of The Economics of Ecosystems and Biodiversity (TEEB)*. Nordic Council of Ministers, Copenhagen

Lee, C. and Han, S-Y. (2002) 'Estimating the use and preservation values of national parks' tourism resources using a contingent valuation method', *Tourism Management*, 23: 531–540

Lovins, H. (2008) 'Rethinking production', in *State of the World 2008: Innovations for a Sustainable Economy*, Worldwatch Institute, www.worldwatch.org/files/pdf/SOW08_chapter_3.pdf, accessed 12 December 2012

MA – Millennium Ecosystem Assessment (2005) '*Ecosystems and Human Well-being: Biodiversity Synthesis*', World Resources Institute, Washington, DC, http://www.millenniumassessment.org/documents/document.354.aspx.pdf, accessed 7 December 2012

Mayer, M., Müller, M., Woltering, M., Arnegger, J. und Job, H. (2010) 'The Economic Impact of Tourism in Six German National Parks', *Landscape and Urban Planning*: 97(2): 73-82

Mazza, L., Bennett, G., De Nocker, L., Gantioler, S., Losarcos, L., Margerison, C., Kaphengst, T., McConville, A., Rayment, M., ten Brink, P., Tucker, G. and van Diggelen, R. (2011) '*Green Infrastructure Implementation and Efficiency*', final report for the European Commission, DG Environment, Institute for European Environmental Policy, Brussels and London, http://www.endseurope.com/docs/120314a.pdf, accessed 7 December 2012

Mitchell, R. and Popham, F. (2008) 'Effect of exposure to natural environment on health inequalities: an observational population study', *The Lancet*: 372(9650): 1655-1660

O'Garra, T. (2007) *Estimating the total economic value of the Navakavu LMMA (Locally Managed Marine Area) in Vitu (sic) Levu island (Fiji)*, Coral Reef Initiatives for the Pacific, School of Marine Studies Faculty of Islands and Oceans, the University of the South Pacific, http://www.crisponline.net/Portals/1/PDF/C2A2_ECONOMIC_REPORT_Final.pdf, accessed 7 December 2012

Pascual, U., Muradian, R., Brander, L., Gomez-Baggethun, E., Martin-Lopez, B., Verma, M., Armsworth, P., Christie, M., Cornelissen, H., Eppink, F., Farley, J., Loomis, J., Pearson, L., Perrings, C. and Polasky, S. (2010) 'The Economics of Valuing Ecosystem Services and Biodiversity', in Kumar, P. (ed.) *The Economics of Ecosystems and Biodiversity (TEEB) Ecological and Economic Foundations*, Earthscan, London and Washington, DC

Proper, K. and van Mechelen, W. (2007) *Effectiveness and economic impact of worksite interventions to promote physical activity and healthy diet*, background paper prepared for the WHO/WEF Joint Event on Preventing Non-communicable Diseases in the Workplace Dalian/ China, http://www.who.int/dietphysicalactivity/Proper_K.pdf, accessed 7 December 2012

Ramsar and United Nations World Tourism Organization (UNWTO) (2012) *Destination wetlands: supporting sustainable tourism*, Secretariat of the Ramsar Convention on Wetlands, Gland, Switzerland & UNWTO, Madrid, Spain, http://www.ramsar.org/pdf/cop11/tourism-publication/Ramsar_UNWTO_tourism_E_Sept2012.pdf, accessed 7 December 2012

Sandel, M. J. (2012) *What Money Can't Buy: The Moral Limits of Markets*, Farrar, Straus and Giroux, New York

Stolton, S., Mansourian, S. and Dudley, N. (2010) *Valuing protected areas*, the International Bank for Reconstruction and Development, Washington, DC

Stolton, S. and Dudley, N. (2010) *Arguments for Protection Vital Sites: The contribution of protected areas to human health*, WWF and Equilibrium Research, http://assets.panda.org/downloads/vital_sites.pdf, accessed 4 June 2013

Tapsuwan, S., Ingram, G. and Brennan, D. (2007) *Valuing Urban Wetlands of the Gnangara Mound: A Hedonic Property Price Approach in Western Australia*, CSIRO Water for a Healthy Country National Research Flagship, Canberra, http://www.clw.csiro.au/publications/waterforahealthycountry/2007/wfhc_ValueUrbanWetlands.pdf, accessed 7 December 2012

ten Brink, P., Mazza, T., Badura, T. and Kettunen, M. (2012) *Nature and its Role in the Transition to a Green Economy: A contribution to Rio + 20* (executive summary), http://www.teebtest.org/wp-content/uploads/2012/10/TEEBnature-greeneconomy.pdf accessed 4 June 2013

Twerefou, D. K. and Ababio, D. K. A. (2012) 'An economic valuation of the Kakum National Park: An individual travel cost approach', *African Journal of Environmental Science and Technology* 6(4): 199-207

UK NEA – National Ecosystem Assessment (2011) *The UK National Ecosystem Assessment: Synthesis of the Key Findings*, UNEP-WCMC, Cambridge, http://uknea.unep-wcmc.org, accessed 18 June 2012

UN-HABITAT (2012) *State of the World's Cities Report 2012/2013: Prosperity of Cities*, United Nations Human Settlements Programme, Nairobi, Kenya, http://www.preventionweb.net/files/28495_urbanprosperity.pdf, accessed 7 December 2012

van Nierop, E. A., Alben, S. and Brenner, M. P. (2008) 'How bumps on whale flippers delay stall: An aerodynamic model', *Physical Review Letters*, 100: 054502

WHO (2007) *Steps to health: a European framework to promote physical activity for health*, World Health Organisation Regional Office for Europe, Copenhagen, http://www.euro.who.int/__data/assets/pdf_file/0020/101684/E90191.pdf, accessed 7 December 2012

ANNEX 5

Step-by-step guidance: tourism, recreation and other cultural benefits

Gantioler, S. and D'Amato, D.

1 Introduction

This step-by-step guidance provides more detailed information on how to apply a number of key methods (outlined in Chapter 8) to assess the socio-economic value of cultural services and related goods. Common approaches include 1) market price methods to determine the economic impact resulting from tourism and recreation, 2) revealed preference methods, such as the travel cost approach and hedonic pricing, to capture recreational and amenity values, and 3) stated preference methods, such as contingent valuation, to make visible values of other cultural services and related goods. The focus of this guidance is in particular on the following methods: generated income and economic impact for market values (Section 2 below), and travel costs and contingent valuation for non-market values (Section 3 below).

2 Market values: generated income and economic impact

The market price method estimates the economic value of ecosystem services and related goods bought and sold in commercial markets. It observes the preferences and the willingness to pay for a service or good at the price offered. With regard to tourism and recreation, the number of visitors to PAs is linked to direct sales of products and services provided to visitors (accommodation, transport, food, events, special equipment, entrance fees, licences, excursions, etc.).

Tourism and recreation: income from entrance fees

Analysis of PA visitor numbers and the entrance fees they are willing to pay to visit a site is a rather straightforward way to capture the socio-economic value of tourism and recreation.

Step 1 – Data collection: The first step consists of finding up-to-date information on the number of visitors and determining the timeframe for analysis. Visitor information should, in principle, cover a full calendar year to guarantee that the analysis captures seasonal fluctuations in visitor flows and avoids, for example, extrapolation of results based on an inadequate evidence base (as one season does not reflect all four seasons). Possible variance in entrance fees needs also to be integrated into the analysis by gathering separate data for different fee categories (Box A5.1).

Step 2 – Analysis: The calculation should be straightforward and consisting of multiplying the entrance fees by the number of visitors. However, multi-level fees might make it necessary to carry out separate analysis for different visitor categories (foreign tourists/national tourists, adults/children/pensioners, etc.) (Box A5.1). If multi-annual data series are available, analysis should be carried out for several subsequent years to reflect the trends in visitor numbers and associated income. Such multi-annual data can also be used to analyse possible correlation between visitor and investment flows over the years.

Information required: Visitor flows are very site-specific and therefore the information used in an analysis should also be site-specific and not extrapolated from other sites. Information sources include visitor logs and/or surveys or, if those are not available, data on entrance checks, ticket sales, etc. Relevant complementary data could also be acquired from third parties such as travel agencies and businesses present in the park. If existing data is not available park gates could be controlled over a defined period of time to gather information on visitor numbers and extrapolated to a year, multiplying it then by the entrance fee.

It is important to note that, while considered as a market value, visitor fees are rarely subject to market rules (see Chapter 8 for discussion on the moral limits of markets). For many PAs there are no entrance fees and where such fees exist they are often set below the price as defined by market rules (i.e. price people would be willing to pay). Consequently, estimating the recreational and touristic value of a site based on entrance fees might actually lead to significantly underestimating the true socio-economic importance of these benefits. This is also because several aspects of these benefits, such as recreation related health benefits, cannot be captured by entrance fees alone. To acquire a more comprehensive picture of the socio-economic value of recreation and tourism, the information on visitors flows and entrance fees is often used as part of the travel cost method or combined with a stated preference approach (see below). If resources for such complementary analyses are not available the caveats of a simple visitor flow–entrance fee analysis should be clearly explained.

Concrete example: Box A5.1

BOX A5.1

Revenue generation from entrance fees and gorilla tracking in Mgahinga Gorilla National Park

The Mountain gorilla (*Gorilla gorilla beringei*) is the rarest subspecies of gorilla with only around 650 individuals. Mountain gorillas have global significance and flagship value for conservation, being one of the rarest of the great apes. In Uganda, gorillas can be seen only in the southwest corner of the country, at Bwindi Impenetrable National Park and Mgahinga Gorilla National Park. Gorilla tourism requires careful habituation of wild gorilla groups to the presence of tourists. Consequently, both the number of tourists and gorilla tourism enterprises need to be carefully controlled to keep disturbance within acceptable bounds. Gorilla tourism also poses a number of risks to conservation including stress caused to individuals in the habituation process, transmission of human diseases through contact with visitors and guards and the risks to habituated groups from poachers, guerrillas and other armed people in the forests. However, the incentives and revenue from tourism has also played an integral role in the survival of species.

Method: Visitor data for Mgahinga Gorilla National Park were collected in 1997–1998 from the park gate records (covering a period of 8 months, extrapolated to 12 months). Park entrance fees and the prices for gorilla tracking varied depending on visitors' residence status (Table A5.1). Income from gorilla tracking was estimated by multiplying the number of trackers (70.5 per cent of visitors) by the sum of park entry fees and tracking fees (taking into account the different fees for visitors' residence status).

Results: The total amount of visitors was estimated to be 1,448 per year. Of these, 97 per cent were foreign non-residents, 2 per cent foreign residents and 1 per cent Ugandan nationals. Over 70 per cent of visits were related to gorilla tracking while mountain climbing (11 per cent) was the next most popular attraction. The data suggest that in 1998 the income from gorilla tracking amounted to US$190,363.

TABLE A5.1 Estimated income from gorilla tracking in 1997–98 in Mgahinga Gorilla National Park

Residence status	*Tracking fee*	*Park entry fee*	*Estimated No. Visitors*	*Number of trackers**	*Income (US$)*
Foreign non-residents (tourists)	175	15	1,409	993	188,670
Foreign residents	150	6	8	5	780
Ugandan nationals	40	1.5	32	22	913
Total	–	–	1,449	1020	190,363

*70.5 per cent of visitors

Source: Adams and Infield (2003)

Tourism and recreation: income from visitor streams

The economic impact of PAs is usually defined as the changes in new economic activities resulting from PA designation and implementation. In relation to tourism and recreation, such impacts relate to the direct spending due to visits to a site (accommodation, food, excursions, tours, etc.), and the subsequent round of impacts this can lead to (procurement of food, cleaning services for hotels, energy supply, etc.). The impact can be assessed by applying different models and approaches, including input-output models reflecting the inter-relationship between activities or using multipliers to estimate how the money flows through the economy, supporting income or jobs directly and indirectly (see Chapters 8 and 10 for further explanation). The different steps below illustrate how such an analysis can be carried out when using multipliers while Box A5.2 provides an example of more complicated modelling.

> *Step 1 – Preparation:* The first preparatory step is to find out whether regional multipliers, such as multipliers for income or employment, have already been developed by previous projects (studies on tourism or culture heritage sites, public statistics, etc.). If such multipliers are available, their suitability for assessing recreation and tourism related benefits should be evaluated. In addition, economic activities and sectors that benefit from tourism expenditure related to PA visits should be determined.
>
> *Step 2 – Data collection:* When working with existing multipliers, the next step includes gathering relevant information on key variables determining the economic impact of a PA. First and foremost, this consists of data on visitor flows to the site. If dedicated surveys are carried out to support the analysis, they should be aimed at providing as sector specific insights as possible, for example determining mean daily expenditures according to the different sectors. In addition, generic visitor surveys should include questions determining the motivations for visits, in order to be able to determine the role the PA plays in attracting visitors to and creating visitor income within a region. Figure A5.1 presents a decision tree that can be applied to determine visitors' affinity to a site (see Box 8.2 in Chapter 8 for more information). Additional demographic data, such as information on gender, household income, visitor type and age, and length of stay, helps to identify socio-economic factors that affect the affinity to a site.
>
> *Step 3 – Analysis:* The visitor spending is calculated by multiplying the number of visitors by the estimated per visitor expenditure in different sectors. Multiplying this value (adjusted for VAT) by the identified direct multipliers results in the estimated direct economic impact of visitor expenditure within the region. By subtracting the direct impact from the visitor spending and using indirect multipliers the indirect impact of the spending can be determined.

Information required: Information related to visitor expenditure is collected via market research studies, government statistics or income and expenditure surveys. Multipliers can be most cost effectively obtained from existing studies undertaken either by private or public institutions. If resources are available, they can determined by carrying out a detailed analysis of inputs to and outputs from different economic sectors in the region.

Concrete example: Box A5.2

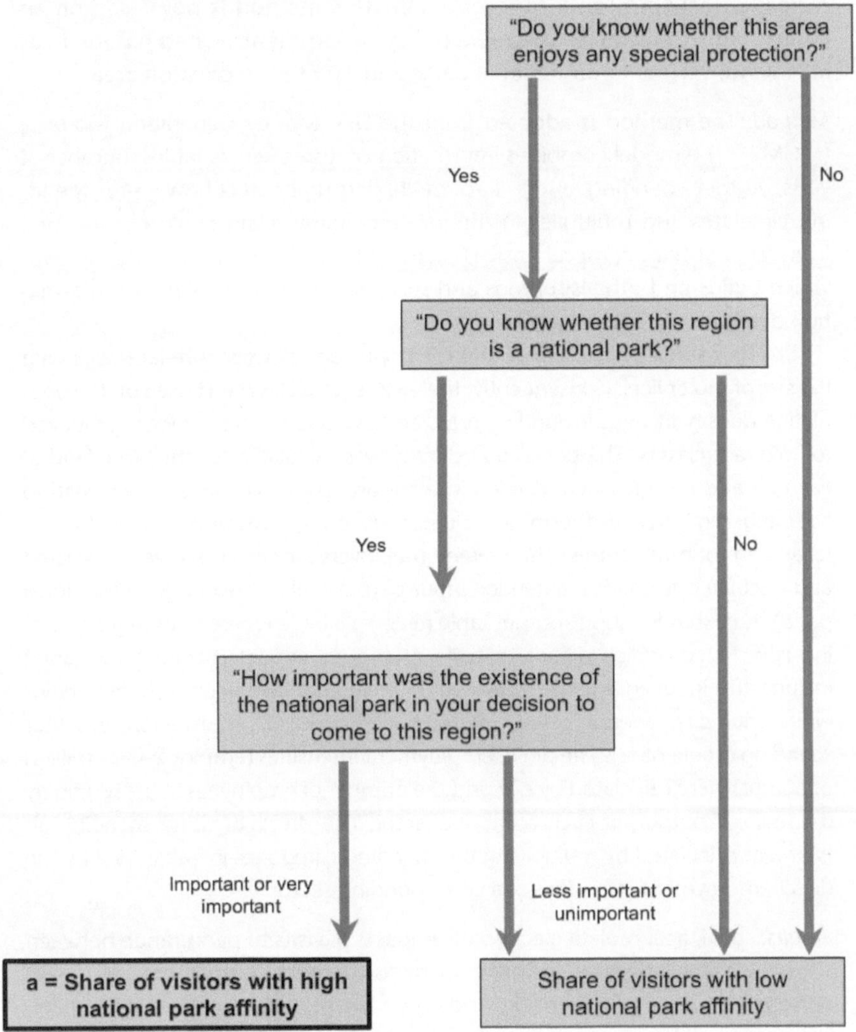

FIGURE A5.1 Approach to determine visitors' affinity to a PA

Source: based on Mayer *et al.* (2010), see Box 8.2 in Chapter 8 for more information

BOX A5.2

Income streams created by tourism at national parks and recreational areas in Finland

In 2009, the Finnish Forest Institute (Metla) and the PA managing agency Metsähallitus Natural Heritage Services developed a national level method for assessing the economic impacts of nature tourism and nature-related recreation activities on local economies. This method is now used on an annual basis to estimate the value of key government owned nature areas in Finland, including 35 national parks and 10 other recreation areas.

Method: The method is adopted from the U.S. Money Generation Model 2 (MGM2). The model combines information on three key variables: number of visits, visitors' spending, and a set of multipliers that reflect how visitor spending circulates and multiplies in the local economy. Data could be obtained from an existing comprehensive, standardised visitor monitoring system, which builds on both visitor logs and surveys, and a national visitor information database (ASTA) by Metsähallitus.

Previous studies have shown that the population density correlates well with the size of multipliers. Consequently, the 45 PAs were classified based on the population density in the surrounding areas and average multipliers were produced for different classes. The population density was calculated for the hinterland of each PA and the sites were divided in three groups: capital area, other built-up area and rural area. A fourth complementary group was also created for PAs located in tourism centres. The average multipliers for each class were extracted and calculated from local extended input-output tables (created for 21 national parks). An extended input-output table refers to a table where household spending is included, which means that the multipliers extracted from these tables include the induced effects of household spending. The economic sectors for which multipliers were created were identified based on information on visitor spending (Table A5.2). The direct employment intensities (employees/€1 million of output) were calculated by dividing the number of employees in the region by the regional output for each economic sector. The effects of visitor spending on PAs were calculated by multiplying the spending categories in Table A5.2 by the direct employment intensities of a corresponding sector.

Results: Local, accumulative economic impacts of visits to parks range between €0.1–€30.6 million/year for different parks, generating an estimated 1-400 person-years of employment depending on the park. The wide ranges reflect, for example, differences in accessibility, landscape and ecosystem features and marketing between the parks. When summing up the benefits at national level, the support of national parks (altogether 37 areas) to local economies amounts to €108.3 million (1,394 man-years)/year. According to the assessment €1 investment in national parks and other key protected areas results in €10 return to local economies.

Source: Huhtala *et al.* (2010) and Kajala (2013)

TABLE A5.2 Spending categories of visitors to Finnish national parks (identified through visitor surveys) and equivalent sectors in national accounting

Spending categories in the visitor survey	*Equivalent economic sector in input-output tables*
Fuel or other purchases from service station	Trade
Costs for local transportation	Transport, storage and communication
Food and other retail shopping	Trade
Cafe and restaurant purchases	Hotels and restaurants
Accommodation	Hotels and restaurants
Organised programme and recreational services	Transport, storage and communication
Other expenses	Average of 'Real estate, renting and business services' and 'Other community, social and personal service activities'

Source: Huhtala *et al.* (2010)

3 Non-market/non-monetary values: revealed and stated preference for cultural services

Recreation values: travel cost method

The travel cost method (revealed preference) is often used to determine the recreational value of PAs. This method assesses the expenses that people are willing to incur or costs they are willing to bear to travel to a site. In its most simple form this refers to actual travel expenses including public transport and flight tickets, costs of petrol for private cars, entry fees and estimated opportunity costs related to time spent (see Chapter 11 for a definition of opportunity costs). Key steps to carry out a simple travel cost method are outlined below, including also some information on more complex approaches.

Step 1 – Preparation: The analysis should start by collecting data relevant to determining the PA's recreational attractiveness. This includes, for example, information on the site's environmental characteristics, most attractive features, available infrastructure and accessibility (availability of different means of transport, etc.). In addition, a scoping analysis of the different types of visitors to the PA should be carried out. For large PAs, the study area should be divided into different zones as travel costs and time-related opportunity costs can vary greatly between different destinations within the site.

Step 2 – Data collection: Data on travel costs are usually collected through surveys (visitor interviews or questionnaires). Key variables covered by such surveys include information on the means of transport, distance travelled, frequency of visits and travel costs incurred. Additional information gathered consists of motivation for the visit, gender distribution and visitors' level of income and education. In order to obtain a representative picture of the visitor flow, the questionnaire can be distributed to (or interviews carried out with) various types of visitors (bikers, walkers, etc.) during different seasons and/or times of the year (high season/low seasons; summer/winter, etc.) using a range of locations within the park.

Step 3 – Analysis: First-cut results could be obtained by multiplying the number of trips by related travel costs and using the acquired background information (means of transport, gender distribution, etc.) as a basis for qualitative interpretation of the results (see Box A5.3). As an additional step, the results on travel costs could be linked to and statistically analysed with the other identified variables, such as type of visitors and different zones within the PA. Such analysis helps to determine interlinkages between costs and other variables highlighting, for example, how travel costs (and in the end the demand for recreational value) reacts to changes in key variables. An increased amount of variables, however, makes the method more resource-intensive and it also requires more skills in statistical analysis and modelling, and as such the support of a dedicated expert might be needed.

The random utility model is a further extension of the travel cost model allowing additional insights on the value of a site to be developed. It tests the effect of changing the quality and quantity of environmental characteristics of a site on the travel costs. More complex forms combine the travel cost method with stated preference approaches, determining visitors' willingness to pay for hypothetical travel costs to visit a site. Such an approach is often applied with regard to entrance fees, to determine the willingness to pay more than the existing prices (consumer surplus).

Information required: As outlined above, key variables for the travel cost method are usually identified via surveys. However, relevant information cannot always be directly acquired from visitors but is calculated based on secondary information. For example, if a visitor has been travelling by car, distances can be determined by asking for the point of departure, and costs can be calculated by multiplying distance with current fuel prices. Train companies can offer special fees in certain countries and thus it is useful to ask the visitor the exact price of the fair to avoid overestimating costs by simply using typical train fairs for the distances. If the mode of transport to the PA is walking or cycling, information on the time spent on travel is more important than associated costs; however, time is also often included as a general cost factor in all travel cost methods (see Box A5.3). Information on environmental characteristics can be retrieved, for example, from site managers.

Concrete example: Box A5.3

BOX A5.3

Travel cost method: the case of Lake Elliðavatn and Lake Vífilsstaðavatn, Iceland

The following study assessed the socio-economic values for education and recreational angling at Lakes Elliðavatn and Vifilsstaðavatn which (partly) belong to the Heiðmörk nature reserve on Iceland.

Method: A total of 269 and 71 anglers were surveyed at Lake Elliðavatn and Lake Vifilsstaðavatn, respectively; 61 and 63 per cent of the responses were usable for the analysis. Information on gender distribution, average age of visitor and main means of transport was collected. In addition, information on the level of education, fishing licences (one-day vs. annual licences) and final use of the retrieved fish (release vs. use) was gathered. Information on the educational value was acquired via questionnaires to schools, aiming to determine time spent by students at the sites and the cost per student.

Results: Based on 2,133 annual trips with an average costs of ISK8,620-ISK12,315 per trip (€54-€77), the recreational value of Lake Elliðavatn was estimated to be ISK19.27–ISK27.15 million (€121,300–€170,960 million). For Lake Vifilsstaðavatn the annual number of visits amounted to 336 with average estimated costs of ISK11,186–ISK11,848 (€70–€774), equivalent to a total recreational value of ISK3 736,124 – ISK3.95 million (€510–€24,900 million). The value of sites in terms of education was estimated at ISK3.81–ISK4.71 million (€24,023 – €29,688 million) for Lake Elliðavatn and at ISK1.97–ISK2.02 million (€12,451–€12,743 million) for Lake Vifilsstaðavatn. These values are likely to be underestimates as they do not take into consideration the use of both lakes by preschools and the University of Iceland.

Source: Jóhannesdóttir (2010)

Amenity values: contingent valuation method

The contingent valuation method (CVM) is also often used in studies evaluating socio-economic values linked to cultural services. It is important to note that CMV is only a stated preference method and therefore it results reflect hypothetical – rather than actual – behaviour and preferences. CVM is usually based on surveys that directly ask people their willingness to pay for the maintenance, use and/or restoration of a certain PA benefit. In some cases, people are also asked about their willingness to accept compensation to give up a certain benefit. Different to CVM, choice modelling (another key stated preference method) asks people to make choices between different options for different alternatives. It creates hypothetical scenarios from which people might choose instead of asking directly about values attributed to a service (see Box A5.4).

Step 1 – Preparation: After determining the scope of the survey (i.e. PA benefits included in the study), relevant information on factors determining the affinity of visitors – or the broader population – to a site needs to be collected. For example, in the case of amenity benefits this refers to information on special environmental characteristics affecting the site's appeal to visitors and/or to the public. In addition, the main beneficiaries who enjoy amenity values should be identified (horse riders, bird watchers, hikers, general public, etc.). In addition, the preferred survey method (interviews vs. questionnaires) and extent of survey (small vs. large sample size) should be determined.

Step 2 – Data collection: Based on the scoping assessment (above), the precise 'willingness to pay' question(s) needs to be defined, including an appropriate indicator (payment 'vehicle') to capture the value of a service. The latter can refer to hypothetical entrance fees, additional travel costs, taxes, or compensations. Example of such a question and indicator of value could be: 'How much would you be willing to pay as a daily, per person entrance fee to a diversified and unique, but restricted high value area in a national park?'

Note that it is not always advisable to initiate CVM assessment with a survey but it might be more appropriate to start by using focus groups to help to define the level of knowledge of the relevant beneficiaries and the kind of background information needed to accompany an eventual survey. Focus groups are also useful to fine-tune the question and indicators used.

Step 3 – Analysis: While some of the results can be synthesised in a qualitative and quantitative manner, CVM commonly results in a statistical analysis that requires expert knowledge and that can be rather resource-intensive.

Information required: Most information is retrieved via surveys, whether by distributing questionnaires or carrying out interviews. Information on the environmental settings of a site can be gained by site managers or administrators.

Concrete example: Box A5.4.

BOX A5.4

Valuation of the existence and amenity value of the Monteverde Cloud Forest Preserve, Costa Rica

The Monteverde Cloud Forest Preserve in Costa Rica comprises 10,000 ha of land, covering both Pacific and Atlantic watersheds and hosting valuable biodiversity. The value of preserving non-market amenities provided by the PA was estimated in order to be able to compare benefits of conservation with that of land conversion.

Method: a closed-end dichotomous choice (yes/no) survey was administered to 575 visitors to the PA, assessing their annual household income, age, gender and level of education. Visitors' willingness to pay (WTP) for preserving biological resources and other amenities provided by the PA were assessed based on a hypothetical scenario of deforestation and conversion of land to pastures and other agricultural uses. Identical English and Spanish surveys were used to collect information from both Costa Rican and non-Costa Rican visitors. Annual per-visitor payments supporting a hypothetical non-profit private conservation organization were selected as a vehicle for the payment, including a single lump-sum payment and recurring annual payments. Four different types of survey forms with different payment sums (bids) were developed and distributed to visitors (one form per visitor). Truncation problems were addressed by including a wide range of bids starting from US$10.00 (with US$10.00 increments) to US$200.00 including 20 bid levels in total. Respondents faced two options: to answer 'no' and face the loss of PA amenity benefits while keeping all of their income or to answer 'yes' and have their income reduced with the resource available in the future. The total existence and amenity value of the reserve was obtained as the mean individual willingness to pay multiplied by the number of visiting tourists per year.

Results: Based on the individual willingness to pay US$121 and a visitor flow of 32,213 persons per year, the total annual value of the existence and amenity of the reserve was estimated to be US$37,517,37. This value was higher than the estimated benefits of land conversion. Despite the generally lower income of Costa Ricans (34 per cent of the non-Costa Ricans' income), their WTP was higher (US$137.41) than WTP of non-Costa Rican visitors (US$118.76), 65 per cent of whom were from the United States. While the existence and amenity value of the PA as such was recognized prior to establishing the reserve, the study demonstrated the extent of this value to a range of national and international beneficiaries, further reinforcing the reasons for maintaining the reserve. Finally, a statistical analysis was used to estimate the influence of demographic parameters on the probability of 'yes' responses indicting that the amount of the bid was the most significant factor determining the response with the probability of a 'yes' response increasing with respondents' income (and education) level.

Source: Echeverria *et al.* (1995)

4 Key issues to be kept in mind

There are several opportunities and risks linked to the application of the different methods, often irrespective of which benefit they are used to valuate. Estimating the value of tourism and recreation can include market price methods that are easy to apply to more complex tools. The methods use observed data of actual consumer preferences and real income and as such are often more easily appreciated in planning and decision-making. However, prices may be distorted and not reflect the actual willingness to pay or benefits, for example because of existing lack of information, monopolies, government intervention, taxes or subsidies. Prices also often describe only the (minimum) amount people are willing to pay for recreation and tourism and therefore are not good metrics for estimating the overall cultural benefits that people obtain from a site.

The travel cost in its simplest form allows the value of cultural services and related goods to be revealed, by looking at the *de facto* market through the lens of related market(s) that could otherwise be very difficult to monetise. By looking at potentially related markets, demand curves for cultural services can be developed. However, more complex and nuanced versions of travel cost methods can be expensive and time consuming, as they build on a large data set and very complex analysis and modelling. It is also very difficult to isolate the value of a particular ecosystem within a PA.

The contingent valuation method (CVM) is a flexible method and as such allows the economic value of benefits to be estimated even in the case when the values are not captured by current markets (amenity values, etc.). On the other hand, CVM is vulnerable to biases influenced by survey design and choice of payment vehicle. It also requires the collection of complex data, supported by sophisticated data analysis and modelling. Careful consideration needs, therefore, to be given to the selection of valuation method and sufficient background information collected to allow an informed and effective decision.

References

Adams, W. M. and Infield, M. (2003) 'Who is on the Gorilla's Payroll? Claims on Tourist Revenue From a Ugandan National Park', *World Development* 31(1): 177–190

Echeverria, J., Hanrahan, M. and Solórzano, R. (1995) 'Valuation of non-priced amenities provided by the biological resources within the Monteverde Cloud Forest Preserve, Costa Rica', *Ecological Economics* 13: 43–52

Huhtala, M., Kajala, L. and Vatanen, E. (2010) *Local economic impacts of national park visitors' spending in Finland: The development process of an estimation method*, Working Papers of the Finnish Forest Research Institute, http://www.metla.fi/julkaisut/workingpapers/2010/mwp149.htm, accessed 23 December 2012

Jóhannesdóttir, H. M. (2010) 'Economic valuation of ecosystem services. The case of lake Elliðavatn and lake Vífilsstaðavatn', in D. M. Kristófersson *Rannsóknir í Félagsvísindum XI*, The social science research institute, Reykjavík

Kajala, L. (2013) 'TEEB Nordic case: Estimating local economic benefits of visitors' spending in protected areas', in M. Kettunen *et al.*, *Socio-economic importance of ecosystem services in the Nordic Countries – Scoping assessment in the context of The Economics of Ecosystems and Biodiversity (TEEB)*, Nordic Council of Ministers, Copenhagen

Mayer, M., Müller, M., Woltering, M., Arnegger, J. and Job, H. (2010) 'The Economic Impact of Tourism in Six German National Parks', *Landscape and Urban Planning*, 97(2): 73–82

9

APPRECIATING THE VALUE OF SUPPORTING SERVICES

D'Amato, D. and Kettunen, M.

9.1 Overview of benefits

Supporting services refer to processes that underpin and drive the functioning of ecosystems (nutrient and biogeochemical cycles, soil formation, primary production, etc.) and support species lifecycles (seed dispersal and species interactions) (Table 9.1). They also include the maintenance and evolution of biodiversity from the genetic to the ecosystem level. These diverse processes, interrelations and interactions form the basis for well-functioning ecosystems, helping to ensure their stability and resilience in a changing world.

Supporting services are different from other ecosystem services and related goods: they provide the basic infrastructure and functioning for all life on Earth and therefore they are a requirement for maintaining the provisioning of all other ecosystem services (i.e. regulating, provisioning and cultural services) (MA 2005, UK NEA 2011). In general, the identification of these fundamental processes as 'services' draws attention to the essential role they play in maintaining human well-being.

The current understanding of supporting services remains rather limited due to the complex and often indirect way they contribute to the availability of other ecosystem services and related goods, especially in the long term (Balmford *et al.*, 2008; UK NEA, 2011). Because of this, a detailed socio-economic assessment of supporting services, particularly in monetary terms, is often not feasible. Furthermore, given that supporting services are an integral part of ecosystems, their value is already reflected in the estimated monetary value of other ecosystem services and goods (Hein and de Groot, 2007). Consequently, the integration of supporting services into the monetary valuation of PA benefits can lead to double counting and overestimating the overall economic value (Chapter 3). It is therefore often more appropriate to assess the socio-economic importance of these services at a qualitative, or when possible, quantitative level if part of a wider assessment

TABLE 9.1 Supporting services and examples of how they underpin other ecosystem services, supporting socio-economic well-being

Ecosystem service	*Examples of benefits*
Nutrient cycling	Nitrogen deposits, influencing water purification and waste treatment
Soil formation	Soil properties, components and processes (nutrient status and pH, etc.) influencing, for example, food production, regulation of soil erosion, nutrient and water cycling
Photosynthesis	Primary production and plants' physio-ecological traits influencing, for example, food and fibre production and carbon storage
Biogeochemical cycles	Water availability influencing, for example, water supply and water regulation
Stability and resilience of ecosystems (maintenance of)	Capacity to recover from disturbances and natural hazards influencing, for example, the quality and quantity of resources, capacity to mitigate the impacts of natural hazards and support the adaptation to climate change
Seed dispersal	Securing species reproduction influencing, for example, food production/security and the existence of culturally important species
Species interactions between trophic levels (maintenance and control of)	Population control by predation or competition influencing, for example, the maintenance of biological control and food production/security
Genetic diversity (maintenance of)	Availability of genetic resources influencing, for example, bioprospecting and agro-biodiversity and supporting pharmaceutical industry and food security
Species diversity (maintenance of)	Species existence influencing food and fibre production, cultural services and the overall ecosystem functioning and resilience
Habitats diversity (maintenance of)	Habitat existence influencing food and fibre production, cultural services, species viability and overall ecosystem functioning and resilience
Unknown/identified ecological processes (maintenance of)	Beneficial processes that are currently not fully understood

Source: own presentation, based on MA (2005), UK NEA (2011), Kettunen *et al.* (2013), Balmford *et al.* (2008)

(see Section 9.2). Monetary values can, however, be included if a specific point regarding the value of supporting services needs to be made that might not be fully appreciated by focusing on the final benefits only.

9.2 Socio-economic importance of benefits

As explained in previous chapters, supporting services are not contributing directly to socio-economic well-being but are an essential prerequisite for it. In other words, without supporting services, the necessary auxiliary inputs that maintain production systems, regulatory processes and cultural values would not exist. For example, photosynthesis contributes to primary production by fixing solar energy and carbon into biomass. This is a basic prerequisite for food production and the development of carbon storage. Similarly, soil processes (Carter 2002, Table 9.2) and biogeochemical cycles underpin water availability and regulation of erosion.

The maintenance of biodiversity is essential, for example, to create opportunities for bioprospecting and to secure agro-biodiversity – that in its turn supports global food security (see Chapters 6-8 and Table 9.2). The existence of attractive or symbolic species that appeal to the public is also essential for cultural services and related goods, as such species often play an important role in creating tourism flows and providing inspiration to people. In Europe, for example, the Natura 2000 Network is the most important conservation tool to protect European threatened species and habitats and has a role in conserving suitable habitats and viable populations of lynx, wolf and bear (Boitani *et al.*, 2012) – three major carnivorous keystone species that also have an important cultural and historic role in Europe (Table 9.2).

Marine nursery habitats, often protected by PAs, have an important role in maintaining population densities and the general stability and resilience of ecosystems by enhancing their capacity to support fisheries. For example, the marine reserves in Baja California (Mexico) show greater population densities and reproductive outputs of the commercially valuable pink abalone (*Haliotis corrugata)* than surrounding unprotected sites (Micheli *et al.*, 2012) (Table 9.2). Similarly, Torre Guaceto Marine Reserve (Italy) has contributed to the re-establishing of essential predatory interactions lost outside the reserve, with community-wide effects (Table 9.2).

Anthropic (i.e. man-made) or environmental pressures (generally more intense outside PAs) drive the loss of abiotic and biotic interactions, weakening ecosystem's ability to function properly. For example, biodiversity loss results in reduced ecosystems' efficiency and productivity (Cardinale *et al.*, 2012). This relationship is non-linear and the change accelerates as biodiversity loss increases. This can result in simplified and impoverished systems with a more limited ability to deliver ecosystem services and related goods. At the same time, there is evidence that biodiversity increases the stability of ecosystem functions through time (Cardinale *et al.*, 2012; Díaz *et al.*, 2005). Therefore, securing the health of supporting services helps to maintain ecosystems' capacities to respond to and recover from stochastic (i.e. occasional and irregular) perturbations (wild fires, flooding, etc.) and human activities (harvesting, deforestation, invasive alien species, climate change, etc.).

Given the above, supporting services represent an irreplaceable social and economic 'safety net' for humans by maintaining genetic resources, soil conditions, and species interactions. These attributes together improve the potential yield of commercially valuable crops, helping to meet food demand while minimising the expansion of cultivated land (Cassman, 1999). For example, genetic improvement of wheat, rice, and maize during the 1960s, achieved by introducing new genes into old varieties, has been identified as one of the factors increasing the productivity of these crops. Well-managed PAs have an important role: they help to preserve supporting services and, as a cascade system, support the economic and welfare values associated with provisioning, regulating, and cultural services inside and outside PAs.

9.3 Estimating the value of benefits

Because supporting services have no direct markets, their economic value can only be assessed through indirect methods. Given the challenges in estimating the socio-economic value of these services, very few attempts at monetary or quantitative valuation exist and only a small proportion of them have been conducted in PAs or at site scale. Therefore, not all of the examples outlined in this chapter are specific to PAs. Nevertheless, they provide useful insights into assessing of the socio-economic value of supporting services.

This section summarises the available methods for estimating the socio-economic importance and value of supporting services, illustrated by a number of concrete examples in Table 9.2 and Boxes 9.1–9.3. An introduction to the different valuation methods mentioned is available in Chapter 3.

Monetary: non-market values

The necessity of trying to estimate the monetary value of supporting services, to support qualitative and quantitative evidence, needs to be carefully considered. This is because it is very difficult to monetise the economic importance of these services and, in particular, only a few methods are available to estimate their (indirect) monetary value.

The replacement cost method has been used to estimate the monetary value of supporting services such as seed dispersal, pollination, and pest regulation. In addition, avoided costs might also be used to value supporting services; for example assessing the costs to fisheries sector (restoration of depleted fish stock) resulting from the loss of system's natural nursery function. In this context, the value of the Eurasian jay (*Garrulus glandarius*) as an acorn disperser in the National Urban Park of Stockholm has been estimated, ranging from SEK35,000 (US$4,900) to SEK160,000 (US$22,500) per pair of jays or SEK15,000 (US$2,100) to SEK67,000 (US$9,400) per hectare. The use of a replacement cost approach excludes any values attributable to jays due to cultural values (aesthetic, ethical) (Hougner *et al.*, 2006) (Box 9.1).

TABLE 9.2 Supporting services and related values

Protected area	*Ecosystem services*	*Examples of benefits/value*	*Source*
Natura 2000 sites, Europe (several ecosystems)	Maintenance of species and habitat diversity	Natura 2000 network is the most important conservation tool to protect European threatened species and habitats and has a role in conserving viable populations of large carnivores in Europe, protecting 15.1 per cent of the suitable habitat for lynx, 17.7 per cent for wolf and 18 per cent for bear. Existence of these species has a role in terms of cultural values (wildlife tourism) and in terms of ecological function (as top predators, they contribute to a number of ecological interactions helping to control populations) (Miller *et al*.,. 2011). The benefits need to be, however, considered in the context of possible costs caused by human–wildlife conflicts (see Chapter 11).	Boitani *et al*., 2012
Reserves in Baja California, Mexico (marine)	Nursery habitats Stability and resilience of ecosystems	The pink abalone *Haliotis corrugata*, a highly valuable but depleted species for commercial fisheries, was recently subject to a severe mortality likely caused by hypoxia (reduced dissolved oxygen) in Baja California (Mexico). The decline of species was observed to be less pronounced within marine reserves. After the peak in mortality, abalone densities were higher within reserves compared to fished areas (for reserves 0.02 (+/-0.004) individuals per m2 or 200 (+/-40) individuals per ha, for fished areas 0.01 (+/-0.001) individuals per m2 or 100 (+/-10) individuals per ha) with greater proportions (approx. 10 per cent) of large individuals persisting inside reserves. The difference in egg production between reserves and fished areas increased in each subsequent year of protection, with estimated reproductive outputs in the reserve ranging from 1.6 to 2.6 times greater than in fished areas.	Micheli *et al*., 2012

Protected area	*Ecosystem services*	*Examples of benefits/value*	*Source*
Torre Guaceto Marine Reserve, Italy (marine)	Trophic interactions	The Torre Guaceto Marine Reserve supports a high density of the sea breams (*Diplodus sargus* and *D. vulgaris*), that are very effective predators of prey sea urchins that can seriously degrade sea grass ecosystems. Consequently, the density of sea urchins (*Paracentrotus lividus* and *Arbacia lixula*) is lower within the reserve than in unprotected locations. As a cascade effect, the seabed inside the marine reserve is covered with a healthy vegetation of seaweed while outside the reserve sea urchins over-consume the weeds causing formation of barrens (bare rocks with encrusting algae). The reserve therefore contributes to re-establishing lost predatory interactions within the marine community.	Guidetti, 2006
Galápagos Islands National Park and Biological marine reserve, Ecuador (several ecosystems)	Nursery habitats Biodiversity as a basis of functioning and resilience of the system	The economic value of the Galápagos refugia functioning as a nursery habitat for fish was estimated to be US$7/ha/year using a benefit transfer approach: based on the similarities of the Dutch Wadden Sea and Galapagos estuarine areas, it was assumed that 10 per cent of the Galapagos fisheries are dependent on the inlets and lagoons of the park. Biodiversity maintenance was considered a necessary precondition to other functions and human activities. The value of this fundamental function was estimated by assuming that 10 per cent of the market value of any activity reliant on the quality of the reserve (including recreation, education and research) could be attributed to well-functioning ecosystems. By using this 'shadow price' method the value of biodiversity maintenance was estimated to be at US$4.9/ha/year and nature protection function was estimated at US$0.55/ha/year.	Ruitenbeek and Cartier, 1999; Original source de Groot, 1992

Continued

Protected area	*Ecosystem services*	*Examples of benefits/value*	*Source*
South Africa (several)	Maintenance of species diversity	Spatial congruence between five ecosystem services (surface water supply, water flow regulation, carbon storage, soil accumulation, and soil retention), species richness, and diversity hotspots was assessed in South African biomes. The grassland and savannah biomes were identified as 'hotspots' for providing all five ecosystem services. There was moderate overlap and a generally positive (however relatively low) correlation between the ecosystem service hotspots and measured species richness and vegetation diversity. Species richness was generally higher than would be by chance in the areas important for water flow regulation and soil accumulation.	Egoh *et al.*, 2009
Not applicable – not a PA specific case study	Soil properties	A review of a wide range of land management studies shows that soil physical and chemical properties. i.e. properties that depend on nutrient cycling and soil formation, have a role in supporting overall soil quality by influencing soil compactibility, friability (ability to break down into smaller pieces), permeability, erodibility, and water and nutrients storage. They also influence decomposition and ecological interactions of soil fauna (Carter, 2002). At the same time, soil fauna and flora, especially in terms of species diversity and functional diversity (diversity of species' niches or functions) affect soil properties and quality (Balmford *et al.*, 2008).	Carter, 2002 Balmford *et al.*, 2008

The production function method, i.e. estimating the contribution of supporting services to the economic value of other services and goods, also provides possibilities for assessing the economic value of supporting services. For example, in the Galápagos Islands, the value of identified supporting services was estimated based on the assumption that these services are a necessary precondition for other benefits derived from the ecosystem (fisheries, tourism and education) and that the scale of which supporting services contribute to the value of other benefits can be broadly determined (shadow price method) (Ruitenbeek and Cartier, 1999) (Table 9.2).

Stated preference methods, such as willingness to pay to avoid the loss of a particular species, are commonly used to provide monetary estimates for benefits that have no market value (see Chapters 6, 7 and 8). However, these methods are mainly suitable for determining the value people attribute to species on aesthetical, cultural, or ethical grounds and rarely capture the benefits related to the role of a species in performing key life supporting functions in ecosystems (Hougner *et al.*, 2006). This is because assessing willingness to pay for such supporting functions requires the respondents to have a good understanding of these rather complex ecological aspects.

Finally, while it might be interesting and/or valuable to obtain an indication of the monetary value of supporting services, it should be kept in mind that these estimates are likely to overlap with the estimated values of other ecosystem services (provisioning, regulating, cultural). Therefore, great care should be taken when integrating the estimated values of supporting services into any aggregated monetary estimates for multiple ecosystem services (Chapter 12). For example, nutrient cycling is essential for forming and maintaining soil's ability to function as carbon storage. Therefore, if the economic value of soil carbon is already integrated into the calculations of an aggregated (monetary) estimate then accounting a monetary value for soil nutrient cycling separately would result in an over-estimation of the overall value of soil (Barry *et al.*, 2011). Note, however, that in order to highlight the importance of soil nutrient cycling it might still be useful to include its estimated monetary value (if available) into the broader socio-economic assessment, even if this value cannot be part of the aggregate (see Chapter 12).

BOX 9.1

The monetary value of a seed dispersal service in Stockholm National Urban Park, Sweden

Composed of 2,700 ha of a mosaic of different ecosystems, the National Urban Park of Stockholm is a large green area located nearby the inner city of Stockholm, Sweden. The park holds one of the largest populations in Europe of giant oaks – *Quercus robur* and *Q. petrea* – two keystone species in hemiboreal forests. Oaks have also played a central role in the historical development of the cultural landscapes of this park for a considerable time.

Method: The seed dispersal service performed by the Eurasian jay (*Garrulus glandarius*) in the Stockholm National Urban Park, Sweden was conducted using a replacement cost analysis. The number of oak trees that resulted from seed dispersal by jays was estimated through an ecological study to determine the costs of replacing this service by human means using two techniques: seeding acorns by humans and planting oak saplings. Seeding acorns was carried out by one or several workers seeding acorns at suitable places in the forest, mimicking jay behaviour.

Planting oak saplings: Cost data for planting of samplings have been obtained from official data, as it is a technique used in the park already. One worker (labour cost SEK2,648 or US$370 per day) can be expected to plant 600 saplings per day, which implies a labour cost of SEK4.41 (US$0.6) per planted sapling. The cost of one sapling can be estimated to SEK10 (US$1.4), which implies a total cost of SEK14.41 (US$2) per planted sapling. With 33,148 saplings to be planted each year, the total costs amount to SEK477,663 (US$66,872) for one year (total cost = total # saplings/supply per day * cost per day) and about SEK 6.7 million (US$938,000) for 14 years.

Seeding acorns: based on a literature review, the labour costs to manually seed 20,000 acorns on one hectare was estimated to SEK2,000 (US$280) per ha for preparation of the ground and SEK5,200 (US$728) per ha for seeding. This implies a labour cost of US$0.05 per acorn (US$1,008/20,000 = US$0.05). However, in the case of the park, 773,458 acorns are in total to be seeded, so cost per acorn needs to be multiplied by a 4.1 factor due to the increased distance between each acorn. The price of acorns is SEK100 (US$14) per kg and, as 1 kg contains about 200 acorns, the price per acorn is SEK0.5 (US$0.07). Hence, the total costs for the first year of seeding equals to SEK109,389 (US$15,314) (total cost = # acorns to be planted * (labour cost/acorn + acorn price). Over a 14-year period this results in costs of SEK1.53 million (US$214,200).

Based on the above, the replacement costs can be estimated as costs/pair of jays in the area or costs/hectare.

Results: Results show that depending upon seeding or planting technique chosen, the replacement cost per pair of jays in the park is SEK35,000 (US$4,900) and SEK160,000 (US$22,500), respectively. Based on the park's aggregated oak forest area, average replacement cost for natural oak forest regeneration by jays is SEK15,000 (US$2,100) to SEK67,000 (US$9,400) per hectare. These are minimum figures that do not account for any values attributable to jays due to aesthetic or ethical considerations. These estimates can help to motivate investments in management strategies that secure critical breeding and foraging habitats of jays, including coniferous forests and jay movement corridors.

Source: Hougner *et al.* (2006)

Non-monetary values

In order to fully appreciate the importance of supporting services to human well-being, it is necessary to better understand the biophysical status and trends of these processes and their contribution to delivering other ecosystem services. Biophysical information can be used as a (proxy) indicator for the status of and trends in supporting services. Examples of such proxies include: overall species diversity and richness, coverage and status of nursery habitats, the number of interactions within ecosystem's food web, and parameters reflecting energy and material flows. It is also of high importance to map the biophysical information on a spatial scale in order to identify the overall spatial availability and 'flow' of supporting services. For example, the Columbretes Island Marine Reserve (Spain) – hosting 18 per cent of the region's lobster habitat – has been estimated to contribute to over 80 per cent of regional lobster egg production. This is likely to be extremely important for local fisheries as 80 to nearly 100 per cent of the total lobster catch comes from areas surrounding the PA (Box 9.2).

Quantifying and mapping the correlation between supporting and other ecosystem services is an important means of highlighting their socio-economic value. While such correlative evidence is often hindered by the lack of relevant biophysical data, some studies have investigated the contribution of supporting services to other services and related goods – including estimating the role of biodiversity in supporting ecosystem services (productivity of systems) (Anderson *et al.*, 2009; Egoh *et al.*, 2009,Tilman *et al.*, 1996; Naeem *et al.*, 1994) (Table 9.2). For example, Costanza *et al.* (2007) used species richness and net primary production as a proxy for, respectively, biodiversity and ecosystem services in order to assess the relationships between biodiversity and ecosystem services in North America at the site and eco-region scales (see Box 9.3). In general, the correlation between biodiversity and other ecosystem services is found to be stronger at the local or regional scale (Díaz *et al.*, 2005), indicating that actions taken to conserve supporting services also protect other ecosystem services at the site and vice versa (Anderson *et al.*, 2009).

BOX 9.2

Quantifying the contribution of Columbretes Island Marine Reserve (Spain) to the fisheries of commercially valuable lobster

The role of Columbretes Island Marine Reserve (Spain) in enhancing regional reproductive output of *Palinurus elephasis*, the most commercially important spiny lobster in the Mediterranean and Northeastern Atlantic, was assessed based on the information on female fecundity and egg production.

Method: A significant part of the lobster fishing takes place in the vicinity of the marine protected area (MPA) (1 to 30 km from the boundary). Female fecundity and egg production of the lobster in the area were estimated through experimental and commercial fishing operations, which recorded data on catch, fishing effort, and female size. Random sampling, simulating typical commercial catch, was conducted by a chartered commercial boat and crew using trammel nets identical to those used in commercial operations. Outside the MPA, data from the commercial fishery was recorded during the annual lobster fishing season. Average female fecundity was estimated based on the size of female lobsters, assuming that their size correlates positively with their fecundity. Trends of female fecundity within and outside the reserve were estimated over time using statistical analysis (linear regression and a power curve). Egg production (number of eggs) for the reserve was estimated based on the mean female fecundity multiplied by the amount of female lobsters caught per area (1 x 1 km). Annual trends in average egg production per area within and outside the reserve were estimated using statistical analysis (linear regression and a power curve). The difference in egg production attributable to the reserve was estimated by comparing the egg production within and outside the reserve.

Results: Mean female fecundity in the MPA increased by 41 per cent over the 10 years of the study and in recent years was more than twice the fecundity of nearby fished areas. At the end of the study, mature females were on average 20 times more abundant, and egg production per unit area was 30 times greater in the MPA than in nearby fished areas. At that time, the MPA, occupying 18 per cent of the regional lobster habitats, was responsible for over 80 per cent of the regional egg production. Overall, after nearly 2 decades of protection, regional egg production was estimated to be six times greater than it would have been without the establishment of MPA. This significant increase in egg production is likely to positively affect 'spillover' of adult lobsters from the MPA to benefit the regional lobster fishery.

Source: Díaz *et al.* (2011)

BOX 9.3

Quantifying the contribution of biodiversity to ecosystem services at site-bases and at eco-regional scale in North America

The role of biodiversity in supporting ecosystem services was investigated both at the site and eco-region scales in North America.

Method: Net primary production (NPP) was used as a proxy for the overall supply of ecosystem services, while species richness (vascular plants) was used as a proxy for biodiversity. For the site scale of analysis, an extensive literature search was performed to obtain observations on NPP at several sites – from temperate forests, to tundra to high mountain meadows – belonging to the Long-Term Ecological Research (LTER) and Forest Service research. In addition to biodiversity, factors such as temperature, precipitation and soil organic matter content (obtained from available data from observatories and database) were also considered in explaining variations in ecosystem functions and services across sites.

For the eco-region scale analysis, North America was divided into 116 eco-regions for which data on biodiversity was assembled for several taxa (vascular plant, tree species, snails, butterflies, birds, and mammals), geophysical characteristics, and habitat threats. NPP values were obtained using satellite imagery coupled with parameters derived from the Biome-BGC – a tool for ecosystem modelling that estimates storage and flux of carbon, nitrogen and water. NPP values for agricultural areas were removed from the analysis because it was expected that high fertilizer and irrigation inputs would alter the analysis of the biodiversity-NPP relationship. The relationship between biodiversity and ecosystem services was tested both at the site and eco-region scales using multiple regression analysis.

Results: At the site scale, 57 per cent of the variation in NPP was correlated with variation in biodiversity after effects of temperature and precipitation were accounted for. At the eco-region scale, 3 temperature ranges were found to influence NPP: there was a negative correlation between the two at low temperatures, medium correlation at neutral temperatures, and positive correlation at high temperatures (26 per cent correlation for all).

Source: Costanza *et al.* (2007)

References

Anderson, B. J., Armsworth, P. R., Eigenbrod, F., Thomas, C. D., Gillings, S., Heinemeyer, A., Roy, D. B. and Gaston, K. J. (2009) 'Spatial covariance between biodiversity and other ecosystem service priorities', *Journal of Applied Ecology*, 46: 888–896

Balmford, A., Rodrigues, A. S. L., Walpole, M., ten Brink, P., Kettunen, M., Braat, L. and de Groot, R. (2008) *The economics of ecosystem and biodiversity: scoping the science*, Report for the European Commission, Cambridge, http://ec.europa.eu/environment/nature/biodiversity/economics/pdf/scoping_science_report.pdf, accessed 1 December 2012

Barry, L., Paragahawewa, U., Yao, R. and Turner, J. (2011) *Valuing Avoided Soil Erosion by Considering Private and Public Net Benefits*, Tahuna Conference Centre, Nelson, New Zealand, http://ageconsearch.umn.edu/bitstream/115512/2/Microsoft%20Word%20-%20Valuing%20Avoided%20Soil%20Erosion%20by%20Considering%20Private%20and%20Public%20Net%20Benefits%20_NZARES%20final_%5B1%5D.pdf, accessed 1 December 2012

Boitani, L., Santini, L. and Rondinini, C. (2012) *The role of Natura 2000 in conserving the large carnivores of Europe*, Zoological Society of London, http://static.zsl.org/files/luigi-boitani-the-role-of-natura2000-in-conserving-the-large-carnivores-of-europe-2204.pdf, accessed 1 December 2012

Cardinale, B. J., Duffy, J. E., Gonzalez, A., Hooper, D. U., Perrings, C., Venail, P., Narwani, A., Mace, G. M., Tilman, D., Wardle, D. A., Kinzig, A. P., Daily, G. C., Loreau, M., Grace, J. B., Larigauderie, A., Srivastava, D. S. and Naeem, S. (2012) 'Biodiversity loss and its impact on humanity', *Nature*, 486: 59–67

Carter, M. R. (2002) 'Soil Quality for Sustainable Land Management: Organic Matter and Aggregation Interactions that Maintain Soil Functions', *Agronomy Journal*, 94: 38-47

Cassman, K. G. (1999) 'Ecological intensification of cereal production systems: Yield potential, soil quality, and precision agriculture', *Proceedings of the National Academy of Sciences USA*, 96: 5952–5959

Costanza, R., Fisher, B., Mulder, K., Liu, S. and Christopher, T. (2007) 'Biodiversity and ecosystem services: A multi-scale empirical study of the relationship between species richness and net primary production', *Ecological Economics*, 61: 478-491

de Groot, R. S. (1992) *Functions of nature: evaluation of nature in environmental planning, management and decision making*, Wolters-Noordhoff, Amsterdam

Díaz, D., Mallol, S., Parma, A. M. and Goñi, R. (2011) 'Decadal trend in lobster reproductive output from a temperate marine protected area', *Marine Ecology Progress Series*, 433: 149–157

Díaz, S., Tilman, D. Fargione, J., Chapin, F. S., Dirzo, R., Kitzberger, T., Gemmill, B., Zobel, M., Vilá, M., Mitchell, C., Wilby, A., Daily, G. C., Galetti, M., Laurance, W. F., Pretty, J., Naylor, R. L., Power, A. and Harvel, D. (2005) 'Biodiversity regulation of ecosystem services', in R. M. Hassan *et al.* (eds) *Ecosystems and Human Well-being: Current State and Trends*, Island Press, Washington, DC

Egoh, B., Reyers, B., Rouget, M., Bode, M. and Richardson, D. M. (2009) 'Spatial congruence between biodiversity and ecosystem services in South Africa', *Biological Conservation* 142: 553–562

Guidetti, P. (2006) 'Marine reserves re-establish lost predatory interactions and cause community changes in rocky reefs', *Ecological Applications*, 16(3): 963-976

Hein, L. and de Groot, R. S. (2007) 'A framework for the Economic Valuation of Land Use Change', in Pertsova, C. A. *Ecological Economics Research Trends*: 105-126, http://books.google.be/books?id=GCkU1p_6HNwC&pg=PA110&lpg=PA110&dq=Supporting+services++double+counting++Hein+et+al.+2006&source=bl&ots=D_DZKI4fRs&

sig=YM_Ud-GhS_RbnzMYVnk02lIdWEU&hl=en&sa=X&ei=lOC9UNPUI8SHhQe83IEY&redir_esc=y#v=snippet&q=supporting%20service&f=false, accessed 1 December 2012

Hougner, C., Colding, K. and Söderqvist, T. (2006) 'Economic valuation of a seed dispersal service in the Stockholm National Urban Park, Sweden', *Ecological Economics*, 59: 364 – 374, http://www.sciencedirect.com/science/article/pii/S0921800905005124, accessed 5 December 2012

Kettunen, M., Vihervaara, P., Kinnunen, S., D'Amato, D., Badura, T., Argimon, M. and ten Brink, P. (2013) *Socio-economic importance of ecosystem services in the Nordic Countries – Synthesis in the context of The Economics of Ecosystems and Biodiversity (TEEB)*, Nordic Council of Ministers, Copenhagen

MA – Millennium Ecosystem Assessment (2005) *Ecosystems and Human Well-being: Biodiversity Synthesis*, World Resources Institute, Washington, DC, http://www.maweb.org/documents/document.354.aspx.pdf, accessed 1 October 2012

Micheli, F., Saenz-Arroyo, A., Greenley, A., Vazquez, L., Espinoza Montes, J. A., Rossetto, M. and De Leo, G. A. (2012) 'Evidence That Marine Reserves Enhance Resilience to Climatic Impacts', *PLoS ONE* 7(7): 1-8

Naeem, S., Thompson, L. J., Lawler, S. P., Lawton, J. H. and Woodfin, R. M. (1994) 'Declining biodiversity can alter the performance of ecosystems', *Nature*, 368: 734-737

Ruitenbeek, J. and Cartier, C. (1999) 'Marine System Valuation: An Application to Coral Reef Systems in the Developing Tropics', *Issues in Applied Coral Reef Biodiversity Valuation: Results for Montego Bay, Jamaica*, World Bank Research Committee Project RPO# 682-22, Final Report

Tilman, D., Wedin, D. and Knops, J. (1996) 'Productivity and sustainability influenced by biodiversity in grassland ecosystems', *Nature*, 379: 718-720

UK NEA – National Ecosystem Assessment (2011) *The UK National Ecosystem Assessment: Synthesis of the Key Findings*, UNEP-WCMC, Cambridge, http://uknea.unep-wcmc.org, accessed 18 June 2012

10

WIDER SOCIO-ECONOMIC BENEFITS

Gantioler, S. and ten Brink, P.

10.1 Overview of benefits

Chapters 6 to 9 above outline how to estimate the socio-economic value of different ecosystem services and related goods provided by PAs. In addition to these individual benefits, PAs also provide a number of wider socio-economic benefits that are difficult to attribute to one single ecosystem service but rather result from a range of them and benefit a number of sectors simultaneously (Gantioler *et al.*, 2010). These benefits refer to the overall contribution of PAs to local and regional development, including the development of rural areas. They also consist of PAs' direct and indirect support to income and employment (Table 10.1). The identification and consideration of these wider benefits complements the benefits outlined in Chapters 6 to 9.

TABLE 10.1 Wider socio-economic benefits

Wider socio-economic benefits	*Examples*
Investment-induced support to local and regional development, including rural development, poverty alleviation and reduction	Direct and indirect/induced economic impact of investment in management of PAs contributing to local and regional economic activity, supporting inter alia (potentially vulnerable) rural populations, and contributing to poverty alleviation and reduction.
Direct and indirect employment	Direct employment opportunities created by management and conservation activities linked to a PA, both on and off-site; employment supported off site linked to PA related activity and spending; jobs supported or created in sectors directly benefitting from ecosystem services (see also Chapters 6 and 8)

Source: own presentation as used in Gantioler *et al.*, 2010

The previous chapters outlined how the different benefits provided by well-functioning ecosystems within PAs can have positive socio-economic impacts on local communities and regions (food production, branded local products, attracting tourism and associated revenue, lower costs of water provision or climate regulation, etc.). Investment in PAs supports local communities also more directly by creating economic and employment opportunities related to the management of the site and, consequently, resulting in spending generated by PA employees and volunteers in a region.

As also highlighted in the previous chapters, investment in the protection and management of PAs, as well as related management measures, has positive impacts across a range of economic sectors, including those that profit from the provision of goods (fisheries, non-timber forest products, etc.), those that are dependent on beneficial ecological processes for continued cost-effective production of goods and services (pollination, water and energy supply, etc.), those that rely on nature to attract visitors to a region (recreation opportunities, amenity benefits, etc.). Furthermore, investment in PAs can lead to increasing broader sectoral investments and/or economic activity by making the region an attractive, high-quality location for new businesses and companies.

The direct impacts above can lead to a share of the income flowing also to a number of other sectors, i.e. sectors providing goods and services to the ones directly profiting from benefits from PAs. This is known as indirect expenditure and income. For example, tourists going to a site tend to spend more on accommodation and sustenance than in entrance fees and on site purchases. Therefore, onsite expenditure is often only a fraction of the benefits related to the role of PAs in attracting tourists. Furthermore, a share of the expenditure on accommodation and restaurants will go to the areas of economy supporting these sectors such as food and beverage, energy and insurance. As a result, the monies will circulate in the local and regional economy creating a cascade of beneficiaries, as each supplier of goods and services in turn purchases goods and services from other sectors. To illustrate this more explicitly, money spent by tourists to visit a PA will go first to entrance fees, shop purchases, hotel, restaurants and travel. This direct expenditure then generally leads – after wages, taxes and profits are subtracted – to a further round of expenditure on inputs (food from supermarket for restaurant, energy for transport, etc.) which in turn leads to further rounds of payments (supermarket purchases from farmers, farmers buying fertiliser or equipment, etc.). With each step there is a benefit, albeit each time smaller, supporting income and potentially also employment. The ratio of the direct expenditure or direct employment and the indirect expenditure or employment is known as the economic multiplier. Furthermore, increased income across the sectors may also help to generate further spending by people working within these 'spin-off income' sectors (also known as the induced impact) as they have more income and economic confidence hence spend – rather than save – money.

The PA related income (above) within/into the local and regional economies can have several outcomes, including investment in education, increased salaries, increased savings and supporting budget balance or employment. Employment opportunities can be linked to the sector-specific utilisation of ecosystem services and related goods (more jobs in tourism, fisheries, agriculture, beverages sector, etc.) (see also Chapters 6 and 8). Additional employment opportunities can also be directly linked to the general management of PAs and related conservation and restoration activities (onsite or/and off-site).

It needs to be noted that the strength of the linkages between PAs and employment might vary between developed and developing countries (Nunes *et al.*, 2011). In developing countries a large proportion of the employment is linked to primary sectors such as fisheries, agriculture or forestry. Therefore, the overall level of employment opportunities in these countries depends strongly on the quality and functioning of ecosystems. In developed countries a smaller portion of employment is directly linked to the provisioning of goods whereas tourism and recreation sectors are an important source of new job opportunities (see Chapter 8). In addition, the income flowing into a region from PAs also affects the quality of jobs provided. In Europe, for example, the flow of income from PAs into regional economies offers new and skilled employment opportunities while in developing economies related jobs are still often of poor quality and low paid subsistence jobs in the primary industries (see example of Kibale National Park, Uganda, in Table 10.2) (Nunes *et al.*, 2011).

The way the investment in and management of PAs occurs and the manner in which spending generated by PAs is directed into the economy play a key role in determining the wider socio-economic benefits of PAs. With due targeting the investment in and management of PAs will support the availability of ecosystem services and related goods, create employment opportunities and contribute to the overall sustainable development of local communities and regions on a long-term basis, including by supporting the welfare of (vulnerable) rural populations. It can also contribute to poverty alleviation (ten Brink *et al.*, 2012; Russi *et al.*, 2013).

10.2 Socio-economic importance of benefits

The following section provides an overview of the wider socio-economic benefits associated with PAs, including an overview of concrete examples (Table 10.2). This section focuses on: employment, regional and local development, and poverty reduction.

In general, it needs to be kept in mind that employment as such/on its own should not be considered to convey economic value in terms of profit generated. It is an indicator that can reflect the social and welfare importance of PAs and that has economic impacts through the purchasing power of a population (see Section 10.1 above). From a financial perspective, however, direct employment and salaries of PA staff are considered costs, for example costs to the local authority paying for the wages and management. Therefore, on the one hand employment and wages of PA staff increase the revenue streams within local communities (economic perspective) and, on the other hand, also contribute to the costs of implementing

PAs, whether to public authorities or non-governmental organisations (financial perspective). The aim of this section is to highlight how local communities can socially and economically benefit from job opportunities created by PAs, underlining at the same time the strong welfare impact of any investment. Chapter 11 will focus on the cost aspects of PAs.

Regional development and poverty reduction

As described above, different sources of income (public investment, visitor spending) can generate income flowing into local communities and regions (salaries, employee spending). For example, it has been estimated that national parks and other key nature reserves in Finland (altogether 37 sites) provided up to €10 in return for €1 of public investment in their management. When aggregating the benefits at national level, the support of PAs to local economies amounted to €108.3 million (1,394 man-years) per year in 2011 (Kajala, 2013) (see Table 10.2). The public investment itself in conserving or restoring PAs can also be used as a proxy to estimate the value a site generates regarding its importance for a region's development. For example, funding of around €10 million flew to the Amvrakikos National Park in Greece to maintain employment opportunities in the fisheries sector (Nunes *et al.*, 2011).

The investment in and income generated by PAs also impacts the general quality of life of local communities and regions. It can affect housing conditions and the possibility of acquiring luxury items (Box 10.2). It also relates to the level of education that can be afforded and to health benefits acquired. All these different components are decisive factors when it comes to regional development and the alleviation and reduction of poverty. There are, of course, also concerns that the designation of an area as PA might actually reduce local income and hinder its economic development due to restricted access to natural resources (see Chapter 11). This is considered to be particularly the case in developing countries, where the dependency on primary provisioning services is often higher. However, several studies have analysed the impacts of PAs on the development of local communities to regions and they generally conclude that PAs usually provide more benefits than costs, especially at regional and national level (Kettunen *et al.*, 2011) (see Chapter 12).

For example, results from a study of national parks in Costa Rica and Thailand suggest that protection of nature within high poverty areas, on average, reduced poverty while also reducing deforestation, particularly regarding sites with intermediate distance from cities (Ferraro *et al.*, 2011) (see Table 10.2). In the Karoo National Park of South Africa, between 65 and 70 per cent of respondents from the community survey indicated that the park improved the community's general quality of life and the appearance of the area, and led to increased tourism and pride of the residents (Saayman *et al.*, 2009) (see Table 10.2). Another study investigated the links between marine protected areas and poverty reduction on Fiji, the Solomon Islands, Indonesia and the Philippines. The results showed substantive positive impacts with regard to income, housing conditions, education, social inclusion and work inclusion of women (Leisher *et al.*, 2007) (see Box 10.2).

However, the studies have also emphasised that the approaches adopted to designation and management are decisive regarding PAs' impact on poverty alleviation and reduction (Saayman *et al.*, 2009; Ferraro *et al.*, 2011; Schmitz *et al.*, 2012; MacKenzie, 2012). Key enabling factors include, for example, the creation and targeting of employment opportunities to communities in the vicinity of PAs, supported by well-defined property rights. Distance to major cities has been noted to have both negative and positive impacts. The former results from increasing nearby opportunities to substitute goods generated within a PA, leading to a general 'devaluation' of a PA. The latter is linked to the accessibility to infrastructure that plays, for example, an important role in attracting visitors and revenue to the area (see Chapter 8).

Furthermore, in both developing and developed countries the governance of PAs is central: an increased stakeholder dialogue and the involvement of local communities in the designation process and management activities have been identified as key measures to increase the positive social and economic impact of PAs on local communities and at regional level.

Employment

There are different ways in which investment into PAs potentially contributes to the creation and maintenance of employment opportunities, and as such has a positive impact on local and/or regional economies and society. Jobs can be directly supported by the management of a PA, including its conservation and restoration activities. With regard to the European Natura 2000 PA network, full-time equivalent (FTE) jobs needed in maintaining the network have been estimated to annually amount to 6,728 with an overall average salary of €28,000 across the EU Member States (17 Member States, Gantioler *et al.*, 2010). Another study estimated that €1 billion of investment per annum in the Natura 2000 network will directly generate 17,900 FTE per year (GHK, 2011). In the Tortuguero National Park (Costa Rica), 235 individuals possessed licences to conduct marine turtle tours in the park in 2002. Most of these guides were residents of local communities, with women making up 20 per cent of the guides in 1999 (Troëng and Drews, 2004) (see Table 10.2).

Direct employment is also generated by sectors immediately benefitting from ecosystem services provided by PAs. This includes sectors such as agriculture, forestry or fisheries (provisioning services, see Chapter 6), water supply (regulating services, see Chapter 7) or tourism and recreation (cultural services, see Chapter 8). Furthermore, PAs support indirect employment through jobs created in sectors profiting from the development of the above mentioned primary dependent sectors (e.g. companies that deliver services to hotels).

For example, tourism and recreation activities related to the European Natura 2000 PA network have been estimated to support between 4.5 and 8 million FTE jobs, and between 800,000 and 2 million FTE jobs if only visitors with high affinity

TABLE 10.2 Examples of economic and welfare values of PAs related to wider socio-economic benefits

Protected area	*Benefits (ecosystem services and related goods)*	*Estimated socio-economic value of benefits*	*Method*	*Source*
Natura 2000 network of protected areas (EU)	Employment	The study aimed at estimating the economic value of the benefits provided by the tourism and recreation sector (see Chapter 8) and the employment supported by the EU Natura 2000 network of protected areas. The economic impact was determined by compiling and calculating data on visitor numbers and spending, scaling up data at EU level and calculating direct and indirect economic impacts of visitor spending by using input/output models. According to the results, tourism and recreation activities in Natura 2000 support between 4.5 and 8 million Full-Time Equivalent (FTE) jobs, and between 800,000 and 2 million FTE jobs if only visitors with high affinity to Natura 2000 are considered. It is estimated that a total of 12 million direct and indirect FTE jobs are supported by the network. It is to be noted though that the results are likely to overestimate the employment impacts due to the lack of primary data and the subsequent need for basing the assessment on extrapolation and scaling up data from limited sources to the EU level.	Income equivalent and quantitative assessment	Bio Intelligence Service, 2011

Continued

Protected area	*Benefits (ecosystem services and related goods)*	*Estimated socio-economic value of benefits*	*Method*	*Source*
Kibale National Park, Uganda	Regional development and employment	The study looked into the benefits (and losses) accrued to 25 communities adjacent to a PA. Considered benefits included park-based employment, tourism revenue and distribution of the revenue between stakeholders, presence of integrated conservation and development projects, and a number of agreements allowing access to resources within the park. Losses were thought to be caused by park-protected animals raiding crops and preying on livestock. Information was collected via focus group discussions and household surveys, the latter being designed to integrate a range of different household conditions within the area. In addition, interviews with enterprises providing park-based employment were conducted, providing information on the number of employees, average pay rate and home village of employees. This included sectors such as park management (Uganda Wildlife Authority (UWA), tourism, research and activities carried out in the context of a carbon offset project. The results of the study indicate that park-based enterprises employ over 800 people resulting in annual salaries of US$570,839 for residents surrounding the PA, with the highest income flowing from tourism, followed by the carbon offset project and research, and least from UWA. The value of park-based employment to study villages ranged from nothing to over US$27,000 per year. Employees came from up to 15 km from the park.	Quantitative, statistical and qualitative assessments; salary costs	MacKenzie, 2012

Protected area	*Benefits (ecosystem services and related goods)*	*Estimated socio-economic value of benefits*	*Method*	*Source*
Costa Rica and Thailand, several protected areas	Poverty reduction	The study explored the impacts (positive and negative) of the designation of PAs on poverty reduction and alleviation in Costa Rica and Thailand. The study was based on data on poverty variables including national census data on household characteristics and assets. Countries were divided into sub-districts which were defined as protected if at least 10 per cent of their total area was covered by PAs. The overall conservation status of sub-districts was determined based on their estimated (and graded) level of avoided deforestation. Additional variables such as the distance to cities were also included in the assessment. A baseline was defined and a statistical analysis carried out. Results suggest that the protection of areas associated with high poverty has, on average, reduced poverty while also reducing deforestation, particularly for those districts with intermediate distance from cities. It is assumed that the positive impacts can be attributed to factors such as profits gained from close-by tourism and the availability of related infrastructure, such as for transport, accommodation, etc., as well as sites now protected by PA having been of little agricultural value (i.e. low opportunity costs). The analysis points to trade-offs if decision-makers focus on achieving high levels of poverty reduction and avoided deforestation, particularly in relation to carbon off-setting projects.	Quantitative, statistic and qualitative assessments	Ferraro *et al.*, 2011

Continued

Protected area	*Benefits (ecosystem services and related goods)*	*Estimated socio-economic value of benefits*	*Method*	*Source*
Tortuguero National Park, Costa Rica, (marine, coastal)	Employment and regional development	On the Caribbean coast, the Tortuguero National Park hosts one of the largest green turtle (*Chelonia mydas*) rookeries in the world. In 2002, visitors spent US$6.7 million in the area, including on food, accommodation, souvenirs, transport and also turtle observation. In 2002 there were 235 tour guides with licences to conduct marine turtle tours in the park. In 1999, local guides from Tortuguero village (526 inhabitants) conducted 72 per cent of all turtle tours. Women made up 20 per cent of local tour guides in 1999. In addition, the park has supported small businesses, such as hotels and tour operators. Tourism in Tortuguero was estimated to provide employment to 265 people. The non-consumptive use of turtles has also proved to provide higher gross revenues than the consumptive use, such as the collection of eggs in other areas, resulting in a higher income also at national level.		Troëng and Drews, 2004

Protected area	*Benefits (ecosystem services and related goods)*	*Estimated socio-economic value of benefits*	*Method*	*Source*
Karoo National Park, South Africa	Regional development	Three surveys canvassing the views of visitors, businesses and local communities were carried out to determine the socio-economic impact of Karoo National Park (NP). The purpose of this research was to determine the economic impact of the Karoo NP on the local economy, impact of tourism business development in the Karoo district, and how the park affects the communities. Results indicated that visitors spent on average between ZAR570.92 (US$65.6) and ZAR882.58 (US$101.4) per group, depending on the accommodation type, with 70 per cent of the money spent in local shops and restaurants. Considering the total number of visitors, total expenditures were estimated to be approximately ZAR10.5m (US$1.2m) in 2007. The economic impact was determined by using matrices capturing the changes of demand (tourist group) on the output (economic impact), and defining according multipliers. Several businesses emphasised that they rely heavily on the NP for their revenue; however, only 4 per cent of the businesses indicated that they completely owed their existence to the park. Most of the jobs created by expenditure on PAs were related to the retail sector (employment of 308 people). 98 per cent of the workers live in the area close by the NP, indicating positive socio-economic impacts to the local community. Between 65 and 70 per cent of respondents from the community survey indicated that the park improved the community's general quality of life; the appearance of the area improved, which led to increased tourism and pride of the residents.	Quantitative and qualitative assessments	Saayman *et al.*, 2009

Continued

Protected area	*Benefits (ecosystem services and related goods)*	*Estimated socio-economic value of benefits*	*Method*	*Source*
Several key government owned nature areas, national parks and recreational areas, Finland	Visitors spending Direct employment	In 2011 the Finnish national parks and other key protected areas were visited around three million times. Based on the assessment, local, accumulative economic impacts of visits to parks ranged between €0.1–30.6 million/year for different parks, generating an estimated 1-400 person-years of employment depending on the park (Note: the wide range is caused by significant differences in the characteristics between different sites, including size). When summing up the benefits at national level, the support of national parks (altogether 37 areas) to local economies amounts to €108.3 million (1,394 man-years)/year. According to the assessment €1 investment in national parks and other key protected areas results in €10 return to local economies. The assessment was carried out using an economic model based on three key variables: number of visits, visitors' spending, and a set of multipliers that reflect how visitor spending circulates and multiplies in the local economy. A comprehensive, standardised visitor monitoring system, including both visitor logs and surveys, provided the data for the calculations.	Quantitative assessment and modelling	Kajala, 2013

to PAs were considered. It was also estimated that a total of 12 million direct and indirect FTE jobs are supported by the network (Bio Intelligence Service, 2011) (see Table 10.2). In the Kibale National Park (Uganda) park-based enterprises employed over 800 people resulting in annual salaries of US$570,839 for residents surrounding the PA, with the highest income flowing from tourism (MacKenzie, 2012) (see Table 10.2). Several studies considered employment related to tourism to PAs as highly relevant, particularly in developing countries as offering opportunities of higher income compared to other sources (Ferraro *et al.*, 2011; Saayman *et al.*, 2009; MacKenzie, 2012).

In general, PA-related employment is of wider socio-economic importance for a number of reasons. First, such employment can be available in fairly remote areas where the overall employment opportunities are scarcer than in towns and cities (RSPB, 2011). For example, part-time and/or seasonal employment linked to PAs (PA management activities, seasonal jobs in tourism, etc.) can provide important 'top up' employment opportunities which supplement other jobs and subsistence activities allowing people to stay in rural areas. PAs also often have programmes to train local people in PA management thus building capacity as well as providing employment. In several places around the world trained PA staff are often the ones with natural resource management expertise in rural areas. Such 'resource people' may become increasingly important in the future when, for example, traditional management practices become less effective due to climate change.

10.3 Estimating the value of benefits

This section summarises the most commonly used methods for estimating the socio-economic value of wider socio-economic benefits, supported by a number of concrete examples in Table 10.2 above and Boxes 10.1 and 10.2 below.

Monetary values

The role of PAs in supporting regional development can be based on the estimated income provided to and retained in an area, whether by direct investment flows (funds for site management, etc.) or via PA visitors. As regards the latter, this can be captured by estimating the amount of spending provided by visitors with high affinity to PAs – as in comparison to general tourists – into different sectors within the region (accommodation, outdoor equipment retail and excursions, etc.). For example, a study was carried out to assess the economic impact of six national parks in Germany (Mayer *et al.*, 2010), determining the level of visitor affinity to national parks by comprehensive surveys (see Box 8.1 in Chapter 8 and Box 10.1 below). Based on calculations that divide the identified total regional income (from tourist spending) by the regional average income per capita, the authors determined that, hypothetically, 4,282 people would be able to live off the income supported by PAs (total income equivalent). Based on the study, roughly 50 per cent of the income was actually retained in the region (see Box 10.1).

The income flowing to a region can also be estimated by assessing the spending of site managers on local and regional services for the management of the PA. Information on market prices can be used to calculate the extent and value of the PAs. The collection of relevant data can be based on surveys, official statistics or existing studies. In addition, site employees and volunteers can support local economies via their spending on local products and services (food, rent, etc.). For example, local restaurants, hotels and shops can significantly benefit from the expenditure of the staff working in PAs. If exact data on expenditures are not available, the amount of spending from site employees and volunteers could be calculated as a percentage of local activity turnover. This can be achieved by roughly looking into the share of PA employees' contribution to a certain sectorial turnover (e.g. 10 per cent, 20 per cent or 50 per cent of a hotel or shop revenue).

Given the expected lower impact compared to visitor spending, PA and staff spending have been less often applied. However, their advantage is that they are relatively easy to carry out, with less need of data and labour. In the absence of information on PA and/or staff spending, the level of financing flowing into PAs (effectively investment costs) can be taken as a proxy to reflect the potential spending and the importance attributed to its impacts (Nunes *et al.*, 2011).

Other possible methods to estimate the monetary value of PAs in supporting broader socio-economic development include analysing the salaries that have resulted from different forms of PA-related spending and to what extent these salaries have been provided to people living in communities in the vicinity of PAs. This can be calculated by multiplying the number of local people employed by park-based enterprises by the average salary paid by that enterprise. For example, in Kibale National Park overall park-based enterprises provided annual salaries of US$570,839 for residents surrounding the PA (MacKenzie, 2012; see Table 10.2).

BOX 10.1

The total income equivalent generated by six German National Parks

The study looked into the economic impact of tourism related to 6 of 14 German national parks, including Niedersächsisches Wattenmeer, Bayerischer Wald, Eifel, Müritz, Hainich and Kellerwald-Edersee. While a part of the results were already presented in Chapter 8 (Box 8.1), the study also calculated total income equivalents (see methods below) to use as proxy of the likely employment opportunities created by visitor expenditures.

Method: The total regional income created by tourism spending was calculated by multiplying the number of visitors per year with mean daily expenditure per visitor and a defined factor (multiplier) for different sectors. The estimated total regional income was then divided by the region-specific average income per capita. The estimation resulted in a so-called total income equivalent (IE), reflecting the hypothetical number of people who would be able to live off the income provided from tourism expenditure. The actual number of jobs created by tourism would be lower than the total IE, as the income from one job can support more than one person. Finally, the IE was adjusted to determine the employment impact specifically attributable to PAs, by only taking into account spending by visitors with high affinity to national parks (see Chapter 8, Box 8.1).

Results: In 2007, the regional income retained from tourism in all of the analysed national parks corresponded to a total of 32,432 IE (of which 93 per cent was provided by the national park Niedersächsisches Wattenmeer). If visitors with high national park affinity only were considered, the total income equivalent amounted to 4,282 IE (of which 78 per cent was provided by the national park Niedersächsisches Wattenmeer). Results showed that between 32 per cent and 35 per cent of the (gross) visitor spending was retained as direct regional income in all national parks. Another 16 per cent, on average, was converted into indirect regional income. About half of the (gross) tourist revenue was thus converted into regional income.

Source: Mayer *et al.*, 2010

Non-monetary values

The most important methods to determine the non-monetary value of wider socio-economic benefits include quantitative approaches aimed at determining, for example, the number of employment opportunities created, statistical analysis to study the correlation between poverty reduction and improved conservation status, and qualitative methods such as interviews, and focus group discussions.

The assessment of direct, on-site employment resulting from PAs is considered to be a simple and less data intensive method to emphasise the importance of an area with regard to its social impact. It also often forms the basis for developing monetary estimates as outlined above. While the assessment of current on-site employment is relatively easy, the assessment of potential future employment impacts raises some challenges. These future opportunities are often determined on a basis of estimated and/or foreseen future spending on management and conservation activities (Gantioler *et al.*, 2010). The complexity of data requirements for assessments increase if the direct employment within sectors profiting from the ecosystem services and related goods and/or indirect employment are taken into account.

The indirect employment is determined by applying multipliers to the number of direct jobs, reflecting the ability of employment, related income and spending to support further employment within the economy. Multipliers are determined through information on and/or estimation of the number of additional indirect jobs per one direct job created (see Box 10.2). A common standard multiplier is 0.5 (1 job supports 0.5 additional jobs), though generally multipliers specific to an area, available through existing studies and region-specific statistical analysis, should be used. For a closed economy (i.e. money staying in the region) the multiplier would tend to be higher than for an open economy (i.e. money flowing out). Data scarcity can pose a challenge to quantitative analysis and consequently a range of existing assessments – regional assessments in particular – have used information from other sites (see value transfer in Chapter 3).

Questionnaires and surveys represent the most important means to collect relevant information for all the methods outlined above. With regard to employment, additional questions on the location of employees help to determine whether created jobs actually benefit the local communities. They are also important for any statistical analysis assigned to determine what type of correlations exist between a range of variables (proximity to a PA and employment). Compared to other approaches outlined above, statistical analyses are labour intensive and require a significant amount of data and expertise. Such analyses have, for example, been used to study the impact of national park designation on poverty reduction in Costa Rica and Thailand, linking variables such as household characteristics and avoided deforestation (Ferraro *et al.*, 2011; Table 10.2).

Finally, qualitative information and methods (interviews and focus groups) that aim to capture people's perception on what impact PAs have on regional development and poverty reduction provide a valuable source for assessing the broader socio-economic value of PAs. They are often used to form a basis for and/or complement quantitative methods and can represent a useful tool in emphasising the importance of PAs in supporting welfare (Box 10.3).

BOX 10.2

Calculating multipliers based on input–output analysis

Multipliers can also be drawn from input-output analysis, i.e. estimating input into and resulting output from a sector, business, company, etc., or other forms of economic impact assessments (see Chapter 8). Hereby a factor is determined that reflects the extent of employment impact likely to be generated by the income resulting from different sources of spending.

In practice, this first requires determining the direct employment by looking at spending across different sectors at the expenditure point and combining this information with sector-specific labour intensity values (e.g. how many jobs per $US million expenditure). Second, the first round of indirect effect needs to be determined by assessing how much money from the primary beneficiary sectors above flows to subsequent sectors and then, as above, combining this information with sector-specific labour intensities. There is then in turn a second, third, etc. effect, with each subsequent step bringing further but ever smaller benefits to sectors. In general, there is no need to pursue too many rounds as the incremental benefits get very small after the second and third round.

The ratio of the total direct and indirect jobs to the direct jobs gives the total employment multiplier. The difference between the two is the additional indirect jobs, also called the incremental benefits.

The above calculations are usually best done with the help of a dedicated model. Alternatively, if the interactions within the economy are well understood and relatively simple, a more pragmatic approach can be adopted, focusing only on the main sectors, flows and a few rounds of indirect effects.

BOX 10.3

Increased socio-economic opportunities and the level of security and empowerment in local communities around marine protected areas (MPAs)

A qualitative and quantitative assessment was carried out to investigate the links between marine protected areas and poverty reduction in four study sites: Fiji (Navakavu), the Solomon Islands (Arnavon Islands), Indonesia (Bunaken) and the Philippines (Apo Island).

Method: The four study sites were selected based on existing expert knowledge on their contribution to poverty reduction. They were also representative of small, one-community local MPAs (Fiji), medium-sized, multi-community local MPAs (Solomons), big collaboratively managed national MPAs with lots of people (Indonesia), and small, co-managed national MPAs with few people (Philippines). Additional information was collected on housing conditions and household characteristics, including data on number and age of children attending school, owned luxury items, diet, sources of income as well as building material use by households. A qualitative assessment was carried out by using focus group discussions and key informant interviews. Quantitative information was collected using structured household interviews to 1) compare communities within PAs to socio-demographically similar control communities with no connection to PAs, and 2) assess perceived environmental and socio-economic changes over the last 5 to 10 years and assess whether people believed these changes were associated with the establishment/existence of marine protected areas.

In total, 68 people (comprising an economist, a social scientist, and a study manager, supplemented in each site by field workers, translators, interviewers, logistics assistants, and note takers) helped with the fieldwork that included more than 950 household interviews (657 in MPA and 301 Non-MPA for control) and more than 50 focus group discussions and key informant interviews, for a total of approximately 1,100 local people consulted.

Income and income growth, housing, luxury items and fish catch were used as welfare indicators and were analysed with the help of radar charts, illustrating changes over time or difference between communities associated and not associated with PAs.

Results: The MPAs' contributions to poverty alleviation were determined regarding increased level of socio-economic opportunities, empowerment and security in the following areas:

Income, housing and luxury goods: Not surprisingly, the sites with increased income levels (three out of four) over time also showed an increase in the number of household luxury goods such as TVs or motorcycles. Income at the study sites Navakavu and Arnavons were more than doubled compared to control sites. The number of household luxury items in Navakavu was almost triple compared to communities with no connection to PA. Housing conditions also improved at two sites. The other site was yet too young to expect any impact on housing and in the remaining site housing had already improved before the study period.

Education: The study revealed little impact on formal education attributable to the MPA in most sites. Only on Apo Island, increased incomes supported funding for schools. However, in the Arnavon Community Marine Conservation Area, around 70 per cent of the children between 6 and 15 years of age attend school compared to only 38 per cent of non-MPA children, and 48 per cent of MPA respondents have completed secondary school compared to just 17 per cent in the non-MPA areas.

Alternative livelihoods: While fisheries benefits were significant in all MPAs, the greatest positive contribution to household incomes came from new sources of livelihoods, especially tourism. Despite episodes of tourism downturns, in Bunaken and Apo Island shifts from working within primary sectors such as fisheries to the tourism industry approximately doubled people's previous earnings. In both locations, tourism training for local people was carried out by private-sector tourism operators.

Governance mechanisms: In all four study sites, new governance mechanisms were introduced and established for the management of MPAs, involving communities in decision-making. This made the MPAs more responsive to community needs, and thus contributed to poverty reduction and served as forums for addressing other community issues, strengthening local governance and reducing conflicts. More than 50 per cent of respondents in Navakavu and Bunaken declared that there were fewer conflicts between neighbouring communities.

Benefits to women: In all four sites, the MPAs helped empower women economically and in some cases socially. Women are the reef gleaners (food and material collectors) in Navakavu and therefore they benefited from the MPA-induced increase in the availability of shellfish around MPAs. In Bunaken and Apo Island, dive tourism created higher-income job opportunities for women, and residents noted an improvement in women's lives because of MPAs. At three sites between 72 per cent and 84 per cent of respondents stated that a stronger status gained by women in community meetings over time was associated with MPAs. For example, in the Arnavons' women became involved in seaweed farming and the making of traditional clothes to supplement their income and as a consequence they also gained a stronger status in community meetings.

Health: Increased fish catches in Navakavu and Apo Island led to a greater protein intake and a perceived improvement in children's health in particular. In addition, greater incomes from the MPA on Apo Island led to more frequent visits from mainland doctors and funding for a resident midwife. The MPA speedboat was used as the water ambulance in the Arnavons and in Bunaken visitor entry fees funded water supply tanks, public toilets, and washing places in several villages and thus improved public health.

Social cohesion: The MPAs were also associated with strengthened unity and social cohesion, sense of belonging and cultural traditions within the communities. There was increased participation in community meetings according to almost 90 per cent of respondents in Navakavu and Apo Island and 72 and 46 per cent in the Arnavons and Bunaken, respectively.

Source: Leisher *et al.*, 2007

References

BIO Intelligence Service (2011) *Estimating the economic value of the benefits provided by tourism/recreation and employment supported by Natura 2000*, final report prepared for the European Commssion, http://ec.europa.eu/environment/nature/natura2000/financing/docs/Economic_Benefits_of_Natura_2000_report.pdf, accessed 7 December 2012

Ferraro, P. J., Hanauer, M. M. and Simsb, K. R. E. (2011) 'Conditions associated with protected area success in conservation and poverty reduction', *Proceedings of the National Academy of Sciences. USA*, 108(34): 13913–13918, http://www.pnas.org/content/108/34/13913.full, accessed 7 December 2012

Gantioler, S., Rayment, M., Bassi, S., Kettunen, M., McConville, A., Landgrebe, R., Gerdes, H. and ten Brink, P. (2010) *Costs and Socio-Economic Benefits associated with the Natura 2000 Network*, final report to the European Commission, Institute for European Environmental Policy/GHK/Ecologic, Brussels, http://ec.europa.eu/environment/nature/natura2000/financing/docs/natura2000_costs_benefits.pdf, accessed 7 December 2012

GHK (2011) 'Evaluating the Potential for Green Jobs in the next Multi-annual Financial Framework', report to BirdLife International and Transport and Environment/World Wide Fund for Nature (WWF), Brussels, http://www.birdlife.org/eu/pdfs/Green_Jobs.pdf, accessed 6 December 2012

Kajala, L. (2013) 'TEEB Nordic case: Estimating local economic benefits of visitors' spending in protected areas', in Kettunen *et al.*. *Socio-economic importance of ecosystem services in the Nordic Countries – Scoping assessment in the context of The Economics of Ecosystems and Biodiversity (TEEB)*, Nordic Council of Ministers, Copenhagen

Kettunen, M., Berghöfer, A., Brunner, A., Conner, N., Dudley, N., Ervin, J., Gidda, S. B., Mulongoy, K. J., Pabon, L. and Vakrou, A. (2011) 'Recognising the value of protected areas', in ten Brink (ed.) *The Economics of Ecosystems and Biodiversity in National and International Policy Making*, Earthscan, London and Washington, DC

Leisher, C., van Beukering, P. and Scherl, L. M. (2007) *Nature's Investment Bank: How Marine Protected Areas Contribute to Poverty Reduction*, The Nature Conservancy/WWF International, http://www.nature.org/media/science/mpa_report.pdf, accessed 4 June 2013

MacKenzie, C. A. (2012) 'Accruing benefit or loss from a protected area: Location matters', *Ecological Economics*, 76: 119–129

Mayer, M., Müller, M., Woltering, M., Arnegger, J. and Job, H. (2010) 'The Economic Impact of Tourism in Six German National Parks', *Landscape and Urban Planning*, 97(2): 73-82

Nunes, P. A. L. D., Ding, H., Boteler, B., ten Brink, P., Cottee-Jones, E., Davis, M., Ghermandi, A., Kaphengst, T., Lago, M., McConville, A. J., Naumann, S., Pieterse, M., Rayment, M. and Varma, A. (2011) *The Social Dimension of Biodiversity Policy*, final Report for the European Commission, Venice/Brussels, http://ec.europa.eu/environment/enveco/biodiversity/pdf/Social%20Dimension%20of%20Biodiversity.pdf, accessed 7 December 2012

RSPB - The Royal Society for the Protection of Birds (2011) *Natural Foundations: Conservation and local employment in the UK*, RSPB, Bedfordshire

Russi, D., ten Brink, P., Farmer, A., Badura, T., Coates, D., Förster, J., Kumar, R. and Davidson, N. (2013) *The Economics of Ecosystems and Biodiversity for Water and Wetlands*, UNEP and Ramsar Convention on Wetlands

Saayman, M., Saayman, A. and Ferreira, M. (2009) 'The socio-economic impact of the Karoo National Park', *Koedoe* 51: 1

Schmitz, M. F., Matos, D. G. G., De Aranzabal, I., Ruiz-Labourdette, D. and Pineda, F. D. (2012) 'Effects of a protected area on land-use dynamics and socioeconomic development of local populations', *Biological Conservation*, 149: 122–135

ten Brink, P., Mazza, L., Badura, T., Kettunen, M. and Withana, S. (2012) *Nature and its Role in the Transition to a Green Economy*, TEEB UNEP, http://www.wetlands.org/Portals/0/publications/Policy%20document/TEEBnature-green%20economy%20final.pdf, accessed 21 December 2012

Troëng, S. and Drews, C. (2004) *Money Talks: Economic Aspects of Marine Turtle Use and Conservation*, WWF-International, Gland, Switzerland, http://awsassets.panda.org/downloads/moneytalks.pdf, accessed 7 December 2012

11

COSTS RELATED TO PROTECTED AREAS

Bassi, S. and D'Amato, D.

In order for PAs to provide benefits – both in terms of biodiversity conservation and socio-economic well-being – a number of measures need to be carried out to ensure that PAs are effectively designated and managed. As outlined in Chapter 3, these measures include maintaining or restoring the site, such as the removal of invasive alien species or reintroduction of native species, investing in physical infrastructure to facilitate human access, and activities to monitor PA's ecological status. Some profitable activities within PAs may be stopped or reduced if they have a negative impact on sites' biodiversity and ecosystems, such as certain agriculture practices or building developments. This means that securing the proper functioning of PAs requires costs to be incurred (see Section 3.3), either in terms of time and resources (management costs) or in terms of forgone lucrative activities which are prevented by the PA (opportunity costs). PAs can also be a source of human-wildlife conflicts and/or lead to broader social and welfare costs – even in extreme cases including displacement of local communities – creating inconvenience or damages to local populations. All of these costs need to be taken into account when assessing PAs in order to identify their *net* benefits (see Chapter 12).

Table 11.1 provides an overview of different costs associated with PAs. Most costs are related to management activities, which cover a broad range of tasks needed to ensure an effective protection of PAs (see Section 11.1). Opportunity costs can also be significant and require particular care when calculating, as they may be difficult to assess (see Section 11.2). Finally, human-wildlife conflicts, and wider social and welfare costs may in some cases lead to significant costs (see Section 11.3).

TABLE 11.1 Costs associated with PAs

Costs	*Examples of costs*
Management costs: one-off	• Costs for the finalisation of sites, such as costs of scientific studies, administration, consultation, etc. • Costs for management planning, such as costs for preparing management plans, establishing management bodies, public consultations, etc. • One-off payments of compensation for development rights • Infrastructure costs for the improvement/restoration of habitat and species • Other infrastructure costs contributing to conservation, such as for public access, observatories and kiosks, etc.
Management costs: recurrent	• Running costs of management bodies, such as staff salaries, rent, electricity, etc. • Costs for review/update of management plans • Costs for public communications • Conservation management measures for the maintenance and improvement of habitats and species • Costs for implementation of management schemes and agreements with owners and managers of land or water for following certain prescriptions • Compensation for rights forgone and loss of income • Costs for the periodic maintenance of infrastructure for public access, observatories and kiosks, etc. • Costs of risk management, such as fire prevention and control, managing of flood risk, etc. • Monitoring costs, for example to collect data about species and habitats • Costs for the surveillance of sites, for example to detect illegal activities

Costs	*Examples of costs*
Opportunity costs	• Forgone output through constraints on land management/land use practices when the management of sites reduces output by constraining farming and forestry practices, (reducing stocking rates, loss of access to communal grazing pastures, limited application of pesticides, capping timber harvesting, etc.) • Forgone opportunities for land use change when PA status reduces opportunities to convert land into intensive agriculture or forestry, thereby reducing the output of land management • Forgone development opportunities when the protection of sites prevents the use of land for built development, including housing, industrial, commercial, energy, tourism and infrastructure developments. This may reduce economic output and/or lead to a wider loss of social benefits
Human-wildlife conflict	• Costs related to damages caused by wildlife on private property, such as on infrastructures, crops or cattle • Costs for preventing damages from wildlife, such as fences, patrolling, etc. • Loss of human lives caused by wildlife
Broader social and welfare costs, including displacement	• Cost of relocation, compensations to the affected citizens, cost of building new dwellings, etc. • Intangible costs from social and psychological distress caused by displacement. In some cases these can be quantified if, for example, claimed in court disputes • Costs of losing governance of lands important for social, cultural or religious reasons • Costs of limiting access to resources used for subsistence (fuel wood, medicinal plants, etc.)

Source: own presentation, based on Kettunen *et al.* (2011), Gantioler *et al.* (2010) and Kaphengst *et al.* (2011)

11.1 Costs of management

Ideally, the estimation of management costs should build on information specific to the area under observation, following a 'bottom-up' approach. Relevant data could be obtained from PA management effectiveness assessments, official reports or PA's balance sheets, including information on staff salaries, infrastructure investments, funding applications or grants, and management plans and accounts. In some cases complementary surveys or expert interviews may be needed to collect additional information, for example, on opportunity costs. Should data on a specific PA not be available, information could be acquired on the basis of experiences from other areas with similar characteristics (see Chapter 3 for benefit transfer). In both cases it is important to ensure that the estimated management costs are adjusted to take into consideration any market distortions (subsidies, tax exemptions, etc.).

In other cases it may be possible that site-specific information (either from the PA in question or from similar areas) is not available, but aggregated analysis may have been conducted at regional or national level. For example, costs related to the prevention of environmental risks, such as mitigation of wild fires and flooding, might have not been assessed for individual PAs, but studies on national expenditures on risk prevention might be available. In this case it should be possible to roughly estimate how much of these expenses can be attributed to a given PA following a 'top-down' approach. Example in Boxes 11.1 and 11.2 demonstrates a range of possible approaches used to collect information on PAs cost, depending on data availability.

Costs linked to land management can be generally identified on a per hectare basis. The majority of these costs tend to increase in direct relation with the size of a PA including, for example, surveillance costs, investment in damage prevention and personnel costs (as larger areas require more management staff). Per hectare estimates allow data to be scaled up or down, supporting value transfer from a site to another and/or cost comparisons between sites (see Box 11.3). However, some costs may not be related to PA's size. For example, expenditures like costs of scientific studies and communication campaigns may be the same for small and large PAs alike. In such cases, costs per PA (rather than per hectare) might be the most useful way to estimate and present costs for these specific activities.

BOX 11.1

Dealing with different levels of information availability when estimating costs of PAs in Macaronesia (Canaries Islands, Madeira archipelago and Azores Islands)

Direct costs of PAs in Macaronesia – comprising the Canaries Islands, the Azores Islands and the Madeira archipelago – were estimated using a bottom-up approach. The sites are all part of the European Natura 2000 network of

PAs. The study used three approaches to assess costs, depending on different levels of information availability: value transfer using habitat information from official documents and local data, site-by-site data collection, and surveys.

Method: The study focused on three types of cost: 1) Costs from actions aimed at habitat conservation, including measures for restoration, conservation, maintenance, management plans/land use planning, and basic studies necessary for management; 2) Costs for basic management of sites, including compensation, wardening, preventive measures, and monitoring; 3) Costs of marine sites, including wardening, information , and buoying.

Two time horizons were analysed: Horizon 1 referred to the initial costs of the sites' establishment apportioned in equal parts over a five-year period; and Horizon 2 referred to the years following the establishment period, when the budget had stabilised.

In the Canaries, most of the PAs coincided with sites protected under national/regional legislation, and it was therefore possible to extrapolate costs from reports and other economic documents of national and regional parks. To calculate habitat type conservation costs, each of the national/regional parks considered (22 in total) was assigned a representative habitat and costs were estimated for 6 habitats. Costs related to site management were instead calculated on a site-by-site basis.

In Madeira, no information was available from official documents, hence basic information regarding management and costs (including for marine areas) was obtained through a survey sent to the Services of the National Park of Madeira.

In the Azores, very little information was available on management plans and costs, therefore costs estimations were carried out by a group of experts by building on basic information obtained in the area. Costs of marine sites were estimated from data for the Colubretes Island, scaled down by 12 per cent to reflect lower personnel costs.

Results: For the Canary Islands, the habitat conservation cost for the first five years (Horizon 1) was estimated at €22,292,000 per year. After the first five years (Horizon 2), costs dropped to €11,986,883 per year. Due to the different restoration needs of different habitats, their cost per hectare was highly variable between time horizons. The most cost-intensive habitat to manage was the thermo-Mediterranean and pre-desert scrub, with an estimated cost of €217/ha during the first five years and long-term costs of €32/ha per year. The basic management costs of the sites were €32,628,045 per year during Horizon 1. This estimate did not include land purchase as these costs were considered unfeasible to determine. As costs associated with habitat conservation decreased over time, while protection costs remained constant, total management costs observed in Horizon 2 amounted to €20,311,307 per year.

In Madeira, habitat conservation costs were estimated at €803,541 per year during Horizon 1, while during Horizon 2 they decreased to €329,145. Management costs amounted to €3,175,601 per year during Horizon 1 declining to €1,943,032 per year during Horizon 2 as damage prevention and habitat conservation costs declined over time.

In the Azores, habitats conservation costs were estimated to be €3,190,227 per year under Horizon 1, declining to €711,843 under Horizon 2. Management costs amounted to €4,122,632 during Horizon 1 and €1,550,898 during Horizon 2. The larger difference of the results compared to the Canary Islands and Madeira was due to the fact that habitats in the Azores were generally more seriously deteriorated, resulting in higher restoration and management costs.

For the region as a whole, habitat conservation costs during the first five years amounted to €26,285,786 per year at Horizon 1 and €12,907,971 per year at Horizon 2. Management costs were estimated at €37,896,034 per year at Horizon 1 and €23,805,237 per year at Horizon 2, including for the management of marine protected areas.

Source: Sunyer (2000)

BOX 11.2

Estimating management costs for Coral Sea Conservation Zone (Australia) using statistical and expert-based approaches

A study by Ban *et al.* (2009) estimated the magnitude of potential management costs associated with three different management options for the Coral Sea Conservation Zone in Australia, a marine protected area (MPA) of over 972,000 km^2 that was established in 2009. The zone consisted of a network of no-take reserves, three smaller no-take areas and a single large no-take reserve covering 30, 30 and 40 per cent of the zone, respectively.

Method: The analysis was based on existing information on the costs of marine park management in Australia and overseas. Costs were estimated using two different methods: a statistical approach and an expert-based approach, over a 20-year time horizon.

The statistical approach used multiple regression (a statistic method) to identify a formula to predict the management cost per hectare, based on information from existing MPAs (top-down approach). The model indicated that management cost per hectare decreased when the area of the MPA increased but then started to increase again. In general, per hectare costs were lowest for long-established MPAs and areas that consisted of special purpose zones. The highest costs were associated with MPAs that consisted of large areas designated under IUCN Categories III to VI.

The expert based approach involved identification of management requirements specific to the Coral Sea Conservation Zone, and the estimation of associated costs based on expert interviews and on information on operating costs for existing PAs (bottom-up approach). Costs were estimated under two scenarios: designating 100 per cent or 30 per cent of the site as no-take zone. Several assumptions were necessary to estimate compliance and enforcement costs. For example, detection of potential illegal activity was assumed to be easier in the 100 per cent no-take scenario given the lower risk of infringement to go undetected. Basic monitoring requirements were the same for both scenarios but additional effort was assumed to be required for marine monitoring in the 30 per cent scenario to detect ecological status in different zones.

Results: Both the statistical (top-down) and the expert based (bottom-up) approach indicated that a complete no-take area in the Coral Sea would be less expensive to manage than several smaller no-take areas.

In particular, the statistical approach estimated that yearly management costs ranged from AU$2.7 million per year for the no-take (100 per cent) scenario to AU$17.6 million per year for the multiple marine PAs scenario (30 per cent). Over the course of 20 years, the difference between the most expensive and least expensive management option would be about AU$430 million.

These estimates obtained through the expert based approach were of the same order of magnitude as the statistical estimate, although under the expert based approach the 100 per cent scenario was a bit more costly than the statistical approach (AU$7.8 million/year), while the 30 per cent scenario was a bit less than the statistical estimate (AU$13.4million/year). In more detail, for the 100 per cent no-take area total staff costs were estimated at AU$2.2 million per year and total management actions were AU$5.6 million per year. For the 30 per cent no-take area total staff costs were estimated at AU$3.1 million per year and total management actions were AU$10.3 million per year. Compliance-related activities represented the biggest proportion of management costs.

Source: Ban *et al.* (2009)

BOX 11.3

Synthesis of different estimated per hectare costs within European PAs

Estimated costs of PAs can vary significantly between countries (Table 11.2). These differences can reflect the actual differences in costs between sites and/or networks but they are also likely to be caused by differences in the assumptions, approaches and data used during the analysis. Consequently, estimated costs of individual PAs or PA networks may differ over the years. In order to truly compare costs of different sites and networks, it is therefore important to understand the assumptions behind the different costs estimates.

For example, various studies on the European Union's Natura 2000 network of PAs conducted over different years led to slightly different results. A first report for the European Commission (Markland, 2002) suggested an annual cost of around €107 per hectare per year for the overall network, based on €6.1 billion per year for an area of 57 million hectares. A later study by Gantioler *et al.* (2010) assessed an average value of €63 per hectare per year (see Box 13.3 in Chapter 13). A report for BirdLife International by Stones *et al.* (1999) assumed a central figure of €80 per hectare per year, while a later estimate (BirdLife, 2009) obtained an average cost of €128 per hectare, based on estimates for six Member States (see also Chapter 13).

TABLE 11.2 Estimated costs of European PAs

Country	*Total annual costs (€)*	*Size of Natura 2000 network (ha)*	*Average costs per ha per annum (€)*
Austria	206,905,000	1,228,000	168.48
Bulgaria	143,482,803	3,759,000	38.17
Finland	126,970,710	5,557,291	22.84
Netherlands	208,571,428	1,100,000	189.61
Slovakia	61,647,855	1,377,425	44.75
Spain	2,749,392,658	14,286,090	192.45
Total	**3,496,970,454**	**27,307,806**	**128.06**

Source: BirdLife International (2009) as in Gantioler *et al.* (2010)

11.2 Opportunity costs

Opportunity costs reflect the forgone economic benefits from alternative activities or uses of a resource on a particular site. In the case of PAs, these arise when the management and conservation of a given site impose restrictions on some economically profitable activities, such as intensive agriculture practices, mineral extractions, logging, certain tourism activities, or building developments.

Opportunity costs include costs such as forgone land development opportunities, forgone output through constraints on land management practices and forgone opportunities for land use change (Table 11.1). These can give rise to direct financial costs if they result in real payments and expenditures; for example, where compensation has to be paid to those affected or where land is purchased to establish new PAs. These can be relatively straightforward to identify and can include:

- Purchase of land: this is most likely to take place in situations where there are conflicting pressures and development options. The price of land in this case may be used as a reflection of the returns that might have been expected from alternative uses and that are 'lost' because of the PA establishment.
- Compensation for forgone development rights or ongoing management constraints: this is sometimes paid to landowners when their profitable activities are limited by the PA management. These compensation payments usually reflect the income forgone from constraining the use of the site, and they can therefore be a good reflection of opportunity costs.
- Management agreements: these normally involve payments to landowners when they have to undertake prescribed management practices which help protect a PA. These payments usually reflect the opportunity costs of alternative land management practices.

Besides these direct expenditures, there may also be wider economic opportunity costs to society which remain uncompensated. These are usually more difficult to quantify. In some cases, these can be approximated by the perceived benefits from the most likely alternative use (Kaphengst *et al.*, 2011). However, this can potentially be an unhelpfully open-ended exercise. Because one can think of a wide range of alternative uses of land within PAs (agriculture, building developments, tourism, wood harvesting, etc.) the consideration of these uses can result in a disproportionate amount of opportunity costs. In practice, it may prove to be difficult – and indeed even inappropriate – to try to capture all the possible opportunities lost. Realistically, some alternative activities would simply never take place; for example because of incompatibility with planning regulations, local stakeholders' opposition, or because they would not attract sufficient interest from investors.

Given the above, the assessment of opportunity costs should be limited to actual concrete financial expenditures and the estimation of uncompensated benefits should be attempted only when alternative uses are clearly explicit. These can include, for example, situations where certain activities are constrained by the PA without formal compensation. In these cases, lost revenues could be assessed using market values (for example, of the estimated loss of crop production) or using value transfer methods from similar experiences (Box 11.4). In some cases, contingent valuation methods can also be applied (Box 11.5).

BOX 11.4

Forgone opportunities from timber resources after the establishment of a PA network in New Brunswick, Canada

In 2003, the provincial government of New Brunswick, Canada, established a network of protected natural areas in its largely forested territory. Opportunity costs related to this transition were estimated on the basis of forgone revenues from timber resources across different sectors, namely the logging and wood product sectors and the government (in terms of lost royalty revenues).

Method: Opportunity costs incurred by each sector were estimated in terms of reduced net revenues for an 80-years horizon and for three different management scenarios. The first scenario ('business-as-usual' or BAU) assumed that harvesting was completely stopped once the PA was established. This was compared to two scenarios representing two hypothetical situations before the establishment of the PA, 'No PNA1' and 'No PNA2' (Protected Natural Areas), where the area was assumed to be harvested to different extents.

A wood supply model (simulating forest growth, natural transitions, harvesting and silviculture actions) including revenue and cost components in each sector (harvest, transportation, royalties costs, etc. for logging sector and silviculture expenditures and unit royalty prices for government sector) was used to assess the potential income forgone. The difference in net revenues between the BAU and the other two scenarios was interpreted as the total change in economic rent in these sectors resulting from the establishment of the PAs. A discount rate of 5 per cent over the 80-years planning horizon was used to estimate costs.

Data for unit costs and revenues was collected using a combination of published and unpublished literature sources, along with supplemental data provided from informal interviews with industry members and government officials in the region.

Results: The study estimated that the establishment of the PA (BAU scenario) effectively reduced the projected harvested volume by an average of 273,700 m^3 per year compared to the first scenario (No PNA1), and by 350,100 m^3 per year when compared to the second scenario (No PNA2), over the 80-year period. This corresponded to an estimated reduction in net revenues for the logging sector of US$23 to US$44 million, while government revenue was estimated to be reduced by US$37 to US$55 million – all in net present value (NPV) terms (see Chapter 12 and Annex 6). The total projected opportunity cost of the PA was estimated to range from US$68 to US$110 million.

Source: Wilson *et al.* (2010)

BOX 11.5

Using contingent valuation to assess opportunity costs in Mantadia National Park, Madagascar

Extending over an area of almost 10,000 ha, the Mantadia National Park in Madagascar imposed restrictions on the access to resources. This had a significant impact on local populations, which depended on the forest for their livelihood. Both the opportunity cost method and willingness to pay/accept were used to assess the economic value of lost access to the forest by villagers.

Method: Contingent valuation was used to assess villagers' opportunity costs and willingness to accept a compensation for the lost benefits from the forest. Data was collected through a questionnaire, surveying 351 households of 17 villages. The analysis took into account data on income from forest products and from slash and burn agriculture, the extent of forest area for which access was lost because of the park, and the extent of villager's access to the area within the national park. Income from forest products was estimated based on frequency of harvest and quantity of goods harvested. Agriculture income was estimated from surveys on area planted, crop yields, and annual costs. Price data was obtained from several sources, including interviews and market prices. The questionnaire also inquired on willingness to accept for losing access to sources.

Results: The two methods provided strikingly comparable estimates of costs borne by villages. The opportunity cost method estimated costs at around US$566,000 per year (equivalent to US$91 per household per year), while willingness to accept was about US$673,000 per year (US$108 per household per year).

Source: Kramer *et al.* (1995)

11.3 Other costs

PAs can also involve other costs, such as costs related to human-wildlife conflicts and negative impacts of PAs' on broader welfare (Table 11.1). Human-wildlife costs refer to damage caused by wildlife in areas used for human activities. These may be negligible in several PAs, but they can represent substantial costs in some areas – especially where the livelihood of local populations is strongly reliant on agriculture and farming. This is particularly the case in developing countries, where agricultural production and livestock can be seriously threatened by some carnivores and other large mammals (Box 11.6). Costs assessments in these cases can rely on estimates of the revenues lost, based for instance on the quantities of livestock and crops lost due to local wildlife, and on current market prices for these goods.

BOX 11.6

Monetary assessment of livestock and crop depredation by large mammals in Bhadra Tiger Reserve, India

The extent of material and monetary losses incurred by resident villagers in conflicts with large felines and elephants were estimated in the Bhadra Tiger Reserve. The Reserve is located at the foothills of the Western Ghats Mountains and covers an area of 495 km^2, mostly dominated by moist to dry deciduous forests.

Method: A survey of 86 households owning livestock – accounting for over two thirds of the estimated livestock population – was conducted in a number of villages to gather information on the livestock holding of each respondent and their relative conflicts with wildlife. The data collected included information on the size and composition of holdings, annual sale or purchase of livestock, annual livestock mortality attributed to predation by large carnivores (tiger, leopards, wild dogs), and whether such kills occurred within the village or in the adjoining forests.

In order to gather data on crop depredation by elephants, 81 households owning crops were surveyed to inquire on the amount of agricultural land owned or tilled by a family, estimated annual crop production, and estimated annual crop losses to elephant depredation. Reported losses of livestock or crops were verified with a local field assistant and randomly with other villagers in order to corroborate the correctness of the information provided. The monetary value of losses was estimated on the basis of livestock and paddy market prices.

The effectiveness of existing compensation schemes aimed at offsetting losses caused by wildlife was evaluated based on a stakeholder survey and existing data on compensation claims (total and rewarded) and awarded sums.

Results: Livestock killing by large carnivores seemed to have a significant impact on cattle populations in the reserve. The sampled households attributed a loss of 219 animals to large carnivore predation (mostly tigers and leopards) over three years. The overall loss amounted to 0.9 animals per household per year, for an estimated annual loss of 12 per cent of the average family holding livestock. In monetary terms, losses amounted to US$56 per household per year, equivalent to 16 per cent of the average annual household income in the region. The overall annual monetary value of livestock losses in the villages was about US$5,056.

Regarding crop damages, within three years about 115–119 tons were lost to elephants, at an annual loss of 39 tonnes. In monetary terms, this was equivalent to US$5,404, or 11 per cent of the monetary value of total annual production.

The annual loss per family amounted to US$108, equivalent to approximately 30 per cent of the average annual household income in the region.

Compensation schemes can offer an effective way to minimize costs associated with wildlife. However, in the case of Bhadra Tiger Reserve, the compensations awarded offset only 5 and 14 per cent of the costs of livestock and crop losses, respectively. The compensation schemes were also hampered by delays in processing the claims. This highlights the importance of basing the development of PA policy instruments (in this case compensation schemes) on proper assessments of costs, to guarantee effective management of and public support to PAs (see also Section 15.4).

Source: Madhusudan (2003)

While not a common occurrence today, local communities have been displaced from PAs by conservation projects (Dowie, 2009). Furthermore, even when not resulting in complete displacement the establishment of PAs might lead to social and welfare costs caused by losing the governance of land that has been important for subsistence (fuel wood, medicinal plants, etc.) and/or social, cultural and religious reasons. This can lead to substantial costs in terms of economic losses to local communities (opportunity costs, etc.) and it can also result in the loss of cultural values and decline in general well-being. Such costs are very site-specific and often challenging to estimate. Monetary assessments may not always be possible; however quantitative and qualitative information, such as the number of people displaced and cultural, religious or social values jeopardised, can help to convey the scale of costs.

For example, in the Democratic Republic of Congo the eviction of local populations from the Kahuzi-Biega National Park in the 1970s led to a dramatic decline in their welfare, which was based on hunting and gathering from within the forest (Nelson and Hossack, 2003). Similarly, in Sibuyan Island in the Philippines, the creation of Mount Guiting-Guiting Natural Park affected 1,687 individuals by constraining their ability to collect honey, rattan, vines, medical plants and other non-timber forest products central to their livelihoods (Tongson and Dino, 2004). In addition, PA status can negatively affect intangible and spiritual values, for example by attracting tourism in areas of religious value to local populations. This was for instance the case in Uluru in Central Australia, where tourists would climb the Ayers Rock in violation of the sacredness of the place (James, 2007).

While the negative examples above highlight the possible negative impacts of PA establishment, it should be noted that widely used participatory planning and management practices have helped to minimise the need for a complete prohibition on access and/or displacement, reducing related costs. Furthermore, several cases exist where the establishment of PAs has resulted in a (net) positive impact on the well-being of local communities (see Chapter 10).

11.4 Total costs of protected areas

Once the costs have been identified and calculated, they have to be aggregated to obtain the total cost of PAs. These can be calculated on an annual basis or assessed for a given number of years (see Chapter 12).

When aggregating costs, it is important to identify which costs are one-off (incurred only once or sporadically during the lifetime of the PA), and which costs are recurrent (incurred on a regular basis for the normal functioning of the PA) (see Table 11.1). One-off costs typically include expenses for the finalisation of sites, management planning, investment costs for land purchase, one-off payments of compensation for development rights, and infrastructure costs to support conservation work for public access. Displacement cost and some opportunity costs, such as the cost of precluding building activities, are also incurred just once in the lifetime of the PA (one-off). The costs for human-wildlife conflicts, for example for the loss of cattle due to large carnivores, can happen several times over the years. These costs are typically unpredictable, irregular and vary in their size and can, therefore, be considered as a number of separate one-off costs.

Recurrent costs are usually associated with regular activities, such as scheduled activities and measures to maintain or improve sites. These include costs for recurrent management planning and for habitat management and monitoring, and opportunity costs related to periodic (annual) compensations for forgone activities.

When calculating the total management costs of PAs across a number of years, one-off costs should be considered only once whereas recurrent costs will have to be multiplied by the number of times they are incurred within the period under analysis. On the other hand, when calculating costs per year recurrent (annual) costs should be included once whereas the one-off costs need to be considered on an annuity value basis, i.e. one-off costs attributed 'pro quota' to each year (see Box 11.7). For example, when the total one-off costs of a PA are €20,000,00 and recurrent costs €5,000,000 per year the total costs for a 20-year period result in €120,000,000 (€5,000,000 x 20 years + €20,000,000) whereas the annual costs during the period would be €6,000,000 (€5,000,000 + €20,000,000/20).

More information on aggregating costs (and benefits), including consideration of changes in net present value over time (discounting), is provided in Chapter 12 and Annex 6. In general, while costs are an undeniable part of well-managed and effective PAs existing evidence shows that costs are often smaller than the overall benefits. A comprehensive assessment of benefits, along with the guidance given in Chapters 6 to 10 and 12, is therefore a useful tool to demonstrate that an investment in nature protection – while crucial for safeguarding to biodiversity - can also be beneficial from a socio-economic point of view.

BOX 11.7

One-off costs in the context of estimating the costs of the EU Natura 2000 PA network

A study by Gantioler *et al.* (2010) on the costs of managing the EU network of protected areas (Natura 2000) provided insights as to the possible approaches for annualising one-off costs for the purpose of developing annual cost estimates.

In general, a period of six years was suggested to be used to develop an annuity of one-off costs. This period was considered to reflect the likely timeframe of a one-off investment. Six years was also the average timeframe chosen by different individual European countries for assessing one-off costs.

The study also noted that considering costs within a hypothetical 'lifespan' of PAs allows for more realistic comparisons to be made between two key conservation strategies (land purchase vs. establishing management contracts with landowners). In terms of costs calculations, this means that there is a need to carry out the assessments within longer timeframes, for example, by using time periods of 20 or 30 years with a 4 per cent real discount rate (see Annex 6).

While the 'six-year approach' gives good insights on the investments needed over a realistic period of time, the 'lifetime approach' helps to ensure that different conservation strategies are, in general, compared on a more equal footing. The two approaches can complement one another, providing a more comprehensive information base for decision-making.

Source: Gantioler *et al.* (2010), see Box 13.3 in Chapter 13 for further explanation of the study

References

Ban, N. C., Adams, A. and Pressey, R. L. (2009) *Marine protected area management costs: an analysis of options for the Coral Sea*, a report prepared for the Protect Our Coral Sea campaign, Australian Research Council Centre of Excellence for Coral Reef Studies, James Cook University

BirdLife International (2009) *Financing Natura 2000: Assessment of funding needs and availability of funding from EU funds*, final composite report, http://www.birdlife.org/eu/pdfs/N2000_Final_composite_report_09.pdf, accessed 12 December 2012

Dowie, M. (2009) *Conservation Refugees*, MIT Press, Cambridge, MA and London

James, S. (2007) 'Constructing the climb: Visitor decision-making at Uluru', *Geographical Research*, 45:398–407

Kaphengst, T., Bassi, S., Davis, M., Gardner, S., Herbert, S., Mazza, L., Pieterse, M. and Rayment, M. (2011) *Taking into account opportunity costs when assessing costs of biodiversity and ecosystem action*, a final report for the European Commission, http://ec.europa.eu/environment/enveco/biodiversity/pdf/OpportunityCostsOfBiodiversityAndEcosystemAction.pdf, accessed 17 December 2012

Kettunen, M., Berghöfer, A., Brunner, A., Conner, N., Dudley, N., Ervin, J., Gidda, S. B., Mulongoy, K. J., Pabon, L. and Vakrou, A. (2011) 'Recognising the value of protected areas', in P. ten Brink (ed.) *The Economics of Ecosystems and Biodiversity in National and International Policy Making*, Earthscan, London

Kramer, R. A., Sharma, N. and Munasinghe, M. (1995) *Valuing Tropical Forests: Methodology and Case Study of Madagascar*, the World Bank, Washington, DC

Madhusudan, M. D. (2003) 'Living Amidst Large Wildlife: Livestock and Crop Depredation by Large Mammals in the Interior Villages of Bhadra Tiger Reserve, South India', *Environmental Management*, 31(4): 466–475

Markland, J. (2002) *Final Report on Financing Natura 2000. Working Group on Article 8 of the Habitats Directive*, EU Commission Working Group on Article 8 of the Habitats Directive, http://www.eu-natur.de/attach/9/Abschlussbericht_Artikel_8_AG_en.pdf, accessed 17 December 2012

Nelson, J. and Hossack, L. (2003) *From Principle to Practise: Indigenous People and the Environment: Issues and Linkages*, Livelihoods and Institutions Group, Natural Resources Institute, University of Greenwich, London

Stones, T., Harley, D., Rose, L., Lasen-Diaz, C., Rayment, M. and Trash, M. (1999) *The Cost of Managing the Natura 2000 Network*, Report for RSPB and BirdLife International, RSPB, Sandy

Sunyer, C. (2000) *Estimation of costs of the Natura 2000 Network in the Macaronesian Region*, Terra Technical Report 2, Madrid, http://www.terracentro.org/Publicaciones/Serietecnica/Serie%20tecnica2%20(en).pdf, accessed 17 December 2012

Tongson, E. and Dino, G. (2004) 'Indigenous peoples and protected areas: The case of the Sibuyan Mangyan Tagabukid, Philippines', in McShane and Wells (eds) *Getting Biodiversity Projects to Work: Towards More Effective Conservation and Development*, Columbia University Press, New York

Wilson, J. J., Lantz, V. A. and MacLean, D. A. (2010) 'A benefit–cost analysis of establishing protected natural areas in New Brunswick, Canada', *Policy and Economics*, 12: 94–103

12

ASSESSING NET BENEFITS

Site level

ten Brink, P. and Kettunen, M.

An assessment of multiple/total benefits provided by a PA will most likely make use of a range of valuation methods, depending on the benefits under review. As the previous chapters illustrate, assessments may focus on estimating the socio-economic value of different – tangible and intangible – benefits provided by PAs, such as provisioning of goods or protection against environmental hazards. In socio-economic terms, these benefits can result in real market revenue, avoided costs or increased personal well-being and cultural importance. With due data, some issues are amenable to monetary valuation, others are more typically represented in quantitative or qualitative terms (ten Brink *et al.*, 2011a; White *et al.*, 2011; Pascual *et al.*, 2010). This mosaic of benefit types, beneficiaries, methodologies and indicators of value does not always render itself to direct comparisons, creating challenges for developing a realistic overall picture of the benefits at site level. The challenges and how to overcome them are discussed in detail in Section 12.1.

Having an accurate appreciation of benefits and their availability in terms of space and time (who benefits from which ecosystem service, at what location and over what timescale) will be invaluable for site managers, local authorities and governments, as well as citizens and concerned businesses. In the case of remote and relatively undisturbed areas, many of the benefits will 'flow' from the ecosystems independent of PA designation and without need of significant intervention and costs. In other cases PA designation and management will be essential to avoid the degradation of ecosystems and related loss of services and goods due to pressures on the site. Furthermore, PAs will also involve a certain amount of costs, from on-site and off-site investments that can enhance the level of benefits (Chapters 3 and 11) to opportunity costs associated with the site's existence or eventual restriction on the use of land or resources. It will be important to understand the full picture of the benefits and costs associated with a site, including the socio-economic implications of different land use and management choices, in order to develop an overall estimate for the net value of the site. The issue of comparing costs and benefits is discussed in detail in Section 12.2.

The mixture and socio-economic value of benefits depends on management and investment choices, as well as the trade-offs between and (choice of) prioritisation of ecosystems services and related goods (Chapter 2). Management and investment decisions on the site can lead to an increase in the availability of certain services and related benefits. A formal PA designation may help increase the level of financial flows to support these management and investment decisions (Chapter 15). Furthermore, the designation of a site as a PA, accompanied with rules/regulations to guarantee the protection, can mitigate the degradation of the site, its ecosystems and its ability to provide ecosystem services. These issues relate to the question of the added value of PAs' designation addressed in Section 12.3.

12.1 Assessing aggregated benefits

Building on qualitative, quantitative and monetary data

Qualitative analysis forms an important starting point for net benefit assessments. It helps to avoid creating an imbalanced overview of benefits, focusing too much on the services and goods for which ample (monetary) evidence is available to the detriment of those with less existing information. The scoping assessment introduced in Chapter 4 (e.g. Table 4.2 and Figure 4.1) provides a means for obtaining a first-cut aggregate picture of the (scale of) benefits at site level, including the flow of benefits to different stakeholders. This overall picture can be further complemented by a map that spatially illustrates the connection between the site and stakeholders benefiting from it. Such an aggregate picture – based on qualitative indicators and quantified perceptions of importance – can suitably reflect the understanding provided by existing studies on the importance and value of different services.

An economic (monetary) aggregated assessment of PA benefits uses insights from quantitative and monetary assessments to complement and further specify the qualitative overview. As a general rule, an aggregated economic assessment of PA benefits should aim to cover as wide a range of ecosystem services and related goods as possible, in light of data availability. As highlighted in Chapters 6 to 10, only some of the ecosystem services and related goods can generally be assessed in monetary terms and, consequently, only a subset of services can be monetised and combined in total economic estimates value (TEV) (see below). Therefore, the final aggregated picture of the benefits of a PA will be based on the total economic estimate, supported by a wider qualitative and quantitative synthesis. It is essential to note all assumptions and all areas where a lack of data has not allowed analysis. The socio-economic analysis should also always be presented in the wider context, making relevant links between the functions, interactions and benefits of a PA to society and the economy. Ultimately, the final picture of the total benefits will be a type of multi-criteria benefits assessment, though will not necessarily need to make use of formal multi-criteria assessment (MCA) techniques. While monetary numbers may be good at attracting headlines and decision-makers' attention (see Chapter 15) the broader, more nuanced picture is needed to ensure concrete uptake of results to support evidence-based decision-making and governance of PAs.

When developing an aggregated benefit assessment it is also essential to be clear as to what type of economic (monetary) value the different individual benefits have been 'assigned' with. The appeal of economic (monetary) analyses of PA benefits is that they result in a common unit for valuation, facilitating comparability between and allowing aggregation of otherwise very different benefits. The monetary analyses, however, make use of different tools which provide different metrics for value. For example, market prices and avoided costs reflect real economic values (existing or potential) whereas monetary estimates based on people's stated preferences capture the appreciation of benefits in terms of broader well-being (see Chapter 14). Consequently, any aggregated assessment should differentiate between and/or be transparent about what elements it consists of, to avoid any misunderstanding of the final outcome.

Further to the above, an estimated economic value may not mean the same to all stakeholders, given different income levels, affordability and dependency. It is important to reflect these differences when developing an aggregate assessment of PA benefits by complementing the estimated site-level monetary estimate with information that illustrates the different importance of benefits to different parties. For example, PAs offer different benefits (and hence monetary values) to companies, society and to the economy as a whole. The values also vary across different geographic scales (local, regional, national, global). Furthermore, the values will mean different things for stakeholders depending, for example, on their income groups within society or the size of group/household/business benefiting. Adjusting values by using purchasing price parities (e.g. by weighting by PPP/GDP) can help. In addition, indicators such as 'benefits as percentage of household income', 'cost savings as a percentage of company turnover' or 'operating profit' help to reflect the distribution of benefits within the overall assessment. The case study of Muthurajawela wetlands (Box 12.1) illustrates how to make good use of quantitative indicators of value to form an aggregated picture of PA benefits.

Finally, when using existing information as a starting point for an assessment care needs to be taken not to aggregate numbers from studies that seek to answer different questions and are therefore not comparable to one another. Some existing studies focus on incremental benefits of specific actions on PAs (e.g. increase in PA's water purification capacity due to restoration), whereas some focus on the total availability of ecosystem services and related goods (e.g. PA's total water purification capacity). There is no easy fix to this, including no simple multiplier to convert an incremental benefits number to a total value or vice versa, as the relationship between incremental and overall benefits is site-specific and changes over time. The best practical solution is to present an overall aggregate picture of the results and place the numbers side by side in due context, clearly stating which estimates relate to incremental and which total values. When resources are available it is also useful to complement the collection of existing studies assessment with an own overall assessment. This can also help to double check the validity of existing values and to determine, for example, whether they are likely to be under- or overestimates (see also Chapter 13).

BOX 12.1

Estimating the aggregate benefits of Muthurajawela wetland, Sri Lanka

The assessment of the economic value of Muthurajawela wetland, Sri Lanka covered a wide range of ecosystem services and used a mix of valuation tools (Figure 12.1). While not specifically focused on PAs, it provides a relevant concrete example on how to assess the aggregated value of benefits derived from a PA.

Methods used in the context of the study included: market prices for fisheries and agriculture, visitor travel costs for leisure, service replacement costs for flood attenuation, mitigation expenditures for waste treatment, avoided expenditures for water supply and recharge, and avoided damage costs for incremental carbon sequestration. The use of common units (total value per year and value per hectare per year) to capture the estimated economic values of different ecosystem services allowed the aggregation of benefits at the site level, despite the use of different methods.

The total economic value of the wetland was estimated at Rs726.5 million per year or Rs0.24 million/ha/year (US$8.1 million a year and around US$2,640 per hectare/year). Muthurajawela wetland is an urban wetland, which explains why the flood attenuation and waste water treatment had the highest estimated economic value.

While interpreting the results, the study clearly highlighted that there was insufficient data to capture the economic (monetary) value of several identified important benefits: health benefits, landscape values, socio-economic importance of plant based handicrafts, value of downstream fisheries, and options/existence values. Similarly, the economic value of climate regulation only covered the annual carbon sequestration of existing wetlands, completely neglecting their value as a stock of stored carbon, potentially at risk of emissions release from wetland degradation.

To address these data deficits, a good use of quantitative indicators of value was made to communicate the broader, social benefits. The results showed that 30,000 people benefit directly and indirectly from the wetland, including 11,600 supported in downstream fisheries activities (Negombo fishing households), 15,000 via recreation and tourism, 6,300 benefiting from freshwater supplies. The study also noted that not only do 5,000 people, but also over 100 industries benefit from flood attenuation, making the business case.

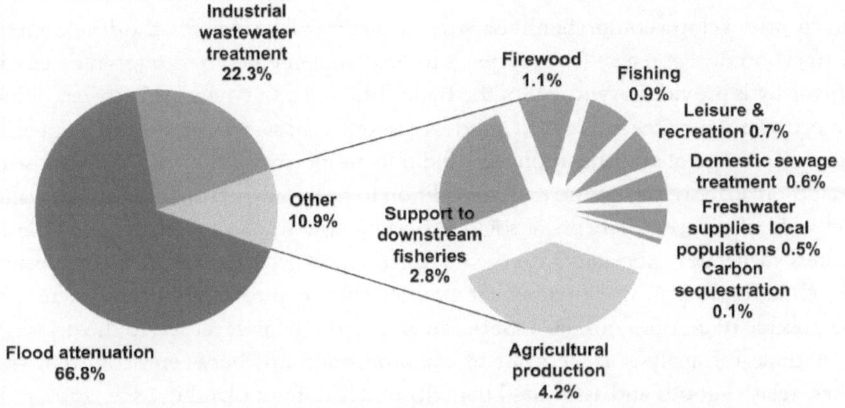

FIGURE 12.1 Aggregated benefits of Muthurajawela wetland, Sri Lanka, based on an economic assessment

Source: redrawn from data in Emerton and Kekulandala (2003)

Economic aggregate and the risk of double counting

As highlighted in earlier chapters, ecosystem services are often interconnected and it is therefore important to ensure that double counting of their economic (monetary) value is avoided during aggregation. A practical way to address this challenge is by understanding the interlinkages between different ecosystem services and related goods (see Chapter 3), identifying whether the same issue is addressed more than once in different places and making sure that no double values for one benefit 'enter' the calculation of an economic aggregate. However, it is to be kept in mind that values left out from the economic aggregate can still be reflected in the context of the broader aggregated picture of the site.

For example, the benefits associated with game hunting to hunters (provisioning of meat and recreational benefits) are interlinked, often to an extent that makes it difficult to distinguish between the two. A simple solution to avoid double counting is simply to count the benefits only at one point. In the Hoge Veluwe Forest PA (the Netherlands) recreational hunting is allowed. While the hunters have recreational benefits, the hunted meat remains park property and is sold by the park. Consequently, the socio-economic values related to the provisioning of game and the recreational value from hunting are distinct and potential double counting between these services is avoided. A total of 300 animals are shot each year, yielding around 10,000 kg of gross animal weight, of which around half is sold as meat. The net revenue for the park related to the sale of hunted animals is around €50,000/year (Hein, 2011). This gives an indicator of value, and avoids double counting, though arguably the estimate is quite conservative as the recreational benefits are not really covered.

Carrying out a trade-off analysis to complement aggregation

Finally, to develop a comprehensive aggregate picture of the benefits, the development of an economic aggregate (above) needs to be complemented by a trade-off analysis. This analysis provides a synthesis of the trade-offs (and also synergies) between different ecosystem services and related goods across different stakeholders (see Chapter 2), based on a range of available monetary and non-monetary indicators. The purpose of a trade-off analysis is to ensure that, in addition to providing information on the value and scale of different benefits, a socio-economic assessment of a PA is also able to address challenges associated with the management of ecosystem services. For example, choosing to prioritise one benefit over another requires a transparent picture of the possible trade-offs, looking at both the wins and the losses related to the decision.

A trade-off analysis can consist of assessing trade-offs between ecosystem services, related goods and associated benefits under different land use scenarios and/or it can focus on trade-offs associated with the adoption of different management measures on the site (Section 12.2). For example, it is important to establish whether an adoption of a certain management practice, aimed at increasing the supply of an ecosystem service, leads to a decline in the availability of and/or benefits associated with another ecosystem services.

Figure 12.2 below presents an example of a monetary trade-off assessment across land-use decisions in the Leuser National Park, Indonesia. The study focused on assessing the net benefits of conservation by comparing the availability and economic value of different ecosystem services and related goods under a conservation scenario with their availability and value under a deforestation scenario (land conversion for agriculture and timber production). The analysis indicated that conservation leads to higher benefits overall, but with clear trade-offs between benefits related to water supply, flood prevention, tourism, carbon sequestration, hydroelectricity and biodiversity conservation on the one hand, and losses to agricultural and timber production on the other. The study from Leuser National Park shows how an aggregate analysis of PA benefits can be developed in a way that it highlights the overall economic benefits while also providing sufficient detail on trade-offs between the benefits.

12.2 Comparing benefits with costs

The aggregated assessment of benefits provides by a PA (above) is, of course, only one half of the picture and a comparison between benefits and costs is needed to create an understanding on the net benefits associated with a site. Such comparisons will help to clarify the final total benefits experienced by different stakeholders, determine whether conservation as a whole is in societies' interest and, finally, assess the overall cost-benefit ratio of decisions (management measures, investment in PAs, etc.). In the case PA designation, a comparison of benefits and costs will also offer insights into who might support the designation and associated conservation and who might oppose it. This is critical for understanding whether and on which basis conservation objectives can be met.

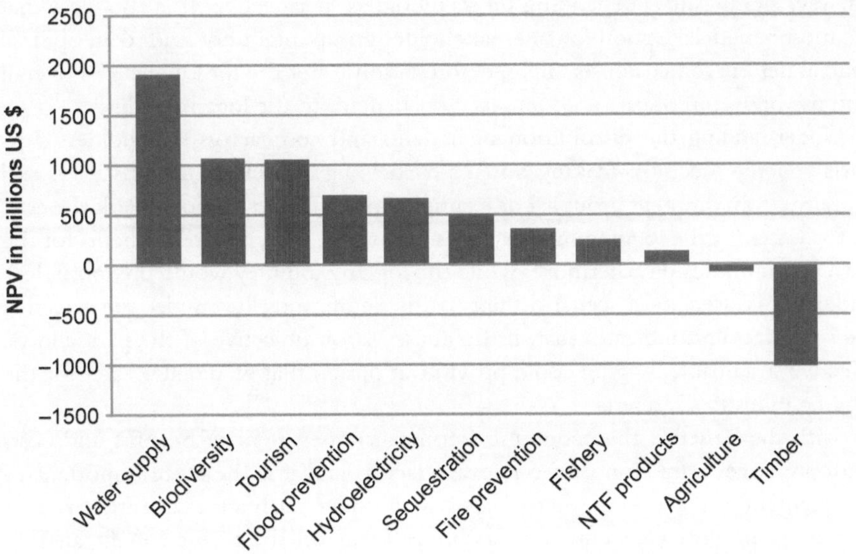

FIGURE 12.2 Net benefits of conservation vs. deforestation in the Leuser demonstrating the trade-offs across different ecosystem services (net present value over period 2000 to 2030, using a 0 per cent discount rate)

Source: redrawn with data from van Beukering *et al.* (2003)

Developing a picture of costs and benefits

The comparison of costs and benefits should be closely aligned with the identified objectives of the overall socio-economic assessment (Chapter 5). Typical contexts for which the comparisons of benefits and costs are valuable include, for example, the designation of a PA and associated conservation vs. alternative uses of land and nature, decisions as regards land use options within existing PAs, and assessments exploring the outcomes of specific measures such as restoration and land purchase to increase PA buffer zones or establish land corridors connecting PAs (ten Brink *et al.*, 2011; Arcadis *et al.*, 2011).

In general, the assessment of benefits and costs needs to build on the identification of impacts on different stakeholders, assessing which of the stakeholders are likely to be significantly affected and what the different impacts of different measures are (i.e. identify 'winners and losers'). For the key stakeholders, an economic assessment of the benefits and costs can usefully be carried out, estimating the economic value of ecosystem services retained and/or lost.

For example, in the context of the Leuser National Park the potential benefits for different stakeholders under three different land use options (conservation, selective use and deforestation) were estimated (Figure 12.3) (van Beukering *et al.*, 2003). The economic analysis shows that the economically 'best' land use option can vary significantly, depending on stakeholders' perspective. It is also clear that the most beneficial option for one stakeholder group might not lead to an optimal general net outcome. For example, deforestation is not the most beneficial overall land use option in the area, despite its clear benefits to the logging industry.

Understanding the distribution of benefits and costs across stakeholders supports concrete decision-making on PAs by helping to identify incentives for and pressures from different groups. For example, in the Leuser National Park the estimated losses for the logging industry are smaller than the estimated benefits for the local communities. Furthermore, while the logging industry would obviously lose under the conservation scenario their loss is significantly less under the selective use scenario. This indicates that, if the conservation objective of the park allows, selective sustainable logging could provide an option that would also cater for the logging industry's interests.

As with the benefits, the economic (monetary) assessment of benefits and costs can only cover a fraction of the relevant factors and it is therefore important to complement any economic net analysis with wider qualitative, quantitative and monetary insights. Consequently, a simple focus on strict cost-benefit analysis (CBA) risks losing a part of the bigger picture (see Box 12.2). The results of an economic assessment will always depend on the geographic coverage of an assessment and it is therefore important to specify what areas have been covered by an economic assessment and what still needs to be captured through other indicators. Economic assessments are also sensitive to the choice of timescale and there are risks of overlooking or underrepresenting benefits or costs over time (see further below and Annex 6).

In general, having qualitative and quantitative comparisons helps form a more nuanced – and convincing – picture on benefits and costs. For example, in the case of the Leuser National Park the economic analysis (Figure 12.3) was complemented by information on the level of water supplied by the PA to households and industry in million m3 per year across the three scenarios over the period to 2030. Under the deforestation scenario, water supply falls from around 240 million m3 to around 80 m3, while under the conservation scenario water supply rises to nearly 500 million m3. Finally, a systematic qualitative comparison of the distribution of benefits and costs – even on a very general level – is a helpful starting point for understanding the real-life implications and possibilities related to the socio-economic role of PAs (Table 12.1), for example the geographic scale of measures required to engage different beneficiaries and compensate for costs.

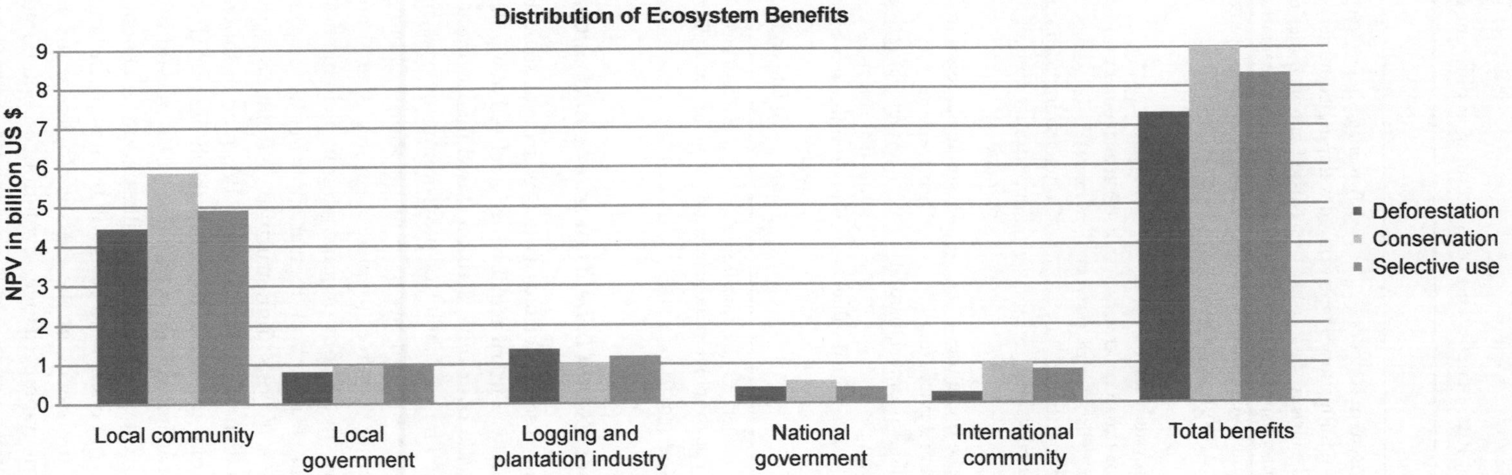

FIGURE 12.3 Understanding the 'best' option for land used within the Leuser National Park, from different stakeholder perspectives

Source: redrawn with data from van Beukering *et al.* (2003)

TABLE 12.1 Examples of PAs' benefits and costs accruing at different scales

	Benefits	*Costs*
Global	Dispersed ecosystem services (e.g. climate change mitigation/adaptation) Nature-based tourism Genetic resources supporting pharmaceutical development Global cultural, existence and option values	PA management* (global transfers to developing countries) Missed opportunities from alternative development programmes* (global transfers to developing countries)
National	Dispersed ecosystem services (e.g. clean water for urban centres, agriculture or hydroelectric power) Nature-based tourism National cultural values	Land purchase * PA management (in national PA systems) * Compensation for forgone activities* Opportunity costs of forgone tax revenue
Local	Consumptive resource uses Local ecosystem services (e.g. pollination, disease control, natural hazard mitigation) Local cultural and spiritual values	Restricted access to resources Displacement PA management (private land owners, municipal lands) Opportunity costs of forgone economic activities Human-wildlife conflict

* These cost categories are in effect transfer costs from the local to national level, or from the national to international level. Section 8.3 provides more information on these and related options.

Source: based on Kettunen *et al.* (2011)

Issues of timescale: costs and benefits over time and calculating NPVs

Naturally, benefits and costs associated with PAs are not static but change over time. A number of benefits – in particular benefits related to regulating and cultural service – tend to increase over time with increased management and investment following designation. The availability of provisioning services and related goods varies depending on the PA and resource in question exhibiting, for example, an increase for recovering resources (depleted fish stocks, etc.) and a decrease for resources with new use restrictions or improving implementation. On the other hand, the costs generally feature both regular annual costs (ongoing management, etc.) and upfront one-off costs (infrastructure, restoration activities, etc.) (see Chapters 3 and 11). In general, costs related to PAs tend to show a typical profile of high upfront costs (one-off investment to establish the site) that level out in the long term (regular management costs with some *ad hoc* one-off investments).

Figures 12.4 and 12.5 below provide an illustration of benefits and costs over years. The actual levels of benefits and costs will be site-specific and their evolution will depend both on the trends in the PA's ecological status as well as socio-economic trends (demographic change, evolution of household incomes, new economic assets and infrastructures, etc.) and interlinkages between them.

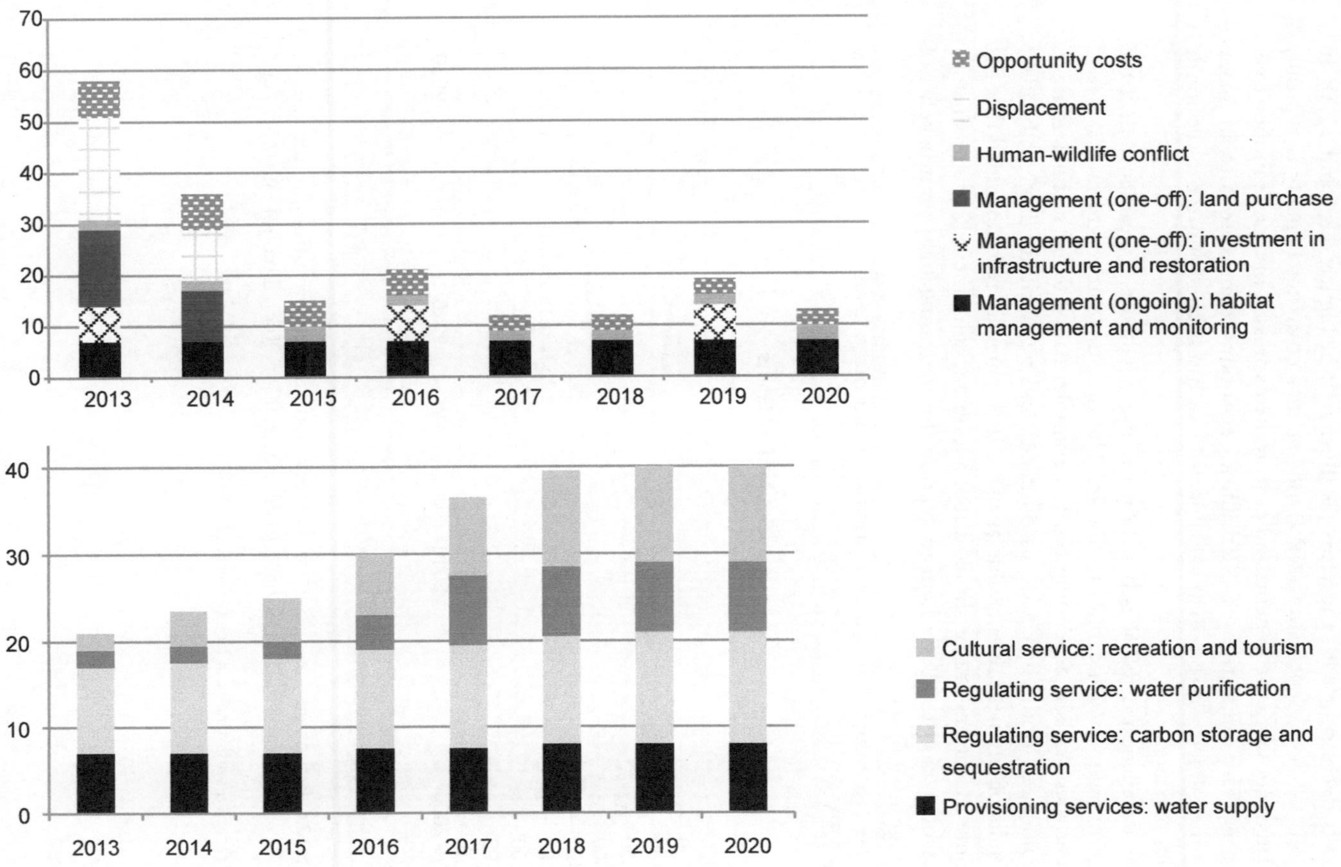

FIGURES 12.4 AND 12.5 Illustrative distribution of PA benefits and costs over time

Source: own presentation

In the context of economic analysis, the costs and benefits are traditionally presented as a flow of costs and benefits over time, often with benefits increasing and costs decreasing with the years (as illustrated in Figures 12.4 and 12.5). In addition, a net present value (NPV) of benefits/costs/net benefits is typically calculated, allowing the benefits and costs to be expressed in today's money over the years. The calculation of NPV allows economic comparisons to be made over time which can be an invaluable tool to assess the long-term economic realities of a site.

The calculation of NPV is based on the use of discount rates that reflect the change in value over a certain timescale. The discount rates and timescales used vary between studies, depending on, for example, national and sectoral contexts and guidelines. Both choices of discount rate and timescale fundamentally affect the results of any net present value analysis. It is generally recommended that the economic valuations of PAs should adopt a long enough timescale (with a matching discount rate) to allow them to capture the increasing/long-term benefits of sites (Figure 12.6) (Gowdy *et al.*, 2010).

Please see Annex 6 for further discussion and more detailed guidance on how to use NPV.

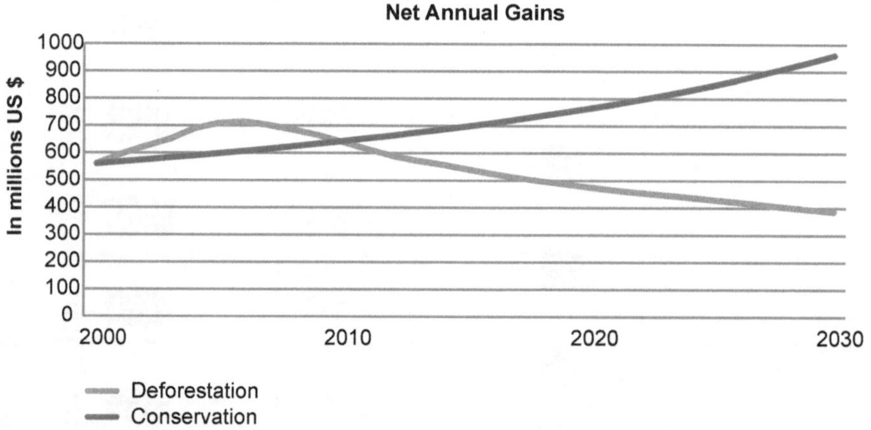

FIGURE 12.6 Estimated net gains of conservation and deforestation in the Leuser National Park, Indonesia (scenarios 2000 to 2030, discount rate 0 per cent)

Source: redrawn with data from Beukering *et al.* (2002)

BOX 12.2

Comparing traditional cost-benefit analysis (CBA) with a broader assessment of benefits and costs

There are different interpretations as to what constitutes cost-benefit analysis. The approach can be applied as narrow mathematical calculations comparing costs and benefits over time, looking only at economic variables, determining whether the benefit-cost ratio is above or below 1 and whether there is a net present benefit or net present cost (i.e. a classic CBA analysis). Alternatively, cost-benefit analysis can be carried out in a wider setting that may use economic metrics, but also non-economic ones where they help offer insights on benefits and costs.

The strengths and weaknesses of the two CBA approaches partly depend on what assessment is being carried out. As noted above, it is important to understand the range of benefits and costs of any decision, and some of these will be amenable to being represented in monetary terms. Furthermore, some of those translatable into monetary terms have different meanings (real costs, cash in hand, avoided damages, well-being benefits). In general, a good assessment of costs and benefits identifies who the winners and losers are, the associated costs and benefits of decisions to the 'winners and losers', and creates a composite picture that combines monetary, quantitative and qualitative insights ideally presented in a spatial context.

The strength of a classic CBA is that it offers a clear and easily understandable framework for assessment using tools that have been accepted by decision-makers and academics. It can also provide solid answers to policy questions or investment decisions. One key weakness of a classic CBA, however, is that it risks overlooking or underrepresenting certain issues. This weakness can be addressed in part by ensuring that a classic CBA is a part of a wider assessment that integrates also other, non-monetary information. This can lead towards more of a multi-criteria assessment (MCA) whereby different metrics are chosen to represent importance or value to complement the economic one. Any assessment seeking to define an optimal solution needs to take into account different factors of importance. This in turn has its own complications as a choice is needed as to what relative weight to attribute to the different criteria.

A further potential weakness of a classic CBA is that some marginal costs and benefits may be non-linear, and the use of a constant marginal value may lead to an overly simple analysis that results in certain issues being underrepresented. This can be addressed by using non-marginal values, such as using carbon prices that rise over time or increasing the marginal value of water provision or recreation over time where demand is projected to increase. Care is needed in choosing non-marginal rates, as they need to be able to withstand scrutiny, and be more appropriate than constant values. One cannot simply transfer unit values of benefits from one context to another, for example from an area where a benefit is available in abundance to areas of scarcity and inelastic demand. In some cases values will change not only non-linearly over time but have a step change in values, for example, if ecological thresholds or social tipping points are passed. This leads to a fundamental challenge to a classic CBA not only functionally (How to address the non-linearity?) but also fundamentally (How to it integrate 'infinite values'?) (Pascual *et al.*, 2010).

Overall, if the cost-benefit assessment is reduced to a stream of costs and benefits in a spreadsheet (or model) and only the monetary values are looked at (i.e. as in Figures 12.4 and 12.5) and/or if the assumptions used are oversimplifications, then there is a significant risk that the aggregate figures underrepresent the rich reality of any site. It can still be quite valuable in illustrating the implications of the choices, but it represents only a partial view. In this case a relatively simple, yet accurate assessment of costs and benefits, including useful monetary, quantitative and qualitative indicators, can provide a better aggregate picture.

12.3 Considering the added value of protected area designation and management

Designating a site as a PA can lead to an added value to the overall site through a signalling effect, as it makes explicit that the area is of natural value and is being (adequately) managed. This may support increased recreation and tourism (see Chapter 8), help attract funding, and increase site management and investments. Some of the benefits associated with the area will of course be maintained even without designation, apart from extreme cases of full conversion of the area to accommodate urban sprawl.

The specific added value of PA designation will depend on a site's ecosystems, its connection to social and economic systems, the pressures on the ecosystem and the way these are affected by the designation and other associated measures, including management and investment decisions. It will also depend on the supply/demand balance for services in the region in light of other PAs (see Chapter 13), wider green infrastructure, and man-made alternatives.

Figure 12.7 presents a schematic for the main components of PA designation over time. The specific added value of a PA designation is represented by two elements: the value of additional measures (restorations, management, etc.) and the benefits of avoided degradation. This gives the incremental benefits. The total value of benefits from the site also includes those benefits that would still remain in the absence of a PA. Most existing studies in practice focus on all three of these benefits together, as they look at the total benefits and socio-economic values of PAs. Some studies, however, focus on incremental changes, particularly if they are assessing the costs and benefit of specific measures.

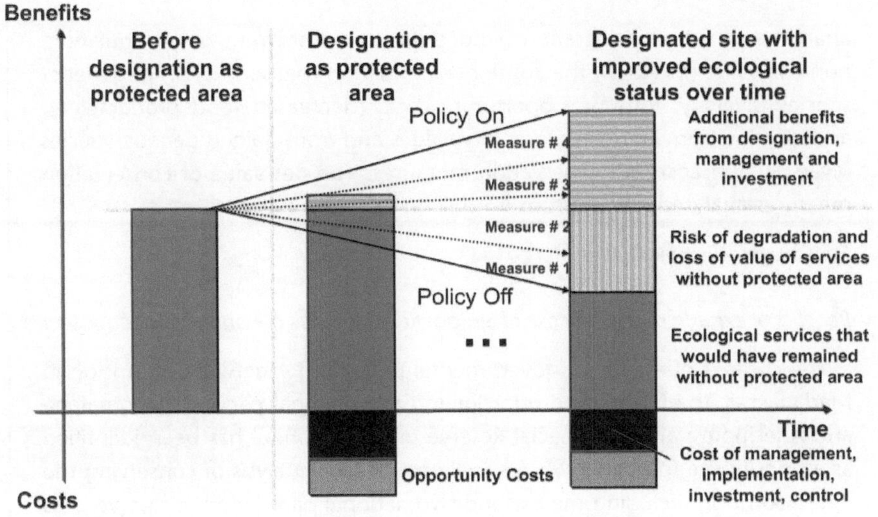

FIGURE 12.7 Incremental and total benefits from PAs (an illustration). 'Policy on' scenario can include a range of different management measures, including restoration

Source: own presentation based on Kettunen *et al.* (2011)

The figure also presents the costs and the opportunity costs. The total new benefits of the area would be equal to the full set of benefits minus the cost (presented as a yearly value or as net present values using different discount rates, as noted above).

The incremental value of PA designation is equal to the incremental benefits from designation, management and investment and the avoided losses from degradation (i.e. excluding the benefits that would accrue anyway) minus the cost associated to the functioning of the PA. The total value of the benefits from the services flowing from the site, (including those that would have accrued anyway) is also a useful value to calculate and use to present the net benefits of the PA (see Chapter 15 on how to present the results). Below are some examples of valuations focusing on the incremental benefits of policy action (i.e. benefits relating to a policy decision) and of specific measures.

BOX 12.3

Case examples of incremental costs or benefits assessments

Net benefits of conservation measures – a case of Kalkense Meersen PA, Belgium

A cost-benefit analysis in Kalkense Meersen Natura 2000 site explored the net benefit associated to conservation measures aimed at restoring the original river landscape, by means of wetlands and estuarine habitats restoration. The analysis took into consideration avoided damage costs from water regulation, non-use value (based on the contingent valuation method, CVM) of genetic/ species diversity, forgone opportunity costs (decreased food productivity), added value from landscape/amenity values and from gained genetic/species diversity, and costs of conservation measures. The net value of conservation measures analysed ranged from €5 million to €19.4 million.

Source: Arcadis Belgium *et al.* (2011)

Benefits of expanding PA – a case of Special Reserve of Pic d'Ivohibe in Madagascar

In the context of a general governmental policy of PA network expansion in Madagascar, the forested corridor joining the national parks of Ranomafana and Andringitra and the Special Reserve of Pic d'Ivohibe has been identified as a priority site for a new PA. A social cost-benefit analysis of conserving the entire corridor, including the existing two national parks and the reserve, was carried out in terms of impact on local livelihoods, timber production, irrigation, carbon sequestration, ecotourism, and international non-use values of biodiversity. The global net present value of conserving the corridor was estimated as large and positive, with a mid-range estimate of over US$330 million (range of US$22 million to US$642 million). The mid-range net present value of costs to forest frontier communities was estimated at over US$1,400 per average household (with a range of US$196 to US$2,610). These costs are distributed unequally among communities and stakeholders. At the national level, the conservation of the corridor has a slightly positive net present value of US$13 million, but given the range of uncertainties (+/– 83 million), this value is rather insignificant.

Source: Hockley and Razafindralambo (2006)

Achieving benefits by designation, participation and benefits sharing – a case of Tubbataha Reefs Marine Protected Area

The Tubbataha Reefs are one of the largest true coral formations in the Philippines, lying at the centre of the Sulu Sea. In the late 1980s intensification of fishing and the use of destructive fishing methods threatened the reefs, which host a multitude of species and genetic diversity. In response to the growing threats, the Tubbataha Reef National Marine Park (TRNMP) was declared in 1988. However, designating the area was insufficient to solve the problem, as interests were divided between those pushing for a fishing ban within the park and those claiming their rights to extract resources in the park. In 1999 agreement was reached on a transition solution to improve the balance of costs and benefits of conservation between the stakeholders. A user fee system for divers was introduced in 2000 and the collected entrance fees were distributed, including to cover compensation payments to local fishermen for their lost access to the park. Fishermen agreed to respect the no-take zone.

Subsequent local monitoring of biophysical indicators showed Tubbataha had a higher fish biomass level than other offshore reefs. Also, fish biomass in the nearby reef Jessie Beazly has doubled since 2000. This, to a large extent, could be attributed to its proximity to Tubbataha (Dygico, 2006). Reef health, fish biomass and densities also improved or stabilised. Live coral cover stabilised at 40 per cent from 1999 to 2003 and reached 50 per cent in 2004.

Fish catches by fishermen near the marine PAs increased from 10kg/day to 15-20kg/day for the period 1999-2004 (Todd and Nunez, 2004) and there was a considerable increase in living standards from 2000 to 2004 in Cagayancillo (Tongson and Cola, 2007).

Source: Sabater and Ledesma, 2004

References

Arcadis Belgium, EFTEC and ECNC (2011) *Recognizing Natura 2000 benefits and demonstrating the Economic benefits of conservation measures - Development of a Tool for Valuing Conservation Measures*, report for the European Commission, http://ec.europa.eu/environment/nature/natura2000/financing/docs/Recognizing_Natura2000_benefits.pdf, accessed 26 December 2012

Dygico, M. (2006) *Tubbataha reefs: A marine protected area that works.* WWF-Philippines, Quezon City, Philippines

Emerton, L. and Kekulandala, L. D. C. B. (2003) *Assessment of the Economic Value of Muthurajawela Wetland. Occ. Pap. IUCN, Sri Lanka*, http://www.iucn.org/about/work/programmes/water/resources/wp_resources_publications/wp_resources_economics/?1154/Assessment-of-the-economic-value-of-Muthurajawela-wetland, accessed 22 December 2012

Gowdy, J., Howarth, R. B. and Tisdel, C. (2010) 'Discounting, ethics, and options for maintaining biodiversity and ecosystem integrity', in P. Kumar (ed.) *The Economics of Ecosystems and Biodiversity (TEEB) Ecological and Economic Foundations*, Earthscan, London and Washington, DC

Hein, L. (2011) 'Economic benefits generated by Protected Areas: the case of the Hoge Veluwe forest, the Netherlands', *Ecology and Society*, 16(2): 13

Hockley, N. K. and Razafindralambo, R. (2006) *A Social Cost-Benefit Analysis of Conserving the Ranomafana-Andringitra-Pic d'Ivohibe Corridor in Madagascar*, University of Wales, Bangor and Conservation International, Madagascar, http://pdf.usaid.gov/pdf_docs/PNADI193.pdf, accessed 26 December 2012

Kettunen, M., Berghöfer, A., Brunner, A., Conner, N., Dudley, N., Ervin, J., Gidda, S. B., Mulongoy, K. J., Pabon, L. and Vakrou, A. (2011) 'Recognising the value of protected areas', in P. ten Brink (ed.) *The Economics of Ecosystems and Biodiversity in National and International Policy Making*, Earthscan, London and Washington, DC

Pascual, U., Gómez-Baggethun, E., Brander, L., Martín-López, B., Muradian, R. and Verma, M. (2010) *The economics of valuing ecosystem services and biodiversity*, in P. Kumar (ed.) *The Economics of Ecosystems and Biodiversity (TEEB) Ecological and Economic Foundations*, Earthscan, London and Washington, DC

Sabater, M. and Ledesma, M. (2004*) Project monitoring report, World Wide Fund for Nature Philippines* (WWF), Quezon City, Philippines

ten Brink, P., Badura, T., Bassi, S., Daly, E., Dickie, I., Ding, H., Gantioler, S., Gerdes, H., Hart, K., Kettunen, M., Lago, M., Lang, S., Markandya, A., Mazza, L., Nunes, P. A. L. D., Pieterse, M., Rayment, M. and Tinch, R. (2011) *Estimating the Overall Economic Value of the Benefits provided by the Natura 2000 Network*, final report to the European Commission, Institute for European Environmental Policy/GHK/Ecologic, Brussels, http://www.ieep.eu/publications/2012/06/estimating-the-overall-economic-value-of-the-benefits-provided-by-the-natura-2000-network, accessed 12 December 2012

Todd, D. and Nunez, E. (2004) *GEFME study of the nature and role of local benefits in GEF program areas: The case of Tubbataha Reef National Marine Park, Sulu Sea, Philippines*, Working Document, http://www.thegef.org/gef/sites/thegef.org/files/documents/Local_Benefits-Case_Study-Philippines-Biodiv.pdf, accessed 26 December 2012

Tongson, E. and Cola, R. (2007) 'Negotiating stakeholder agreements for conservation: The case of Tubbataha Reefs, Philippines', *Science Diliman*, 19(1): 47–63

van Beukering, P. J. H., Cesar, H. S. J and Janssen, M. A. (2002) 'Economic valuation of the Leuser Ecosystem in Sumatra', *ASEAN Biodiversity*, 2(2):17–24

van Beukering, P. J. H., Cesar, H. S. J. and Janssen, M. A. (2003) 'Economic valuation of the Leuser National Park on Sumatra, Indonesia', *Ecological Economics*, 44: 43-62

White, S., ten Brink, P., Hayashi, K., Liekens, I., Ninan, K., Meire, P., Shine, C., Simmons, B., Tinch, R. and Wielgus, J. (2011) 'Recognising the value of biodiversity: new approaches to policy assessment', in P. ten Brink (ed.) *The Economics of Ecosystems and Biodiversity in National and International Policy Making*, Earthscan, London and Washington, DC

ANNEX 6

Calculating net present values (NPVs) and the effect of different discount rates

ten Brink, P.

Discounting is the practice of attaching a lower weight to values (like costs and benefits) that arise in the future. For example, a discount rate of 4 per cent means that €100 this year is worth €104 next year and €100 next year is worth €96 in today's terms ('present value'). This is due to both practical and psychological reasons: money can be invested to make a return in future years and people are 'impatient' and typically prefer consumption (or benefits) today rather than tomorrow. A high discount rate would suggest that there are either high returns to be made (or perceived as available) or high preference for short term consumption.

The reduction in the value such as in the above example may not appear significant as it focuses on the short term, but a 4 per cent discount rate applied over 50 years would lead to €100 in 50 years' time being valued only as one-seventh of that in net present value (NPV) terms. A 10 per cent discount rate would lead to it being less than 1 per cent of the present value. Here the future is effectively absent in the calculations. On the other hand applying a 1 per cent discount rate leads to the value in 50 years' time being worth around 60 per cent of todays' value, so there is less erosion of value over time. See Figure A6.1 for the effect of different discount rate scenarios over time.

The choice of discount rate is usually influenced by national and sectoral contexts. It can also build on 'normal practice' and/or national guidelines. Similarly, the choice of timescale used in the analysis can vary significantly. Some take a 20-year timescale, others longer. Both choices of discount rate and timescale fundamentally affect the results of any net present value analysis.

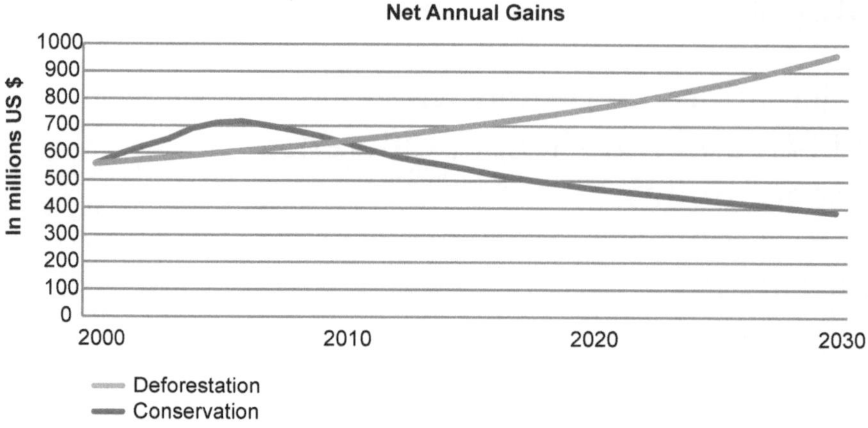

FIGURE A6.1 Impact of different discount rates on future values when discounted to the present (based on $US10 million flow of benefits per year)

Source: own presentation

Examples of discount rates and timescales used in practice to estimate the economic value of ecosystem services and related goods are outlined below. Note that not all of these examples are related to PAs.

- A study estimating the benefits of the conservation of Finnish old-growth forests (Kniivilä *et al.*, 2002) used both 3 per cent and 5 per cent discount rates, leading to significant differences in values. Net present values of the benefits from conservation of old-growth forests were €186 million ($US175 million) at 3 per cent discounting, and €106 million ($US100 million) at 5 per cent.
- A 4 per cent discount rate and 30 year timescale was used in the Leuser National Park case in Indonesia (van Beukering, 2002, 2003).
- The assessment of the benefits and costs linked to the restoration of Skjern River in Denmark used 3, 5 and 7 per cent discount rates (Dubgaard, 2004). The net present value was significantly positive at 3 per cent (DKK228 million or US$36 million), still positive although lower at 5 per cent (DKK67 million or US$8 million) and around zero at 7 per cent (Dubgaard, 2004 in White *et al.*, 2011)
- A 12 per cent discount rate was used to assess the benefits of non-timber forest products to tribal communities in the Rajiv Gandhi National Park (Ninan, 2007; Ninan *et al.*, 2007a, 2007b).

Net present values can be used to aggregate the estimated value of ecosystem services and related goods over time, deriving a single number for the benefits over time. For a comprehensive picture, it is recommended to calculate net present values for single services, complemented by an overall net present value that combines the values of different services and/or goods into a single aggregate. These calculations can best be done using a range of discount rates, for example using high discount factors based on financial markets (interest rates used on treasury bonds, bank transactions, etc.) and lower discount rates which would allow future generations to be included in the calculation. This is both an issue of ethics and maths, as a lower discount rate would allow longer term benefits (i.e. those to future generations) to be better represented in the calculation.

Applying a range of discount rates helps to illustrate the importance of timescale in the calculations and the importance of assumptions in the analysis. This can help inform discussions of both the meaning of the results (i.e. clarifying what part of a benefits or cost stream has a major impact on the analysis) and robustness of the results (as sensitivities to different factors can be explored).

While the traditional use of discount rates has been to use a static figure (such as 4 per cent over the assessment timescale), there is an increasing call for the use of discount rates that decline over time (a 'hyperbolic discount rate'), to ensure that the future is not underrepresented or disregarded in the analysis. For example, a discount rate could gradually decrease from 4 per cent to zero over time, or tend to a fixed rate like 1 per cent. For a more detailed discussion see for example Chapter 6 of TEEB Ecological and Economic Foundations (Gowdy *et al.*, 2010).

BOX A6.1

Calculating net present values

To calculate a net present value the following steps are needed:

Step 1: Estimate the availability of benefits and/or costs over time by choosing 1) the costs and benefits necessary to the assessment and 2) the most relevant timescale relevant for the analysis.

Step 2: It is best to use values in constant terms (i.e. stripping out inflation). Alternatively one can have everything explicit in the calculations, using real prices (i.e. what would be seen in the market), then add in inflation projections, and then convert the real prices to constant prices by stripping out the inflation component.

Step 3: Choose the discount rate(s) to be used in the analysis, ensuring that 1) they will not lead to key issues being underrepresented; and 2) they are understandable and acceptable. Ideally ensure a sufficiently wide range, including both the discount generally used in the country and sector (this helps acceptability of the results) and lower discount rates that allow future costs and benefits to be represented in the analysis.

Step 4: Apply the discount rate to the costs and benefits – this creates a discounted flow of costs and benefits. Today's value of a benefit in five years' time is:

$$\frac{B_5}{(1 + d)^5}$$

Where B_5 is the value of the benefit at year 5, and d is the discount rate.

Step 5: Sum up the calculated annual discounted values for different years to derive net present value for the benefits at the end of the timescale.

To illustrate the effect of net present value rates of 10, 4, 1 and 0 per cent (the latter being the undiscounted cash flow), they have been applied to an illustrative case where new designation is accompanied by new investment over the short term (e.g. restoration costs) which leads to increasing benefits over the long term. The timescale considered is 50 years. Whether a decision is seen as offering net benefits overall will often depend on the discount rate (see Table A6.1).

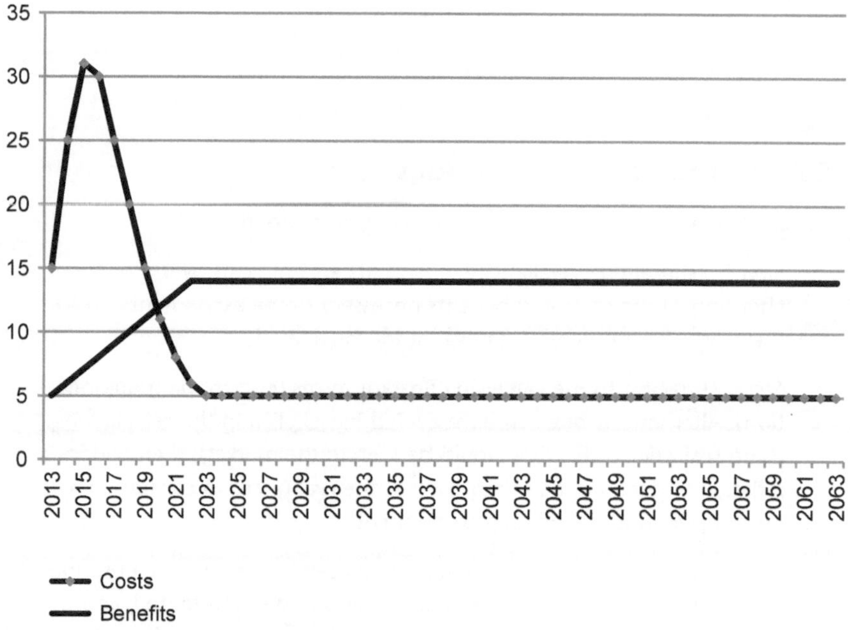

FIGURE A6.2 Illustrative flow of costs and benfits of a protected area, with investment following designation

Source: own presentation

TABLE A6.1 Net present value of costs and benefits over 50-year period

Discount rate	*10%*	*4%*	*1%*	*0%*
Costs	157	233	333	391
Benefits	117	274	519	669
Net benefits	-39	41	186	278

Source: own presentation

References

Dubgaard, A. (2004) 'Cost-benefit analysis of wetland restoration', *Journal of Water and Land Development*, 8: 87–102

Gowdy, J., Howarth, R. B. and Tisdel, C. (2010) 'Discounting, ethics, and options for maintaining biodiversity and ecosystem integrity', in P. Kumar (ed.) *The Economics of Ecosystems and Biodiversity: Ecological and Economic Foundations*, Earthscan, London

Kniivilä, M., Ovaskainen, V. and Saastamoinen, O. (2002) 'Costs and benefits of forest conservation: regional and local comparisons in Eastern Finland', *Journal of Forest Economics* 8: 131–150

Ninan, K. N. (2007) *The Economics of Biodiversity Conservation: Valuation in Tropical Forest Ecosystems*, Earthscan, London

Ninan, K. N., Jyothis, S., Babu, P. and Ramakrishnappa, V. (2007a) 'Nagarhole – The Context of Tribal Villages Located Within and Near a National Park', in K. N. Ninan (ed.) *The Economics of Biodiversity Conservation: Valuation in Tropical Forest Ecosystems*, Earthscan, London

Ninan, K. N., Jyothis, S., Babu, P. and Ramakrishnappa, V. (2007b) 'Uttar Kannada: The context of agricultural cum pastoral villages located within and near a wildlife sanctuary', in K. N. Ninan (ed.) *The Economics of Biodiversity Conservation: Valuation in Tropical Forest Ecosystems*, Earthscan, London

van Beukering, P. J. H., Cesar, H. S. J. and Janssen, M. A. (2002) 'Economic valuation of the Leuser Ecosystem in Sumatra. In: Conservation Dividends?', *ASEAN Biodiversity*, 2(2): 17-24

van Beukering, P. J. H., Cesar, H. S. J. and Janssen, M. A. (2003) 'Economic valuation of the Leuser National Park on Sumatra, Indonesia', *Ecological Economics*, 44: 43-62

White, S., ten Brink, P., Hayashi, K., Liekens, I., Ninan, K., Meire, P., Shine, C., Simmons, B., Tinch, R. and Wielgus, J. (2011) 'Recognising the value of biodiversity: new approaches to policy assessment', in P. ten Brink (ed.) *The Economics of Ecosystems and Biodiversity in National and International Policy Making*, Earthscan, London and Washington, DC

13

ASSESSING NET BENEFITS

Multiple sites

ten Brink, P. and Kettunen, M.

Chapter 12 presented issues relating to developing an overall picture of the values and costs of PAs at the site level, including the range of challenges related to developing such assessments and how to address them. This chapter focuses on aggregating benefits across a range of sites for a region or country, focusing on issues such as data availability and the choice of method to develop multi-site aggregates. Furthermore, the aggregation of benefits across multiple sites needs to consider the complementarity and/or competition between benefits across sites. This chapter explores the above issues offering guidance on what the challenges to developing multi-site aggregates are (Section 13.1) and how to develop aggregate values in practice with case examples to illustrate practical solutions (Section 13.2). The chapter is complemented by Chapter 14 on how to interpret the results.

The issue of aggregation of costs across sites, while not easy, is somewhat less complicated than the aggregation of benefits. Consequently, this chapter will focus primarily on the benefit aggregates. However, a case example is also given on cost aggregation using different approaches of scaling up. Finally, given the scope and target audience for this book, developing aggregate estimates on global benefits and/or costs falls outside the scope of the chapter.

13.1 Issues to reflect when aggregating values for multiple sites

As illustrated in Chapter 12, there are a number of challenges in deriving an estimate for the value of PAs at site level while ensuring that the estimate captures all relevant benefits and beneficiaries, takes into account changes over time, and provides robust results. For developing an aggregate picture across a range of sites further challenges have to be considered.

Two approaches can be used to derive aggregates for multiple sites: 1) a bottom-up approach that scales up the estimated value of ecosystem services and related goods from a set of individual sites to a wider set of sites, and 2) a top-down approach that builds on detailed, region-wide and spatially explicit geophysical data to derive aggregate values. The bottom-up approach is the most often used, building generally on scaling up per hectare values from a set of sites for which existing data is available and extrapolating to the wider study area. Both methods still suffer from data availability (lack of data in suitable form and/or with suitable geographic coverage, etc.) and require a number of assumptions to be used as a basis for aggregation. Section 13.2 below provides more detailed information on the application of the bottom-up approach. The top-down geophysical approach has already proven valuable for studies on carbon storage and increasing efforts at spatial mapping should lead to further promising applications in the future. However, due to its rather technical nature and yet limited application it is considered to fall outside the scope of this book.

Deriving multi-site aggregates for the current or past benefits is not easy as it requires deep understanding of original data (methods used, representativeness of actual values, possible interlinkages with other sites, etc.). It is yet more difficult if projecting into the future, for example, when looking at how increasing funding for management will lead to changes in the level of benefits. While a subset of benefits can be added across sites and/or scaled up directly, not all benefits that accrue on a site level will lead to a simple 'summed-up' total benefit at the level of multiple sites. There can be complementarity and/or competition between benefits associated with different ecosystem services and related goods (Table 13.1).

To address this challenge, it is vital to understand the spatial interrelations between the provision of ecosystem services, related goods and their beneficiaries. It is also important to understand the relationship between supply and demand to inform consideration on the marginal values of benefits and consequently the values on a site and values of the aggregate for the region. When the benefits from different sites are complementary (non-rival) it is possible to simply sum up the estimated value of current and future benefits across multiple sites. In other cases the assessment of benefits from multiple sites will be a type of zero-sum game where foreseen site-specific increases in benefits cannot be added across several sites in a region. In yet other cases markets and/or demand for a benefit might be growing, leading to (partial) complementary effects across sites.

For example, the carbon stored or sequestered in one site is not affected by how much is stored or sequestered in another and the values of the carbon stored in different sites can be added up to form an aggregate value for multiple sites. This is also because the change in the supply of carbon sequestration and/or storage does not generally affect the marginal price of carbon. The same general rule applies for most regulating services, assuming that there is enough demand. However, as soon as the supply of benefits is greater than the demand there can be competition and price effects. For example, if a number of PAs support water supply then the marginal unit value of the water-related ecosystem services (and price of water

TABLE 13.1 General socio-economic linkages between benefits provided by multiple PAs

Benefit	*Socio-economic value in the case of multiple sites*	*Explanation*
Provisioning services and related goods	Possibly competitive	In economic terms, increased availability of a resource generally diminishes its relative value. On the other hand, increased availability helps to guarantee that a resource can be used in a sustainable manner. Development and branding of 'PA friendly' products might diminish the market share of similar products produced on other sites in the same region. On the other hand, having a critical mass of products from across sites in the same region can help increase overall brand recognition and demand.
Regulating services and related goods	Likely to be complementary, with few possible exceptions	Regulating services and related goods (pollination, regulation of water quality and flow, mitigation of natural hazards) are relatively fixed spatially (i.e. dependent on the ecological characteristics and hydrology of the site). Therefore, they also have a spatially fixed set of beneficiaries and are less likely to suffer from competition with similar services at other sites. However, if there is more capacity than needed for the demand (for example enough wetland area to maintain water quality or mitigate flooding) then the marginal value of an additional hectare of land providing regulating services can go down. As for climate change mitigation, any local addition to carbon stock and/or sequestration complements the overall national and global carbon stock.
Cultural: recreation and tourism	Complementary and/or competitive	Encouraging ecotourism and recreation at one site may reduce visitors and the value of ecotourism at a neighbouring site. On the other hand, increasing tourism and recreational opportunities at multiple sites within the same region might attract more tourism to the area as a whole (seen as an opportunity to enjoy several sites during one visit).

Benefit	*Socio-economic value in the case of multiple sites*	*Explanation*
Cultural: broader cultural benefits	Complementary and/or competitive	As above
Broader local/ regional socio-economic benefits	Possibly competitive and/or complementary	Multiple sites might face competition in terms of available investments. However, multiple PAs might also be a factor drawing in further investment, i.e. being an advantage for a region.
Long-term resilience/ insurance	Complementary	Multiple PAs increase biodiversity and the quality of ecosystems at the regional level, increasing overall resilience.

Source: own presentation, based on Kettunen *et al.* (2009) and Gantioler *et al*: (2010)

provided) will be lower. On the other hand, in some cases the interactions between sites – and resulting complementarity – might mean that multiple sites are required for the benefits to be realised. For example, more than one PA might be needed to effectively control flooding and natural hazards in the area. Similarly, the viability of natural enemies for pests (pest control) might require the availability of multiple PAs supporting a range of different habitats (e.g. top predators, migrating species).

Furthermore, the development and branding of products linked to PAs may diminish the market share of similar products produced on other sites in the same region. Similarly, the availability of genetic material, species and/or habitat used to inspire biomimicry or develop new pharmaceutical products on one site might make any similar benefits on other sites redundant. Encouraging ecotourism and recreation at one site may lead to reducing visitors and the value of ecotourism at a neighbouring site. On the other hand, increasing tourism and recreational opportunities at multiple sites within the same region might also attract more tourism to the area as it would be seen as an opportunity to enjoy several sites during one visit. Furthermore, brand recognition and increased sales for one product may also lead to increased interest in other similar products (i.e. away from non-local environmentally branded products) and an overall growing market.

Given the above, one should not take for granted that the socio-economic importance of, and obtained values for, different benefits can be simply added up across sites. The more sites there are in the area the greater is the risk of individual sites 'competing' over the provisioning of some ecosystem services and related goods. In general, when the availability of a certain potential benefit is in excess of demand then one site can affect the socio-economic value of benefits on another site. As a rule of thumb, if the 'demand for' a service or benefit is larger than the potential supply from multiple sites then it is possible to add up the values of the benefits across sites.

13.2 Assessing aggregated benefits of multiple sites by scaling up

A fully accurate aggregated picture of the socio-economic value of multiple sites should ideally be based on site-specific information across the sites in the study area. Unfortunately, however, there are generally few studies available and obtaining the required and/or similar level of information for all sites is often beyond the time and resources available. Consequently, it is often more feasible to consider basing the assessment on already existing estimates from a similar site (or group of sites) and using these estimates as a basis for scaling up and extrapolating benefits to different geographic locations (see Chapter 3 for value transfer). In many cases applying value transfer methods can give values for other sites that are arguably defensible estimates that help illustrate and communicate the value of a site (White *et al.*, 2011). However, due consideration should be given to ensure that value estimates from one site can be used as a proxy for the same benefit on another site.

The robustness of scaling-up outputs reflects the robustness of the inputs (see Chapters 3 and 14). Any scaling-up aggregate that builds on existing studies (via value transfer) will need to check the reliability and robustness of the benefits estimates employed and to ensure as far as possible that the scaled-up values are not compromised due to the use of unreliable or flawed estimates of individual benefits. While using peer-reviewed journal articles as a basis for scaling up reduces the risk for errors it is not guaranteed that the values used are robust and policy relevant. In addition, the availability of academic papers is still rather limited. The best feasible solution is therefore to exercise judgement regarding the choice of estimates used as a basis for scaling up and ideally also explain the choices. All sources used need to be clearly referenced, in order to provide a transparent assessment and data trail.

It is important to underline that if the existing information base is insufficient to carry out a robust scaling-up valuation then only an experimental valuation should be carried out, to help to identify the data/information needs for a subsequent generation of assessments (Box 13.2 below).

Value transfer and scaling up gives rise to some important methodological challenges, such as how to integrate extraordinarily high benefit estimates for particular sites and how to fill in gaps in available evidence. As regards the former, it is common practice to not make use of 'outliers', i.e. studies that have very high values that are considered one-off (not typical) and whose inclusion could unduly influence the results. In the case of outliers it is important to state that they have not been taken into account and the reason behind the choice (see also discussion on robustness in Chapter 14).

An ideal approach for filling the gaps and scaling up is to do a thorough meta-analysis (an analysis of many similar types of study to determine relationships) that helps to develop a (statistically significant) benefits production function, i.e. an equation that describes the level of derived benefit and its relationship to a range of drivers of value. This benefits production function should not be confused with the production function presented in earlier Chapters (Chapter 3). A benefits production function presents in mathematical form the relationship between benefits and factors driving the benefits (explanatory variables) that can then be correlated to the other PAs in the study region and then aggregated (Box 13.1). The relationship, developed on the basis of a set of sites for which there are studies, can then be used to estimate the values for those sites not yet covered but for which good data exists on the drivers of value. While the application of benefit production functions is foreseen to be an important tool in the future only a few studies are yet available (Box 13.1) (de Groot *et al.*, 2012).

Consequently, a more frequently used and pragmatic approach to scaling up is to build on calculated average (or median) per hectare values (Box 13.2). This is far from perfect as while the value of sites are related proportionally to area for some services (many provisioning and regulating services such as food provision, water supply, carbon storage, etc.) for others the relationship is not linear (cultural services such as scientific knowledge, tourism, etc.). Per hectare values are applied to the area under study, to estimate the overall benefits of multiple sites as a whole. Based on ten Brink *et al.* (2011) the different methodological approaches for up-scaling per hectare benefits include:

- Using mean value of benefits per hectare: taking a simple mean from available benefits estimates and up-scaling
- Using mean value of benefits per hectare, excluding outliers: omitting exceptionally high benefits estimates for individual sites, which are likely to be unrepresentative and will unduly skew the benefits estimates if included
- Using median value of benefits per hectare: where several benefits estimates are available, the median may provide a more representative value than the mean, particularly if the latter is biased by one or more exceptional estimates
- Using GDP adjusted benefits per hectare: as differences in national income per head can be expected to influence the value of benefits, and many of the available estimates relate to the more affluent countries, adjusting values by GDP per capita may provide more representative estimates. Care is needed where beneficiaries are from a mix of geographic scales)
- Using stratifying benefits estimates: there are good reasons to expect that the available estimates will not relate to a representative cross section of PAs. For example, studies may have tended to focus on sites which are more valuable or which deliver particular benefits that are more easily quantified. This issue could potentially be addressed by grouping existing estimates according to different site characteristics, such as by habitat or according to whether sites can be expected to deliver benefits of low, medium or high value because of their location, characteristics and profile.

None of the above per hectare approaches are perfect but are able to provide an illustrative value that can help communicate the importance of multiple PAs in a regional and/or national context (see Chapter 14 on interpreting the results).

BOX 13.1

Benefit production functions: a future way forward for assessing benefits of multiple sites

An increasing number of comparable, site-based benefits valuations for PAs allows a 'meta-analysis' to be carried out, forming a basis for deriving a 'benefits production function' that shows the relationship between the estimated socio-economic value and site variables affecting the value (see bullet list below). The benefits production function could then be used as a 'value transfer function' to extrapolate to other existing PAs in the region for which no site-specific valuation has been done. In practice, separate benefits production functions would need to be developed for terrestrial sites and for marine sites, given their different drivers of value. Separate functions could also be usefully developed for different subcategories of ecosystems, such as done for inland wetlands by de Groot *et al.* (2012). A benefits production function can also adapted from one region to another if the contexts are not too dissimilar.

The wide variation of per hectare values by site, while affected by different methodological approaches used and the scope of the assessment (see Chapters 6 to 10) can also be driven by the potential key factors presented below. These key factors could be used as the basis for developing benefits production functions.

- Habitat type and species, related to their impact on ecosystem functions
- Area, due to its potential impact on the extent of ecosystem services such as carbon capture and storage or flood control
- Conservation status, as a reflection of ecosystem health and linked to resilience
- Uniqueness/rarity, which can influence the scientific value, potential for bioprospecting and tourism/recreation
- Spatial relation to other resources and their abundance, determining the perceived scarcity/abundance of substitutes as well as potential for benefits (see Section 13.1 above)
- Proximity to population and accessibility and population density around sites, as the proximity of beneficiaries can impact the perceived benefits
- Income, linked to ability/willingness to pay
- Prices, to calculate monetary benefits for ecosystem services such as carbon sequestration

A statistically significant benefits production function requires a certain level of site-specific data. While geographic scale impacts the level of studies needed (more needed for larger/more diverse area), a fair basis for analysis would encompass at least 200 high quality, comparable primary valuation studies (around 20 studies per a key factor driving benefits). The required data sources would be less for assessments focusing on only one region or one ecosystem-type in similar climatic and socio-economic settings. As temporal and spatial conditions are important and methods for economic valuation are evolving, some past studies will not be useable in the future and new studies will be needed in the future to allow the development and application of robust benefits production functions.

Source: ten Brink *et al.* (2011)

BOX 13.2

Scaling-up per hectare values to estimate the scale of benefits from multiple sites: estimated benefits of the EU Natura 2000 PA network

In 2011 a first assessment of the benefits provided by the EU Natura 2000 network of protected areas was carried out. The purpose of the assessment was to support the global biodiversity target on increasing the awareness of the values of biodiversity (CBD Aichi Target 1) and also to help to ensure sufficient funding for biodiversity from the EU budget.

The study used three different methods to assess the overall value of the benefits of Natura 2000 sites and to aggregate them across the network:

- An ecosystem service-based approach, i.e. aggregating service by service. For carbon storage monetary aggregate values were derived. For other services composite pictures that relied on quantitative evidence and site-based monetary insights were developed. It is to be noted that only an experimental monetary assessment was made on water quality and no aggregate was attempted for flood control, in both cases due to the site specificity of the values and the lack of sufficient data for a proper analysis.
- A site-based approach, building on data from a number of site-based studies, assessing mean and median per hectare values and scaling up.
- A habitat-based approach where site-based estimates were used to estimate per hectare values for individual habitats, which were then combined with data on extent of habitats at network level, to provide EU wide estimates. This provided a logical basis for up-scaling, as similar habitats are likely to deliver similar types of benefit across the network (although the value of many services varies significantly by location).

The results presented in Table 13.2 below represent the first, indicative estimates for the scale of benefits provided by PAs in Europe. Significantly more data is needed to develop statistically significant analysis (see Chapter 14).

TABLE 13.2 Estimated benefits of EU Natura 2000 network of protected areas

Approach	*Value estimates*	*Assumptions*	*Level of robustness/ usability*
Site based	€223–€314 billion/year	Grossing up from 35 values from 21 studies. GDP adjusted site based.	Best currently possible preliminary indicative value. Use with care; Lots of caveats.
Habitats based	€189–€308 billion/year	Grossing up from 33 study numbers for 7 habitats – coastal, freshwater, heath and sand, grasslands, bogs and mires, forests.	High dependence on studies from few countries.
Carbon sequestration/ storage	Current Stock: €600 billion-€1130 billion for the Natura 2000 network Over next 10 years, there will be an increase of €79-€88 billion in carbon value if ecosystem quality is improved; or €82-€92 billion if there is a 10 per cent increase in forest area	Stock value from carbon storage (living and dead carbon) and CO_2 value range €17.3 and 32/tCO_2 for 2010. Annual sequestration building on sequestration rates. Stock values are gross values.	Relatively robust estimate for the value of the stock of (living) carbon. Result is likely an underestimate of the total value given that sequestration not addressed.

Source: ten Brink *et al.* (2011)

13.3 Estimating the future benefits of multiple sites

Where studies are looking to assess the values of multiple sites in the future a further series of assumptions will need to be applied. The assessor will need to understand what the drivers are to value, how these are likely to evolve in the study context, choose the assumptions and always ensure that the choice and the rationale for the choice (and the data sources) are well documented (see Chapter 14). The following drivers of value should be considered:

- Demography: more people, greater demand, higher value
- Income and education: greater income and education, usually leads to a higher value for certain services
- Site ecological health: improving ecological status (which is likely, with some exceptions, as a result of positive management practices) will lead to greater provision of services
- Climate change: the greater the risk of climate change and actual/likely damage from climate change over time, the higher the carbon price and hence the higher the value of climate regulating services
- Rarity: the existence of rare endemic species/habitats will lead to higher values
- Availability and abundance of similar sites in the region: where there are many sites, the marginal value of a given site will be smaller (given that there are substitutes, see Section 13.1 above); the loss or degradation of sites elsewhere will make the provision of services from the study site more critical, and enhance their value
- Preference and social norms: there may be changes in habits, interests and values that can change the value of benefits over time (e.g. preference towards or away from nature tourism).

It is useful to develop at least two scenarios of values for the future, using different assumptions. One could, for example, have a high and low scenario, with one integrating upper end estimates for population growth, income changes, GDP growth, carbon values, and one integrating lower end estimates. This will lead to a range of values and a reasoned choice is of critical importance to ensure that the value has a chance of representing the future. For example, the value of carbon sequestration can be linked to the marginal avoided damage of climate change, and this marginal value will be a lot higher in a world of 6 degrees C temperature change, than 4 degrees C, than 2 degrees C, the official target (see Annex 4 for further information).

13.4 Comparing costs and benefits at an aggregate scale

Aggregating costs across sites faces similar, though slightly fewer challenges, than aggregation of benefits. Box 13.3 below provides a concrete example of aggregating costs in the context of the EU Natura 2000 network. While assessing what the costs of managing PA networks are in a country is arguably a matter of aggregating the costs of different PAs some challenges need to be taken into consideration when developing multi-site estimates. These include:

- Some available cost estimates are likely to be current costs (a snapshot of what costs are) whereas others might be incremental costs (additional costs related with new designation). Some future cost estimates focus on the costs of achieving good ecological status (costs to achieve the desired outcome) while others focus on actual or planned costs, based on existing budget allocations (real costs).

- A network of sites is more than just a series of PAs and connectivity between sites is required to ensure ecological functioning of the network. Some cost estimates might include management costs related to buffer strips or wider connectivity features.

The methods and challenges for comparing costs and benefits at a site level have already discussed in Chapter 12. The advice given in Chapter 12 also applies in the context of multiple sites, with focus on seeking to compare like with like while being clear and transparent with comparison made.

A number of studies comparing costs and benefits of PAs at national or regional level already exist. While it is not possible to go through the various methods adopted in each of the cases with the space available in this book these examples can offer additional routes for readers to explore. For example, in Scotland, the ecosystems protected by EU Natura 2000 network provide benefits to the Scottish public worth more than three times their associated costs, including direct management and opportunity costs (Jacobs, 2004). In Brazil's Amazon, ecosystem services and related goods from PAs provide national and local benefits worth over 50 per cent more than the return for smallholder farming, including (reduced) ecosystem service flows following conversion (Portela and Rademacher, 2001). The benefits provided by PAs studied in Manaus, Brazil, drew three times more money into the economy at the regional level than extensive cattle ranching, the most likely alternative use for PA land (Amend *et al.*, 2007). In Madagascar, investment in managing the national PA system and providing compensation to local farmers for the opportunity costs of forgone farm expansion would pay for itself and generate an additional return of 50 per cent from tourism revenues, watershed protection and international financial support to biodiversity (Carret and Loyer, 2003). In the EU, the costs of the Natura 2000 network have been estimated to be around €6 billion per year whereas the benefits from the network have been broadly assessed at between €190 and 310 billion per year (Table 13.2 and Box 13.3).

Even though the above comparisons are not exact or precise, the order of magnitude difference is sufficiently clear to make a case for the socio-economic importance of PA networks, creating a valuable communication tool to demonstrate their benefits (see Chapter 15).

BOX 13.3

Aggregated costs of EU Natura 2000 PA network, based on the estimated costs of PA networks at national level

In 2008, the authorities in each EU Member State (27 countries) were asked by the European Commission to complete a questionnaire detailing the costs of implementing and managing the EU-wide Natura 2000 PA network. The questionnaire was accompanied by a written guidance document which provided advice about its completion. Opportunity costs that did not translate into compensation were excluded from the estimates.

The questionnaire asked for information on:

- Size of national networks, in terms of the number and surface area of sites
- One-off costs of implementing national networks, including management, land purchase and infrastructure costs
- Recurrent costs of managing national networks, including management planning and habitat management and monitoring
- Breakdown of costs between different land use types
- Costs of managing new as well as existing sites
- Staff numbers and wage costs involved in managing national networks, including current levels and those anticipated in future.

Three gap-filling estimates were used to account for missing data from two EU Member States:

- Simple area based extrapolation: this assumes that the missing costs are equivalent to the per hectare average for the 25 Member States for which data were available (€63.4/ha)
- Extrapolation based on EU-15 and EU-12 averages: this extrapolates the costs based on the identified average for the 15 established and 12 more recent members of the EU (€69 and €48 per hectare per year respectively)
- Regional based extrapolation: North-Western (NW), Northern (N), Southern (S), and Central (CE) Europe to fill the gaps

The estimated total annual costs for the 25 (of 27) EU Member States for which data were available were €5,101 million per year. The additional costs of the network in the two Member States for which data was missing were estimated at between €378 and €671 million per year, resulting, in an estimated annual total cost for the EU of between €5,479 and €5,772 million per year. Using the average estimates from the three extrapolation methods suggests a total annual cost of managing the Natura 2000 PA network in the EU of €5,658 million per year.

When interpreting the results it is to be noted that the basis for calculating the estimates varied between countries. Most of the input data focused on the current costs or planned costs, while only one country presented an estimate for estimated costs required to reach favourable conservation status of the national network. Furthermore, costs related to the establishment and management of marine sites were absent from the calculations. Consequently, the overall estimated costs are likely to be an underestimate for the financing needed to secure full finalisation and effective management of the network.

Source: Gantioler *et al.* (2010)

References

Amend, M., Gascon, G. and Reid, J. (2007) 'Beneficios economicos locais de areas protegidas na regiao de Manaus, Amazonas' *Megadiversidade*, 3: 60

Carret, J. C. and Loyer, D. (2003) *Madagascar Protected Area Network sustainable financing: economic analysis perspective (Executive summary)*, World Bank, Washington, DC, http://www.conservationfinance.org/upload/library/arquivo20120703174353.pdf, accessed 31 December 2012

Gantioler, S., Rayment, M., Bassi, S., Kettunen, M., McConville, A., Landgrebe, R., Gerdes, H. and ten Brink, P. (2010) *Costs and Socio-Economic Benefits associated with the Natura 2000 Network*, final report to the European Commission, Institute for European Environmental Policy/GHK/Ecologic, http://ec.europa.eu/environment/nature/natura2000/financing/docs/natura2000_costs_benefits.pdf, accessed 31 December 2012

Jacobs (2004) *Environment Group Research Report: An economic assessment of the costs and benefits of Natura 2000 Sites in Scotland*, final report to the Scottish Government, www.scotland.gov.uk/Resource/Doc/47251/0014580.pdf, accessed 31 December 2012

Kettunen, M., Bassi, S., Gantioler, S. and ten Brink, P. (2009) *Assessing Socio-economic Benefits of Natura 2000 – a Toolkit for Practitioners* (September 2009 Edition), Institute for European Environmental Policy (IEEP), Brussels, Belgium, http://ec.europa.eu/environment/nature/natura2000/financing/docs/benefits_toolkit.pdf, accessed 16 January 2013

Portela, R. and Rademacher, I. (2001) 'A dynamic model of patterns of deforestation and their effect on the ability of the Brazilian Amazonia to provide ecosystem services' *Ecological Modelling*, 143:115-146

ten Brink, P., Badura, T., Bassi, S., Daly, E., Dickie, I., Ding, H., Gantioler, S., Gerdes, H., Hart, K., Kettunen, M., Lago, M., Lang, S., Markandya, A., Mazza, L., Nunes, P. A. L. D., Pieterse, M., Rayment, M. and Tinch, R., (2011) *Estimating the Overall Economic Value of the Benefits provided by the Natura 2000 Network*, final report to the European Commission, Institute for European Environmental Policy/GHK/Ecologic, Brussels, http://www.ieep.eu/publications/2012/06/estimating-the-overall-economic-value-of-the-benefits-provided-by-the-natura-2000-network, accessed 12 December 2012

White, S., ten Brink, P., Hayashi, K., Liekens, I., Ninan, K., Meire, P., Shine, C., Simmons, B., Tinch, R. and Wielgus, J. (2011) 'Recognising the value of biodiversity: new approaches to policy assessment', in P. ten Brink (ed.) *The Economics of Ecosystems and Biodiversity in National and International Policy Making*, Earthscan, London and Washington, DC

STEP III

Interpreting, using and communicating the estimates

The purpose of Step III of the assessment guide is to offer guidance on how to interpret and use the results of socio-economic assessments. This step provides general guidance on using the results of valuation(s). In particular, Step III aims to assist in understanding the robustness of value estimates and using the different estimates to form an overall picture of the total socio-economic importance and value of the site (Chapter 14). In addition, advice on communicating the results to different relevant stakeholders and different possibilities for using them to support PA management is provided (Chapter 15).

14

INTERPRETING THE RESULTS OF SOCIO-ECONOMIC ASSESSMENTS

ten Brink, P. and Kettunen, M.

As seen in the previous chapters socio-economic assessments of PAs will provide a mixture of information (monetary, quantitative, qualitative and spatial), derived using a range of data sources and methodologies. This chapter explores what the results of the assessment mean and how they should be interpreted. It also draws attention to how results could be misinterpreted and what those carrying out the assessment might wish to do to minimise this risk.

The results obtained can have different meanings: some benefits have current or potential economic value captured by the markets, whereas others relate to well-being and have different value to different people (Section 14.1). Furthermore, what the results actually tell about the overall benefits associated with a PA depends on the scope of the assessment and the coverage of benefits (Section 14.2). Depending on the methods used and data available, the outcomes of an assessment will also vary in the level of robustness of the results (Section 14.3), confidence in the results (Section 14.4), and whether the final results are fit for purpose given the overall objectives of the assessment (Section 14.5).

14.1 General meaning of the results

An assessment of ecosystem services and related goods provides a picture of the benefits that the economy and society derive from PAs, focusing on the anthropocentric value of sites. Consequently, such an assessment can only ever be a partial picture of the importance of PAs. It is important for any socio-economic assessment to be fundamentally clear that even if it provides a representative picture of the benefits to mankind, it would still not capture the intrinsic value and broader ecological importance of the site.

The results of benefit assessments will present a range of metrics for the importance of PAs. While all these metrics help to capture the socio-economic value of

PAs, the 'meaning' of value will vary depending on the metric and the method. Some benefits will be real economic (market) benefits that can be picked up, for example, in Gross Domestic Product (GDP) accounts and company bottom lines. Other benefits will be real in the sense that they lead to avoided costs. For example, where natural water filtration allows for lower operating costs (lower mechanical filtration needs) this can lead to higher profitability for a company, if there is not a full pass through of the savings to consumers. Higher company profitability will also be felt in GDP statistics.

Some PA benefits are potentially real economic benefits, pending market creation, and hence represent potential assets. For example, carbon storage and sequestration and benefits related to water supply and purification will become concrete economic benefits once a market is set up and transactions start. Payments for ecosystem services (PES) are examples of markets created that turn potential economic benefits into real ones, from the perspective of the economy (Chapter 15; see also ten Brink *et al.*, 2011).

Other benefits are welfare benefits and, while real at a human and social level, will not be picked up by economic metrics such as company profit and loss accounts or GDP. In the above example, if all the benefits of PA-based water purification were to go to consumers, with them paying lower prices for their water due to lower costs of water purification and provision, then there would be only a consumer benefit and no producer benefits. This would result in no direct effect on company bottom lines and GDP. Indirectly, however, higher household disposable income could subsequently have effects on the economy. In reality, cost savings created by natural water purification will benefit both water companies and consumers, with the balance depending on market and regulatory conditions.

Given the differences above, it is essential to be clear as to what type of socio-economic value the assessment relates to. The assessments should differentiate between the values that are real in cash terms, those that are potential real savings or revenues from markets, and those that are real in well-being terms. Care is, of course, needed in communicating such differentiation as some audiences may choose to ignore well-being benefits and focus only on the benefits that lead to tangible economic gains.

The distinction between different values is, however, important as decision-makers and other stakeholders may wish to know whether the identified and assessed benefits of PAs will lead to GDP growth, local and regional development and jobs (see Chapter 10), new markets with economic flows or whether they can help balance budgets and lead to costs savings. Alternatively, some will ask whether, how and to what extent PAs can help with poverty alleviation, creating sustainable livelihoods and improving individual and community well-being. Finally, the conservation community will wish to know the biodiversity value of PAs and how highlighting different socio-economic values of sites can help to safeguard these values (Chapter 15). Answers are needed for each of these questions and those producing the assessments have the responsibility for communicating clearly what the estimated socio-economic values mean in practice.

14.2 'Reading' socio-economic assessments

The answer to the question as to how to interpret different value estimates depends on the objective and scope of the socio-economic assessment (i.e. what questions a study set out to answer) and its purpose and relevance in the decision-making context (i.e. what evidence will it come up with and how does it relate to specific decision questions). Care is needed both in presenting and interpreting the results. The assessors should make it absolutely clear what the study is addressing, what it is not covering, what the methodological caveats are and what the assessed values relate to (Section 13.1 above). In particular, as noted in Chapter 12, regional and/ or national assessments need to be handled with care as any scaling-up exercise introduces an additional level of assumptions and uncertainty affecting the level of precision, robustness and meaning of the results.

It is important to realise that the results of socio-economic assessments might be easily misinterpreted: some readers may look at headlines or take a cursory look at the results, failing to take the whole picture into account. Consequently, the third party should make due efforts to understand what the results mean and, if seeking to build on them, use them correctly. While the assessors cannot be expected to foresee all possible contexts in which the results might be used – and present detailed guides on how to use them – some transparency and reflection will be valuable.

An assessment may, for example, focus on the question of what the incremental local and national benefits are of designating a PA and managing it so that it reaches a desired, good ecological status (see Section 12.3). Alternatively, the study may focus on the aggregate picture and explore what the incremental national benefits are likely to be of completing a well-functioning and coherent PA network (Chapter 13). The results of such assessments will be incremental numbers related to policy action and/or management measures and they will not, therefore, represent the full set of benefits and associated socio-economic values provided by site(s). Similarly, focusing only on national interest and not taking into account global benefits can result in underestimating the full socio-economic value of the ecosystem services and related goods. In general, one should not compare or add up the assessed values for incremental benefits with gross benefits.

While interpreting socio-economic assessments of PAs, it is important also to understand that the limited scope of existing and/or future assessments does not mean that the other benefits are inconsequential or unimportant. As argued throughout this guide, the aim of any socio-economic assessment should be to identify and assess the whole set of socio-economic values of PAs at some level (qualitative, quantitative and monetary) while at the same time not forgetting the intrinsic value of biodiversity. However, existing socio-economic assessments of PAs (monetary assessments in particular) tend to focus on a subset of ecosystem services and related goods, even where they aspire to have a wide coverage. This is inevitable given data constraints and the practical necessity of having a manageable process.

For example, it is easier to look at the benefits of water purification and supply than to assess the benefits related to health, spiritual experiences or potential future utility of genetic resources (bioprospecting). It is also easier to carry out assessments on the basis of current generations identifying local beneficiaries and private sector interests, rather than reflecting the benefits of PAs to future generations, broader communities and/or national and global beneficiaries. For aggregate assessments, it is easier to estimate the socio-economic value of carbon storage and sequestration but for the moment it is nigh impossible to do so for flood control, given the stochastic nature of flooding and limited information about interactions between PAs, wider natural infrastructure and man-made infrastructure.

How people will read the results will depend on their perspectives, the questions they are seeking to address, the decisions they are seeking to take. The different stakeholders and users of the results will each be looking for different types of data and are likely to focus only on part of the picture from their perspective. Assessors should arguably be aware of these needs and perspectives and ensure that the results are presented in a manner that will be both be useful for different audiences and also prevent any misuse of results. Factoring these considerations into the interpretation and presentation of the results will help in the demonstration and communication value of the study (see Chapter 15 for further discussion).

For example, the results of any PA socio-economic assessments can be useful for:

- Funders considering whether to offer funding to cover site management: these readers will seek to understand the importance of the site (biodiversity, ecosystem services and related goods) and the benefits that either already accrue (hence merit protection) or could accrue with due management (hence are worthy of investment);
- Local or regional actors seeking to engender appreciation of PAs in the local and regional socio-economic fabric: these readers seek information on the different beneficiaries, including their spatial distribution and the type and level of benefits accrued;
- Site managers or decision-makers: these readers will be interested in identifying cost-beneficial management activities focusing on specific measures such as restoration for cost-effective carbon storage or investment in infrastructure to encourage recreation and tourism. Decision-makers may also wish to understand the different incentives affecting land use choices, helping to identify the political and economic reality that drives non-compliance with PA regulations (illegal logging, etc.);
- Conservation-minded NGOs and decision-makers: these readers might be seeking to complement arguments of the uniqueness of biodiversity values with insights on the socio-economic values to help create a stronger case to oppose interests in converting PAs to other uses, or arguing for activities to be allowed on site.

14.3 Robustness of estimates

Site-specific estimates

The robustness of site-specific socio-economic values is determined by a range of issues, including the quality of data, scope of assessment, methodologies used and assumptions made. Low quality or out-of-date data can undermine even the most methodologically advanced assessments. As regards assessment scope, choosing too narrow a focus means that some issues will not be covered (see 14.2 above). Finally, assumptions can have major impacts on the outcome of the study – recall, for example, the discussion on discount rate and future generations in Chapter 12 and Annex 6 – and assessors need to be very clear when presenting the results as to the sensitivity of the results to assumptions and what this means.

The following key questions will help to assess the robustness of final results:

- Was the data used obtained from a formally recognised and reliable source and did it reflect reality? If compromises had to be made (filling data gaps based on assumptions, use of value transfer, etc.) as good local data was unavailable, what were the implications on robustness?
- Did the choice of scope lead to a sufficiently comprehensive full picture of benefits associated with the site and were all key stakeholder groups covered (local communities, minorities, future generations, etc.)?
- Were the most suitable and best available methods used and what impacts did the assumptions used in the assessment have on the results, for example, did they lead to any bias that needs to be addressed?

These questions can already usefully be asked up front in doing the assessment (see Chapters 4 and 5), but it is also important to do so after the results have been derived, to enable the assessor to step back and see the values in perspective and subsequently provide a balanced and evidence-based commentary on how robust the results are so as to help readers see the results in context. It is important not to oversell the results and present them together with a summary as to their robustness.

Finally, it is also important to reflect whether the estimated values can be attributed to the site itself or whether they also require the presence of a wider non-protected landscape and/or man-made infrastructure, for example, flood control based on a mixture of PA and wider natural infrastructure, supported by dikes or levees, etc. In the latter case, the interconnections should be clearly presented.

Aggregated multi-site estimates

While the above issues are also relevant to the robustness of aggregated estimates for multiple sites, there are also a range of additional issues that need to be considered, notably what method is used for scaling-up (if a bottom-up approach is used). Therefore, in the case of aggregated values additional care should be taken to present the results in a way that helps readers to fully appreciate the level of robustness of the results (underlying assumptions, level of extrapolation, etc.), what the results mean and what the results are potentially useful for.

As noted in Chapter 13, the robustness of the outputs reflects, inter alia, the robustness of the inputs. It is important that the assessor reflects the quality of the input data, presenting a clear discussion on data sources, their reliability, and appropriateness for the study context and discusses transparently the level of robustness of the derived benefits estimates. If any high profile studies have been dropped from the assessment, it can be helpful to explain why. This will help in the perception of robustness and interpretation of the results and also show that the assessors are not purposefully trying to derive 'big numbers', but rather doing a careful assessment. While it is always best to aim for developing an accurate picture of the benefits, from a perspective of credibility, it might be advisable to err on the side of being conservative as this helps avoid any accusations for vested interests and attempts to intentionally overstate the scale of benefits.

Finally, there is a risk of using outdated data as a basis for aggregation. Care is needed in selecting the base studies and also transparent discussion needed in the assessment as to the choice of base studies and which were included/excluded and why. Future projections incorporating any changes to the status quo will need to explain what changes have been factored into an assessment and what the rational and data sources are behind the choice assumptions (population growth, income changes, GDP growth, carbon values, etc.) Having a careful commentary and data trail, will help readers interpret the results and have confidence in the results.

14.4 Level of confidence

When people talk of the level of confidence in the results, they can either mean it in the general everyday language or, if talking to economists, statisticians and scientists, 'confidence to 95 per cent' or 'confidence to 99 per cent' (i.e. only a 5 per cent or 1 per cent probability of the answer not being right) (e.g. de Groot *et al.*, 2012). Ideally, the authors should make use of any available indicator of confidence, to ensure honest and transparent communication of results to the readers. Similarly, the use of scenarios and sensitivities, backed up with transparent presentation as regards assumptions, methods and data sources can help in creating an understanding as to the level of confidence one can have in the results. The use of peer review processes can also be an invaluable indicator to readers that they can have confidence in the results. An extra level of checking to see whether others agree with the results is good practice. Whether a 'blind copy peer review' process (as in peer-reviewed journals) or a more transparent consultation and review process (useful for general assessments, consultancy reports) is adopted is a choice for those developing the study.

Ideally, the assessors should provide as detailed information as possible on the variables used in the assessment so as to allow others to interpret and build on the results correctly. For example, having information on the number of households, household income and proximity of different population groups to the site provides a valuable evidence base to allow readers to understand the benefits better and eventually project benefits into the future (by adjusting for demographics, household income changes, etc.). On prices, transparent assumptions on carbon prices used can be invaluable in readers being able to interpret whether the carbon sequestration or storage values are high or low, and hence whether the results should be seen as over or underestimates and eventually creating the potential to adjust the results (see Annex 4). As regards a PA's ecological status it is useful to give an indication of the level of conservation and/or degradation of sites.

For aggregate assessments based on scaling up, the results should be generally presented as illustrative estimates to communicate the order of magnitude of the benefits. Only if there is a solid evidence base should assessors suggest that the estimated benefits are more than illustrative. Given the diversity of PAs and their social and economic context, it will be near to impossible to derive high level (e.g. global) bottom-up aggregated assessments that both meet the academic rigour of 95 per cent confidence bounds and fully reflect the issues on the ground. Consequently, it is important not to overstate the case for such valuations.

There are different ways for authors to present the level of confidence in the methods used and results obtained, from using statistical confidence indicators to simply discussing the issue more qualitatively. The following pragmatic approach developed by ten Brink *et al.* (2011) provides a practical way for qualifying the level of confidence in the results and guiding users how to make use of the data:

- Deep green: robust method used with due data that should lead to robust numbers, fine for publication, citation, without need for significant context
- Light green: fairly robust tools leading to illustrative and indicative values, useable with due caveats, transparent presentation of limits and what the numbers mean
- Orange: methods to be used with care, but can lead to illustrative and indicative numbers; results not to be presented out of context of this being a first assessment
- Red: weak, very experimental methods, to explore ways forward, the results are not suitable to be used to support decision-making.

14.5 Wider reflection on the utility of the results

Valuation, whether monetary, quantitative or qualitative, is both useful as a means of increasing understanding on the value of PAs and as a tool to guide decision-making. What is fit for purpose depends on what use will be made of a derived value. In certain circumstances the estimated benefits of conservation and restoration will be in an order of magnitude larger than the costs including value of loss of services from ecosystem conversion. In these cases there is no need for particular precision. If the conclusion is clear that conservation or restoration makes sense, then there is arguably little point (unless the objective is scientific understanding) of paying for a more expensive study to obtain a more precise answer. Similarly, for most 'big policy questions' it generally matters less whether the benefits are four or five times the cost; the key question is whether the benefits are larger than the costs.

However, for some decisions the answer may of course not be clear cut, and if that is the case, further analysis and data may be needed – both biophysical and monetary, to complement the initial analysis. The level of effort should be proportional to the need. When it comes to non-compliance cases, there are merits in the results of the assessment being more precise, but even in these cases, some valuation and some negotiation (i.e. legal and political bargaining) will be needed to obtain a settlement deemed fair and appropriate.

In summary, in any assessment of value it is useful to be clear as to what level of precision is needed for the question the assessment is aiming to illuminate. A robust order of magnitude may be sufficient, or fit for purpose in some cases; however, in marginal cases more precision will be needed. Of course, in all cases effort should be made to be as robust as possible, and to clearly note all data sources, choice of methods, scope and assumptions.

References

de Groot, R., Brander, L., van der Ploeg, S., Costanza, R., Bernard, F, , Braat, L., Christie, M., Crossman, N., Ghermandi, A., Hein, L., Hussain, S., Kumar, P., McVittie, A., Portela, R., Rodriguez, L. C., ten Brink, P. and van Beukering, P. (2012) 'Global estimates of the value of ecosystems and their services in monetary units', *Ecosystem Services*, 1(1):50-61

ten Brink, P., Badura, T., Bassi, S., Daly, E., Dickie, I., Ding, H., Gantioler, S., Gerdes, H., Hart, K., Kettunen, M., Lago, M., Lang, S., Markandya, A., Mazza, L., Nunes, P. A. L. D., Pieterse, M., Rayment, M. and Tinch, R. (2011*) Estimating the Overall Economic Value of the Benefits provided by the Natura 2000 Network*, final report to the European Commission, Institute for European Environmental Policy/GHK/Ecologic, Brussels, http://www.ieep.eu/publications/2012/06/estimating-the-overall-economic-value-of-the-benefits-provided-fby-the-natura-2000-network accessed 12 December 2012

15

USING AND COMMUNICATING THE RESULTS

Sections 15.1–15.3 by Stolton, S. and Dudley, N.
Section 15.4–15.5 by Kettunen, M. and D'Amato, D.

As discussed in Chapter 2, good socio-economic assessments, i.e. assessments based on a range of qualitative, quantitative and monetary evidence, can help to make the case for establishing and appropriately managing PAs. With perhaps the exception of academic studies, assessments should not, however, be considered as an end in themselves: the final success of an assessment should be measured based on the implementation of results. The time and effort put into an assessment can be jeopardised by a lack of planning relating to how results are likely to be used and communicated. Thus, although focusing on these aspects will clearly take place towards the end of an assessment, the uptake and communication of results should be considered throughout the process, ideally including development of a dedicated plan at the start of the assessment process.

A plan for using and communicating the results of an assessment should be structured around a number of pivotal questions including:

1. *What is the reason for doing the assessment (the objectives)?* (see Chapters 4 and 5)
2. *Who is the target audience?* (see Chapters 4 and 5)
3. *What would one wish this audience to learn/understand?*
4. *What information and evidence base helps the audience to form an informed decision or/ and what kind of information is needed to motivate the target audience?*
5. *How can one reach the target audience in an effective way?*
6. *Who will benefit from the assessment?* (see Chapters 4 and 5)
7. *How will these benefits be communicated to the different beneficiaries?*

Elements of this plan are discussed in more detail below.

- *Objectives and targets*: The identified objectives of an assessment might include, for example, aiming to increase government funding for the PA, building on a better understanding of PA benefits; or development of a PA management plan that complements a site's conservation objectives with dedicated measures aimed at maintaining and/or restoring the PA's socio-economic benefits. While these general objectives are useful it is also always beneficial to identify more *targeted goals* to ensure assessment results are taken up by and communicated to the right stakeholders. Broad and inexplicit goals which have no measurable outputs should generally be avoided. Instead goals should be more focused and expressed in a way that is easier to measure and report on. For example, socio-economic assessments supporting the development of PA management plans should include specific outcome oriented targets, such as targets related to the uptake of results by relevant local stakeholders (e.g. development of concrete business ventures). Similarly, the outcomes of raising public awareness should ideally be expressed in terms of targets such as: publishing two articles on PA benefits in local newspapers; including a feature on PA benefits on the PA website; production of leaflets or signage in the PA which explains PA benefits to visitors, etc.

- *Audiences*: Identification of the *target audiences* for communications, preferably during the planning stages of the assessment; along with effective *communications channels* to reach target audiences; and choice of spokespeople and presentations of the assessment results which are of *relevance* to the target audience.
- *Strategy*: Identification of '*windows of opportunity*' for the use and communication of results such as links to specific events, decision-making processes or policy cycles.
- *Resources*: Ensuring appropriate *resources* (personnel and funding) to support the implementation of results (e.g. use of results to inform PA management or business planning) and/or carrying out communications (e.g. presentation of results in relevant committees and working groups, publications or presentations at key events) (Pabon-Zamora *et al.*, 2008).

In the following sections four of the overarching objectives of socio-economic assessments are discussed in more detail: understanding, awareness and advocacy; support to decision-making and management (including identification of appropriate measures and instruments); identifying and addressing social impacts; and mobilising funds. In addition, advice is provided as regards to how to ensure that the outcomes of socio-economic assessments result in synergies between biodiversity conservation and ecosystem services.

15.1 Understanding, awareness and advocacy

Perhaps the most common objective in valuation assessments to date has been to increase the understanding and awareness amongst (target) audiences about ecosystem services and related goods provided and supported by PAs. In addition, assessments often aim to influence policy and resource allocation decisions within political, economic and social systems and institutions. Although the motivation for awareness-raising and advocacy often relates to increasing funds for protection (see Section 15.4) other objectives (see Section 2.6) may be to influence policy/legislative decisions which increase support for PAs (e.g. including PAs in strategies for sustainable development or climate change mitigation) or to increase public support for PAs, particularly in areas where PAs are threatened by competing land-use decisions.

Of all possible assessment 'objectives' awareness-raising and advocacy is the one to be most precise about when developing the assessment. In particular, as noted above, it is important to develop objectives that have measurable targeted goals which allow the implementation of the assessment to be considered. A commonly used set of criteria to develop indicators that help measure the achievement of goals and overall success of a project is contained in the acronym SMART. SMART criteria include that a measure has a *Specific* purpose (i.e. project objective), and is *Measurable* (the process for measuring the achievement of the project objectives should be defined during the assessment planning phase – see Chapters 4 and 5), *Achievable* (this may mean aiming for a series of small 'successes' rather than a vague objective such as changing overall government policy), *Relevant* to the success of the objectives, and *Timely*, i.e. the indicators should be achieved within a predefined and relevant period (this may be quite a time after the actual assessment and projects should ensure that project staff/funding are available to implement the communication strategy and deliver the expected results within a realistic timeframe).

Thus the success in increasing understanding and raising awareness will depend on the quality of the assessment itself and its findings (Chapter 14), but it will also be heavily influenced by the robustness of the plan to use and disseminate the findings and advocate change. As an example, the seven questions outlined above are discussed below for a specific hypothetical awareness-raising example. A case study in Box 15.1 provides an example of how this type of planning can work in reality.

EXAMPLE – Planning for an assessment to identify water resources from PAs with a view to increase support for PAs from parts of government and industry involved in water supply:

1. *Objectives:* developing a better understanding of PA benefits amongst key decision-makers in government and service sectors (particularly water authorities) to increase support for management of the PA, including through better working relationships with government departments and the service sector and by sharing of expertise across sectors.
2. *Target audience:* decision-makers from the service and production sectors (e.g. water and agriculture sectors), any private water companies and the organisations in charge of assigning financial resources (e.g. relevant government departments).
3. *Changed audience perceptions of PAs:* by ensuring that PAs are seen as strategic areas for sustainable development and natural resource management rather than as unproductive areas that cost the government money.

4. *Processes to increase understanding and/or change perceptions:* by explaining the links between conservation and water supply and showing that the water sector depends directly or indirectly on natural resources management, the target audience is encouraged to take positive actions in terms of conservation as a way of maintaining a high quality supply of water, which can also help them increase their prestige, respect, authority and acceptance amongst a wide range of stakeholders.
5. *Contacting target audience:* by ensuring that the right people within the water sector receive capacity-building material; timing the communication strategy so that it does not overlap with other events that will distract the audience (e.g. elections) but is timed when major relevant decisions are likely to be taken (e.g. policy or budget negotiations)
6. *Assessment beneficiaries*: assuming the objective of increased awareness raising is achieved there are likely to be many beneficiaries as a result of the valuation project ranging from PA site managers to local communities who receive more from PA benefits sharing mechanisms: in this case perhaps through payment for ecosystem services schemes. Carrying out a short exercise to identify all the beneficiaries (see Section 4.2) can therefore be an important resource for the communication plan.
7. *Communicating to beneficiaries*: some beneficiaries (e.g. PA managers and staff) may be closely involved in the valuation and thus will not need to be included in the communication strategy; however, in large PAs news of the results might be communicated in staff meetings and newsletters to ensure that the lessons learned from individual site valuations are communicated across PA networks. For other beneficiaries, such as local communities who, for example, might start receiving payments for ecosystems services following acknowledgement from water companies of a PA's role in ensuring municipal drinking water quality, the reasons for undertaking an assessment and the results from the assessment may not be clear; an element of the communication strategy should address this. It is important when working with the private sector or with large public sector service providers that the PA's role in the provision of benefits is fully explained - ideally PA staff and communication departments should work together on the communications strategy.

Communication activities should focus on those actions which are most likely to reach audiences critical for achieving the goals and objectives of the assessment (see previous section). A number of different tools used to build advocacy and develop awareness are outlined in the table below.

TABLE 15.1 Applying the right tools to the right audience for communicating results

Audience	*Interest in PAs*	*Use of the valuation study*	*Communication tools*
Local community	Extractive use, recreational use, harvesting, derived economic benefit (e.g. tourism).	Increase in knowledge about the value of the PA. Demonstrate need for sustainable use of natural resources.	Local outreach, including, community education campaign, community meeting, local news story, local radio. Exchange visits with communities in/near PAs (if specific objectives can be shown to have been achieved in these communities).
Nearby cities and/or towns	Possibly very low interest or even a negative perception of PAs. Lack of awareness of uses and services provided by PA and associated economic benefits.	Increase awareness of PAs' role in the provision of clean water, maintenance of health benefits, disaster mitigation (flood control, avalanche or landslide control, etc.), carbon storage and climate change mitigation, recreation and support for tourism trade, etc.	Local outreach (local news stories in newspapers, radio, TV, etc.), open days, guided tours.
Service sectors (water authorities, tourism boards, etc.)	Possibly low interest with benefits of the PA taken for granted.	Increased understanding that many sectors depend directly or indirectly on natural resources management undertaken by PA.	Site visits, individual meetings and well-focused (innovative) presentations, evidence-based research (well presented summaries with full reports available for technical review).
Local business (fishers, foresters, farmers, hoteliers, etc.)	As above benefits (such as protecting spawning grounds or sources of water) are often taken for granted.	Increased understanding that businesses can depend directly or indirectly on natural resources management undertaken by PA.	Local outreach of evidence-based research including, for example, through presentations at individual meetings, local news stories, site visits focused on specific benefits and research results.

Continued

Audience	*Interest in PAs*	*Use of the valuation study*	*Communication tools*
Politicians and national policy makers	Possibly very low interest. Lack of awareness of uses and services provided and associated economic benefits.	Increase awareness of the economic use of the ecosystem. Describe national and local economic benefits associated with protecting ecosystems and potential costs if ecosystem degrades.	Presentation, maps, policy brief, poll results, site visits, short film, story placement in high profile media.
Local NGOs/ International NGOs	Awareness usually quite high but may not be aware of all benefits.	Provide all parties with the same data on which to reach a consensus about the economic benefits of protected areas.	Policy brief and full report, presentation, side event at regional or international conservation meeting.
Multilateral and bilateral donors	Often focused on specific agendas such as development and climate change adaptation/mitigation	Increase in awareness of the link between PAs, climate change, poverty reduction and social and economic development, etc.	Site visits, policy brief, presentations at high level international meetings, individual meetings, international high profile media, etc.

Source: own presentation, adapted from Pabon-Zamora *et al.* (2008)

BOX 15.1

Funding PAs for their role in provisioning and regulating water: a case study from PAs around Quito, Ecuador

The central principles of the 'payment for environmental services' (PES) approach are that those who provide environmental services should be compensated for doing so and that those who receive the services should pay for their provision.

PES schemes have been particularly successful in making the links between PAs and water. But such schemes will only work if there is clear evidence that the natural ecosystems and conservation management practices in the watershed can help provide a clean, regular supply of water and that downstream service users (e.g. water utility companies, hydropower companies, and other industries) who depend upon these services are willing to pay for their maintenance and persistence (Goldman *et al.*, 2010). Situations where there is a single source of funds (such as a water company) and a single entity to receive funds (such as a village council or a land-owners' cooperative) are the easiest to employ successful PES schemes. Advocacy and awareness raising are a vital component in the development of any PES scheme.

In Ecuador, Quito, about 80 per cent of the population receives their drinking water from three PAs (Reserva Ecológica Cayambe-Coca, Reserva Ecológica Antisana and Cotopaxi National Park). A PES scheme, Fondo para la Protección del Agua (FONAG), enables water users to contribute financially on a voluntary basis to manage these PAs for their role in water services.

Incentives for participation in FONAG vary. FONAG started life in 2000 with US$20,000 from Quito's public water authority, with the aim of avoiding or at least reducing future costs for water treatment and supply by ensuring these functions were provided by the conserved ecosystems, and US$1,000 from the NGO The Nature Conservancy (TNC), whose incentive was the development of long-term financing for the conservation of PAs. An annual commitment of US$45,000 was provided by Quito's power authority, in acknowledgment of the importance of ensuring sufficient water and low levels of sediment flowing into their hydropower grid. The first private contributor was the brewery company Cervecería Nacional who saw that contributing US$6,000 annually to the fund would help secure clean water and minimise production costs; the water bottling company, Tesalia Springs, contributed to the fund for similar reasons. By 2010 the fund had grown to US$10 million, meaning nearly US$1 million annually is now being invested into conservation projects in the PAs and watersheds around Quito (TNC, undated and UNEP, undated).

During the set-up phase of FONAG, lessons learned in effective communication of the value of PAs noted the importance of:

- Using a variety of models and tools (hydrological, ecological, and socio-economic) depending on the target audience to map service flows and clearly demonstrate through scenarios the importance of watershed conservation to the supply of water services.
- Analysing water users businesses and providing users with clear financial details and potential revenue streams associated with watershed protection.
- Carrying out a range of feasibility studies to engage stakeholders, establish finances, set objectives, develop strategies, and prioritise areas for PES investment (Goldman *et al.*, 2010).

Thanks to FONAG, almost 2,000 people are estimated to have received economic benefits associated with watershed management and conservation and over 65,000 ha of Quito's watershed is now under improved management (Canales and Jouravlev, 2012).

15.2 Support to decision-making and management

In most cases, if PA site managers are asked to identify the main values and benefits from their PA they will answer 'biodiversity' *and* 'conservation'; if you ask most advocates of PAs why they support over 12.7 per cent of the world's land and 1.6 per cent (Bertzky *et al.*, 2012) of global oceans under some form of management, the simple answer will be 'biodiversity conservation'. But as this book has shown this is actually just one, albeit the most fundamental, benefit of PAs. This section looks at how the results of advocacy and the increased awareness of PAs' socio-economic benefits can be used in PA decision-making and management.

The assessment of socio-economic benefits can affect two main types of decisions: (i) by providing a contribution to PA system planning and the designation of new sites and (ii) deciding whether the current PA system, and systems supporting the governance of land use and sustainable management of natural resources in general, is adequately protecting a country or region's full range of PA benefits (e.g. putting in place measures to avoid pressures on PAs).

Changing global social conditions mean that many different stakeholders will expect to have a say about whether a PA is created or not and designation will often depend on a complex process of negotiation, trade-offs and agreements. The science of selection is usually heavily influenced by the politics of what is possible or acceptable. It means making choices about the location, size and management objectives and measures of PAs that take into account other needs within the landscape or seascape. The art of PA design is as much about balancing biodiversity and human needs, and finding the points of overlap, as it is about the strict science of assessment and planning. Effective PA networks will increasingly only work if all the benefits and values – both biodiversity values and socio-economic importance – are recognised and support for protection is gained as a result.

The same process is increasingly true with respect to management objectives and governance types in new or existing PAs. The best choices for biodiversity often have to be balanced and traded off with competing demands. Choice of management objectives and associated IUCN category (see Chapter 2) and choice of who manages are both often key steps in gaining acceptance of protection; but balancing human needs with those of biodiversity often involves difficult trade-offs.

PA system planning and designation for different values and benefits

Assessing the benefits and related socio-economic values of PAs can support sound decision-making by helping to compare the costs and benefits of different options when planning and designating PAs. For example, socio-economic assessments can be used to identify synergies between conservation and other benefits and identify which sites provide which socio-economic benefits and which benefits are abundant/scarce at regional level. Assessments can inform the debate between those responsible for PAs and those affected by them, making visible the real trade-offs and social, environmental and economic consequences involved in alternative options. Furthermore, socio-economic valuation can support addressing trade-offs by helping to delineate different areas within a PA and to define their respective roles. In general, biodiversity values play an important role in defining core areas for PAs, whereas buffer zones can be usefully designed from the perspective of maintaining and sustainably using different ecosystem services.

The motivation for selecting and designating particular PAs has varied over the past one hundred years, with initial PAs being chosen principally for their aesthetic, cultural and recreational values while more recently species-driven protection (often charismatic species such as the giant panda or the game species of Africa) has been dominant. As more is understood about the wide range of benefits from natural ecosystems, the more the gaps in existing PA coverage become apparent. Crop wild relatives (CWR), for example, are a critical resource for sustaining future food security. Despite this, many areas important for CWR conservation are not protected, and where CWRs do occur within PAs they are often not surveyed or monitored and no management provision is made for their conservation. A slow realisation of the importance of CWR has led to recent projects funded by the Global Environment Facility (GEF)/World Bank and GEF/United Nations Environment Programme (UNEP) for *in situ* conservation of CWR in the Fertile Crescent (a semi-circle of land covering Palestine, Israel, the western part of Syria, south-eastern Turkey, eastern Iraq and western Iran) and five developing countries (Armenia, Bolivia, Madagascar, Sri Lanka and Uzbekistan) of particular importance to CWR respectively (Hunter *et al.*, 2012).

Adequate protection of values and benefits

Understanding the socio-economic benefits of PAs can help to ensure biodiversity conservation, increase the support of local people to the PAs and provide, as noted in Section 15.1, powerful tools to engage policy and financial support for PA management.

As Chapters 6 to 13 show, the assessment of socio-economic value can be carried out at the level of one or multiple sites, so both options are considered below:

1 *Assessment of multiple sites:*

- Assessments can show synergies across sites which can be exploited in effective management and capacity-building (e.g. by identifying all the PAs in a country which help protect water quality, best practices can be shared and joint advocacy plans developed particularly with water authorities. This is the case with the FONAG in Quito (Box 15.1), which has been the role model for the development of several other water funds in Central and South America).
- Synthesising, or even aggregating, the values of most or all the PAs in a region (see Chapter 13) can create a good evidence base and powerful arguments for their continued funding and management.

2 *Site level assessments should include planning to address the full range of benefits in documents such as the management plan and business plans (see Box 15.3); issues to consider may include:*

- How to prioritise management needs across values. Once a benefit has been identified, systems may need to be developed to understand impacts on the benefit (e.g. the role of forest management in water provision or natural hazards regulation) and to monitor any resource use or impacts that may affect the delivery of the benefit. Management systems can then be developed based on the results of this research and monitoring (see case study in Box 15.2).
- Identification of values that need more management than others. As noted above regarding CWRs, some benefits and associated values have not always been identified as being important in PA management in the past but as our understanding develops, the need to ensure effective management of all ecosystem services and associated values also increases.

Being able to accommodate different points of view on the importance of values and benefits, both biodiversity and socio-economic alike, to different stakeholders is a critical factor in using and communicating the results of an assessment. The much sought-after win-win situations, where local communities and PA managers work towards the same conservation objectives, are increasingly being achieved. The potential for such mutually beneficial results is usually dependent on a thorough understanding and wide discussion of exactly what different interest groups stand to gain and lose. The Sanctuary of Flora Medicinal Plants Orito Ingi Ande, for example, is the only PA in Colombia primarily dedicated to the conservation of medicinal flora, but through the protection of the culturally important plants yoco (*Paullinia yoco*) and yagé (*Banisteriopsis caapi*) the whole ecosystem is protected.

In the words of Juan Lozano, former Minister for the Environment of Colombia:

> The declaration of the Sanctuary Orito Ingi Ande is an important landmark in the history of the National Natural Parks of Colombia. The category of Sanctuary of Flora of Medicinal Plants harmonises the western point of view of the conservation of biodiversity with the traditional integrated management of the Cosmovision of the Kofán People. The preservation and survival of their customs and traditions in the use and management of medicinal plants depends on the conservation of these territories.
>
> (Stolton and Dudley, 2010)

The declaration of a floristic sanctuary is therefore seen as a means of combining the protection of the habitat and the biological resources with that of the traditional medicine systems, thus contributing to recovery of the area's natural, cultural and intangible heritage.

BOX 15.2

Changing PA management to provide benefits to local people: case study from Bwindi Impenetrable National Park, Uganda

Bwindi Impenetrable National Park in southwest Uganda is managed by the Uganda Wildlife Authority (UWA) with the prime purpose of conserving the montane forests and their associated wildlife populations, especially the mountain gorilla (*Gorilla beringei beringei*). Many people live around the PA boundary (about 300 people per km^2). Local use of wild plants for food, medicine, craft making, clothing, building and artwork have played an important role in local livelihoods for centuries. However, when Bwindi was made a National Park in 1991, local people were stopped from harvesting resources from the park. Conflicts arose between UWA and local people, resulting in widespread fires and intense poaching (Bitariho and Barigyira, 2009).

Since 1994, a Multiple Use Programme (MUP) has greatly reduced conflict. Local people are allowed to harvest plants used as medicines and for basket weaving and also to place hives in the forest for beekeeping. They also assist park managers by reporting illegal activities encountered during plant collection.

The MUP's success has been in ensuring adequate assessments of ecosystem services are used as a way of balancing the needs of the local people with the health of the Bwindi forest. Methods used include participatory rural appraisal (interviews and observations to assess needs); herbarium specimen collection (for botanical identification of the plants); and determining the plant resource density and distribution in the PA (to assess ecosystem health).

The result is a change from the initial management, which barred people from accessing the PA. Recommended annual harvest quotas for various plants are included in a Memorandum of Understanding between UWA and local resource user groups. Monitoring of several species of plants used for medicine and craft materials is in place using permanent plots monitored on an annual basis. As a result sustainable harvest levels have been established (Bitariho *et al.*, 2006). This process is ongoing and new plant resources are being monitored and harvestable quotas/quantities developed.

Source: Bitariho *et al.* (2006), Bitariho and Barigyira (2009)

15.3 Identifying and addressing social impacts

Assessment results can also be used to identify and where possible address any social impacts of protected areas. As the case study from Bwindi has shown (Box 15.2), the designation of a PA has sometimes cut off the supply of some ecosystem services to people living within or near the PA. Detrimental social impacts of PAs on local people are nearly always related to provisioning and cultural services, in particular where access to these services is altered due to PA establishment or management regimes. Assessments therefore need to assess the costs to those who help to 'maintain' benefits as well as the overall benefits to societies as a whole.

Assessing ecosystem services and related socio-economic values, understanding their importance to local populations and finding ways to manage management impacts are often underlying reasons for undertaking an assessment. Clearly not all benefits reach people in the same way; nor do they have the same implications for PAs. Local communities in particular are likely to experience both the costs and some of the benefits more intensely and more immediately than the population as a whole.

It is possible to categorise the ways benefits from PAs reach local people as follows:

- Derived from the provisioning services provided by the PA, such as resources collected or harvested directly from the PA including non-timber forest products, fish (see Section 2.3)
- Derived from the regulating services provided by the PA, such as clean water and erosion control (see Section 2.3)
- Cultural services provided, protected or maintained by the PA, such as education, health and well-being (see Section 2.3)
- Other benefits derived directly from the PA, such as jobs in the PA (see Section 2.4)
- Derived indirectly from the protected area, such as hospitals set up thanks to funds raised by the PA or subsidies for the PA and its surroundings
- Empowering and engaging poor people, such as through co-management of the PA.

If costs seriously outweigh benefits then it is important to take steps to address these in an equitable manner. A frequently promoted, general principle is that PAs should at least 'do no harm' to local communities. This sounds easier than it is to achieve in practice; even conservation actions that benefit the vast majority of people still may cause costs to a minority. A more realistic aim may be that there should be no net harm and any losses are subject to agreed and adequate compensation. A full understanding and good communication of costs and benefits to all the affected stakeholders; who 'wins' and who 'loses' and by how much (see Chapter 12); who might deserve compensation and who, on the other hand, might provide this; can together help to ensure that a PA is established and managed in a fair and equitable manner. In particular, this might include that:

- Beneficiaries of PAs understand the actual or potential costs and who is likely to incur these.
- Local stakeholders in particular understand likely costs and options for negotiation and compensation.
- All stakeholders understand potential wider and future benefits that might themselves help compensate costs.
- Everyone knows that this knowledge will be used as a basis for developing mechanisms to ensure an equitable distribution of costs and benefits (such as payment for ecosystem services schemes).

In a world where PAs are under increasing pressure to demonstrate multiple values, managers are therefore faced with a three-step process: 1) ensuring processes are in place to gain a full understanding of the values and who they benefit; 2) developing ways of capitalising on PA benefits which do not undermine the natural values that the area was established to preserve (as in the case study in Box 15.2); and 3) ensuring any benefits that accrue are equitably shared (see case study in Box 15.1).

The case study from Turkey (Box 15.3) looks at the use by PA managers of the PA Benefits Assessment Tool (PA-BAT) (Stolton and Dudley, 2009; see also Chapter 4) to help understand and categorise benefits. The findings were used to capitalise from benefits and ensure equitable sharing of benefits can follow (as noted in the case study).

BOX 15.3

Understanding local peoples' perspectives of benefits and values: a case study from Küre Mountains National Park, Turkey

Turkey's Küre Mountains extend parallel to the Western Black Sea Region for some 250 km. The Küre Mountains National Park (Küre Dağları Milli Parkı – KDMP), was declared in July 2000. The overall vision for the PA is:

Conservation of biological and cultural diversity in Kastamonu Bartın Küre Mountains National Park and Buffer Zone and formation and implementation of a management model based on cooperation for sustainable use of natural and cultural assets in a manner also offering benefits to local communities.

There are almost no settlements in the core area of KDMP but some 20,000 villagers in 123 villages in the buffer zone. The major economic activities in the region are forestry, agriculture, apiculture, wooden handicrafts, weaving, chestnut growing and tourism. In the poorest forest zones per capita income is lower than €400 per year.

The PA Benefits Assessment Tool (PA-BAT) (see description in Section 4.1) was implemented in KDMP in March 2009 (Stolton and Higgins, 2009). At the time the PA did not have a management plan, although a draft plan was produced in 2000. There was also no communication plan, but a draft sustainable tourism development strategy (2007-2013) had been prepared, which put a major emphasis on ecotourism development, including achievement of PAN Park status (i.e. unique European wilderness area) by 2013, as a strategy towards sustainable development. Park management had a strong focus on involving local people in planning initiatives related to the national park and buffer zone, but no systematic assessment of the PA's benefits and values had previously been attempted.

In total three meetings were held around KDMP to assess the values and benefits of the PA. The first public meeting was attended mainly by local Muhtars (leaders of the village) in Bartın. The meeting concentrated on using a simplified version of the PA-BAT based around a PowerPoint presentation and assessed the values (subsistence, economic and potential value) for local people living inside and near the PA. The second meeting was for representatives of park management (national parks and forestry) and local university departments. The group was divided into two working groups who between them completed all the PA-BAT data sheets relevant to the park. The third meeting was also a public meeting of mainly local officials from the forest, water and parks sectors.

Figure 15.1, summarises the results of using the PA-BAT. Each of the relevant values was assessed in the workshops through a simple scoring system (in this case from 0 to 2). The discussion around each of the values was rich and captured in a narrative report. Figure 15.1 clearly shows significant differences in the perceptions of values between local people, managers and service providers. For example, the local community noted the major importance of sacred springs in the PA, while managers, researchers and service providers assessed these as being of minor importance. The importance of permitted traditional agriculture, wild food plants, and medicinal herbs in the buffer zone were also assessed differently between the groups, with managers thinking traditional agriculture was more important than the local people, whilst managers thought collection of medicinal herbs of no importance, whilst local people assessed this as being of major importance.

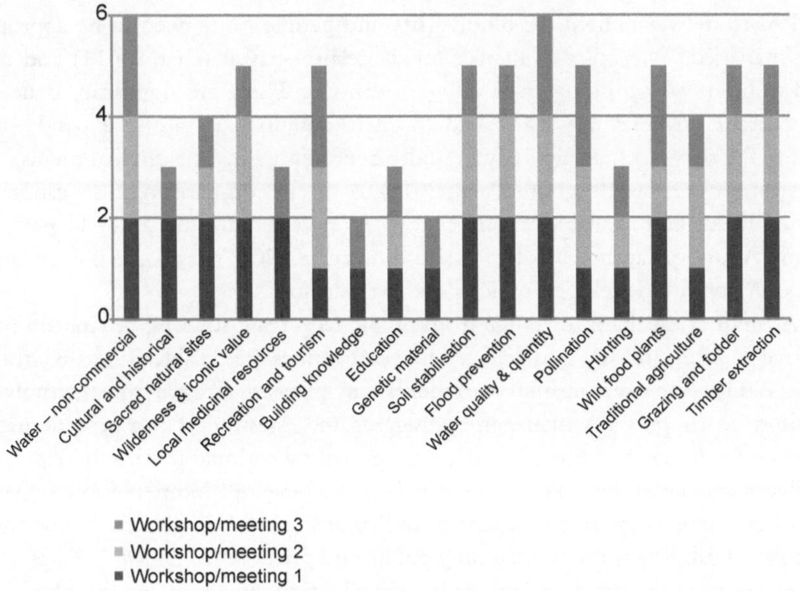

FIGURE 15.1 Quantitative assessment of the perceived benefits and values associated with Küre Mountains National Park, Turkey. The benefits have been identified based on results of three stakeholder meetings/workshops, each ranking them on a scale from 0–2 (no importance – high importance). Building on this data, the total perceived importance of the identified benefits and values in the park has been expressed on a scale of 1-6

Source: Stolton and Higgins (2009)

The PA-BAT assessment of ecosystem services was used as baseline information for a more detailed economic valuation study in the park. The economic value was calculated based on existing market prices and the benefit transfer approach for provisioning services (timber, firewood, traditional grazing, non-timber forest products); carbon market prices for regulating services; and existing recreational fees for cultural services. The total value of the national park was calculated to be approximately US$25 million for 2010.

Highlighting and understanding the different perceptions through the PA-BAT provided useful background information for management planning and served as a basis for the development of a business plan for the KMNP and its buffer zone. This (draft) business plan includes, for example, plans to sustainably increase the production of the identified non-timber forest products (e.g. wooden carved spoons, chestnuts) supported by the establishment of local market places to sell the products.

Source: Stolton and Higgins (2009)

15.4 Mobilising funds

For PAs to deliver benefits to biodiversity and people, they need to be appropriately managed. The costs of management are non-trivial (Chapter 11) and they need to be met by funding from different sources. There are significant concerns that current financial flows are inadequate to sustain – let alone expand – the existing PA networks, particularly in under-represented marine environments (e.g. Emerton *et al.*, 2006; Berghöfer *et al.*, 2012). A recent comparison of the estimated funding needs and available resources for PAs revealed around a 35–40 per cent gap in PA funding at the global level (Kettunen *et al.*, 2011). This gap is even larger in poorer and less developed regions such as Asia and Africa.

PA funding mechanisms range from public to private sources, relying on both external funding inflows and self-generated revenues (see Table 8.2 in Kettunen *et al.*, 2011). Socio-economic assessments can play an integral role in mobilising funds to support PA management: aggregated assessments can provide useful evidence for budgeting funds for PAs at regional or national level, whereas more detailed socio-economic valuations can help to develop compensation schemes and/or encourage concrete investments in PAs at site level. In general, a diversified portfolio of funding sources, including public and private mechanisms, can prevent PAs becoming too reliant on one finance mechanism, and so increasing the long-term sustainability of their management.

Direct investment in PA management (government and donor's budgets, NGO grants and private donations, etc.) continues to form the key means for funding PAs. Socio-economic assessments can strengthen investment in PAs by highlighting multiple (public) benefits provided by the site, creating visibility and political support for conservation (Section 15.1 above) and in this way encouraging financial support to sites. For example, the estimated 10 to 1 benefit-cost ratio of investment in national parks in Finland (Box A5.2 in Annex 5) helped to maintain government investment in the national PA network during a period when the government was generally unsympathetic to conservation (Kajala, 2013). In western Ireland, the economic valuation of public and private benefits (landscape and cultural value, tourism, etc.) provided by the Burren National Park were estimated to be significantly higher than the associated costs, with a 235 per cent minimum rate of return on government support to conservation activities in the area (van Rensburgh *et al.*, 2009). Demonstrating the economic value of these benefits was considered to have played a role in successfully securing future financing to protect the traditional rural landscape in Burren (Kettunen *et al.*, 2011).

The direct support to PAs is increasingly being complemented by new, more innovative funding mechanisms that build on the identification and economic valuation of specific ecosystem services. Such mechanisms are also often aimed at ensuring the equitable sharing of costs and benefits between stakeholders. These mechanisms include, for example, sustainable promotion of new benefits from PAs, building on new partnerships to manage and/or use PAs resources. Furthermore, economic valuations can be used to establish market-based charges for PA benefits (fees, tourism charges, licencing and concessions, payments for ecosystem services, etc.), earmarked to support the management of sites.

A case study from Kaya forests, Kenya, provides a concrete example of identifying new PA benefits and establishing partnerships. The coastal Kaya forests are a gazetted National Monument under the National Museums Act, managed by the local community through a set of traditional rules and regulations. The socio-economic assessment of Kaya forests demonstrated the dependence of local communities on these sacred and culturally important forests also for their livelihood, revealing also unsustainable levels of exploitation (fuel wood, medicinal herbs and building materials, etc.) (Mhando Nyangila, 2006 and Gachanja and Kanyanya, 2004 in Berghöfer *et al.*, 2012). Based on the assessment, local communities approached the national museum of Kenya for management and conservation assistance, in the hope that they might develop sustainable utilisation of the forests' resources. As a result, new sources of revenue were created. For example, the Kaya Kinondo Ecotourism Project was established in partnership with the local communities, using local guides to take visitors through the forests. In 2001, communities around Arabuko Sokoke Forest earned US$37,000 from guiding, beekeeping and butterfly farming.

Public and private payments for ecosystems services (PES) can be used to incentivise the maintenance and sustainable management of ecosystem services, giving those responsible for protecting and/or maintaining benefits an incentive to continue to do so (Boxes 15.4 and 15.5). There are already more than 300 PES programmes in existence (Wunder *et al.*, 2008), providing a good evidence base and starting point for adopting these schemes on PAs. The scope of these existing schemes ranges from local to regional and national, typically targeting maintenance of water quality, improving carbon sequestration, protecting soil or conserving biodiversity. In general, PES schemes offer potential to obtain new funds also for PAs and they can be particularly relevant in situations where trade-offs exist between different land uses by tipping the balance to make conservation more profitable for the landowner/user. In general, PES schemes can offer a flexible way to incentivise and/or finance management of PAs; however, careful design and preparation is needed to ensure that PES schemes are effective and appropriate for local conditions (ten Brink *et al.*, 2011).

At the global level, creating international markets for ecosystem services through offsetting schemes or trading can also provide opportunities for PAs in the future. Transnational and international PES schemes for global public goods are amongst the most prominent recent proposals for financing schemes (Kettunen *et al.*, 2011). For example, the UN Framework for Convention on Climate Change (UNFCCC) proposal for REDD+ (Reducing Emissions from Deforestation and Forest Degradation) aims to create financial value for the carbon stored in forests, offering incentives for developing countries to reduce emissions from forested lands and invest in low-carbon paths to sustainable development. REDD+ aims to go beyond preventing deforestation and forest degradation, integrating the role of conservation, sustainable management of forests and enhancement of forest carbon stocks into the compensation mechanisms, and so providing an international PA-relevant PES scheme (UN-REDD Programme, 2012).

BOX 15.4

Monetary compensation for water provision PES scheme determined through contingent valuation in Paso de Los Caballos River Basin, Nicaragua

Paso de Los Caballos River Basin (740.6 ha) is located in the municipality of San Pedro del Norte, Nicaragua. The main economic activity in the area is agriculture. Problems regarding water quality and quantity led 125 households from San Pedro del Norte to propose and negotiate a PES scheme with the support of local and regional NGOs, which identified priority areas for funding in the upstream basin recharge area. The 125 households created a water committee and reached 5 individual agreements with upstream landowners, covering a total of 39.2 hectares for reforestation and conservation of the prioritised areas.

Each household contributes US$0.31 per month to the scheme while landowners receive a payment of US$26 per hectare per year. As recipients of the payment, landowners commit to a number of sustainable land management practices: avoiding fires before, during and after sowing; developing organic agriculture; conducting soil conservation practices; developing agroforestry systems; promoting regeneration; and committing to preventing livestock from invading areas under PES schemes. The level of monetary compensation was determined based on the information on opportunity costs for farmers and assessing what the providers of payments perceive as a fair compensation.

While not directly related to PAs, the case from Nicaragua provides a good example on how awareness of ecosystem services can help to establish effective 'PES partnerships' between different stakeholders.

Source: Corbera *et al.* (2006)

BOX 15.5

Assessing the feasibility for establishing water-related PES schemes in Shivapuri Nagarjun National Park, Nepal

A feasibility study was carried out in Shivapuri Nagarjun National Park, Nepal, to explore the possibilities for establishing PES schemes to support sustainable management of the site. Shivapuri Nagarjun National Park lies to the north of Nepal's capital city, Kathmandu, and it is well recognised for its rich biodiversity and water-related ecosystem services provided by Sundarijal catchment. The catchment is responsible for maintaining up to one third of the piped water supply in the Kathmandu Valley. The water from the catchment is also used for generating hydroelectricity, irrigating paddy fields, producing bottled water and soft drinks.

The feasibility study indicated that people living in the villages inside the PA are facing socio-economic hardships, having little trust in and being in intense conflict with park authorities. With few livelihood options, local people have recently resorted to making alcohol using fuel wood collected from the PA (both activities prohibited by law). On the other hand, the study estimated the net value of the water-related services of Sundarijal catchment (revenue minus costs of water distribution and electricity generation) to be US$870 per hectare per year. The cost of PA to people living within the PA (damage to crops and livestock by wildlife, limited access to market) was estimated to be US$498 per household per year.

A foreseen PES scheme would collect payments from the stakeholders benefiting from the PA's water supply (water distribution companies, hydropower companies, tourism sector, downstream industries and paddy farmers) and use these payments to incentivise local people to assist in conservation and park management, reducing also the conflict between park authorities and local communities. Given the estimated economic value of water-related benefits, a PES scheme is seen as a viable mechanism to mobilise funds to the Shivapuri Nagarjun National Park while also addressing the current inequitable sharing of costs and benefits.

The following steps are essential to further promotion of PES in protected areas in Nepal:

- review and development of supportive policies for adopting PES mechanisms
- piloting of a PES scheme, and documentation and dissemination of the process and lessons
- development of a regulatory framework for implementing PES, using the experience and lessons from local participation and benefit distribution in protected areas in Nepal

- public awareness building for greater support of PES in PA
- clarification of the roles and responsibilities of multiple stakeholders in protected area and PES management.

Source: Icimod (2011)

15.5 Considering synergies and conflicts with biodiversity conservation

As already discussed in the earlier chapters (e.g. Chapter 2), while there are several synergies between biodiversity and ecosystem services, focusing solely on the latter does not necessarily guarantee positive impacts on the former. Consequently, the impacts of promoting PA benefits on biodiversity should be carefully considered when interpreting, communicating and/or using the results of socio-economic assessments.

For example, in Costa Rica legal egg collection of Olive Ridley Sea Turtles (*Lepidochelys olivacea*) in Ostional Wildlife Refuge supports the local economy and equitability; however, it also provides a framework for the illegal egg trade undermining conservation (Troëng and Drews, 2004 in Chapter 6). In comparison, ecotourism related to green turtles (*Chelonia mydas*) in Tortuguero National Park supports small businesses, such as hotels and tour operators and also acts as a deterrent against the illegal taking of turtles and eggs and facilitates reporting of such activities to the authorities, leading to an increase of 417 per cent in turtle nesting since 1971 (Troëng and Drews, 2004 in Chapter 10).

The relationship between biodiversity, ecosystem services and related goods is often related to the type of services in question. Supporting and regulating services (Chapters 7 and 9) built to a large extent on the maintenance and/or restoration of well-functioning ecosystems and therefore highlighting the socio-economic importance of PAs' in delivering these services generally supports biodiversity conservation. However, some exceptions do exist. For example, restoration of agricultural pastures into wetlands might be advantageous for flood management but detrimental for bird species nesting on the pastures. The maintenance of carbon stocks is usually beneficial for biodiversity, for example, when protecting old forest with a high conservation value. Therefore, any valuation emphasising the economic value of maintaining a PA's existing carbon stocks is usually in line with biodiversity goals. However, in some cases climate measures and conservation objectives may not be fully compatible. For example, there is a growing interest to use reforestation as a means to increase carbon sequestration and mitigate climate change. In this context, it is important to note that not all reforestation activities, such as planting monocultures, are beneficial for biodiversity. Therefore, any attempt to increase carbon sequestration in PAs should clearly consider the related (positive or negative) impacts on biodiversity.

As stated in Chapter 6, demonstrating the socio-economic value of PAs as a source for provisioning services and related goods helps to safeguard biodiversity only if the resource in question is exploited in a sustainable manner, without jeopardising the PA's set conservation goals. In general, estimating the socio-economic value of provisioning services is most directly relevant to biodiversity goals, for example, when it is used to help to demonstrate the importance of habitat/species/genetic variability (value of wild crop relatives and local breeds supporting food and health security, etc.) or highlight the benefits arising from sustainable long-term use of a species/group of species (value of fisheries, etc.).

As shown in Chapter 8, the role of biodiversity in delivering cultural services and related goods is closely connected to human interaction with nature. It relates to the role that the diversity of species, ecosystems and genes has played in shaping a community's knowledge, culture, creativity, inspiration and belief. This can relate to, for example, a single charismatic species which attracts visitors to a site and is also a key part of people's religious beliefs. It can also include the diversity of structures and features that an ecosystem offers, making it attractive to visitors. It can also be just the knowledge of the existence of an exceptional natural area that increases people's affinity to a site. Tourism and recreation to PAs can bring nature closer to people, increase awareness and the willingness to safeguard biodiversity. Affinity to PAs can, however, represent a double-edged sword if no careful consideration is given to how the influx of people and resulting impacts can be managed to avoid degradation of the PA.

Building on the above, the choice of promoting certain socio-economic benefits of PAs can both contribute to and risk jeopardising the achievement of site's biodiversity goals. Supporting the interpretation, communication and use of socio-economic assessments with a co-analysis of ecosystem services and biodiversity 'hot spots' helps to determine where conflicts might occur. Such an assessment can always be carried out on a qualitative level. However, quantitative and spatially explicit analysis require dedicated expertise in GIS and modelling and, as such, are already recommended to be built into the socio-economic assessment themselves (Nelson *et al.*, 2009 in Box 15.6).

BOX 15.6

Assessing synergies and trade-offs between multiple ecosystem services and biodiversity conservation by using the Integrated Valuation of Ecosystem Services and Trade-offs (InVEST)

The Integrated Valuation of Ecosystem Services and Trade-offs (InVEST) is a spatially explicit modelling tool that can be used to acquire information on the biophysical flow of ecosystem services on a landscape level (see Chapter 5).

The tool was used to predict changes in several ecosystem services (water regulation, maintenance of soil quality, carbon sequestration, crops and timber, role of environmental attributes in affecting property value) and biodiversity conservation in Willamette Basin, Oregon, under three land use scenarios involving different grades of development vs. conservation in order to highlight synergies and trade-offs between multiple services and conservation.

Method: The scenarios, evolving around the implementation and/or loosening of current policies for conservation, were 'implemented' on a set of spatially explicit raster grid LU/LC maps (30 m x 30 m grid cells) of the Willamette Basin (10-year intervals from 2000 to 2050).

The biophysical provisioning of ecosystem services was quantified and mapped using proxy indicators: level of dissolved phosphorus, landscape characteristics and soil infiltration, soil erosion rate, carbon content of above and below ground biomass and rate of carbon sequestration. In addition, the social value of carbon storage was assumed at US$43 per Mg of carbon (based on value transfer) with a discount rate of 7 per cent and also assuming that the social cost of carbon will decline at a rate of 5 per cent/year. The value of crops and timber was derived by value transfer and using a discount rate of 7 per cent/year to compute the net present values (NPV) of commodity production across time (see Annex 6 for further explanation of NPV). NPVs were determined as a production function of crop/timber type, soil productivity, level of irrigation/forest rotation time, crop/timber prices and production costs. The amount of actual and potential habitat area provided for a species was used to determine the capacity of each LU/LC map to support a suite of 24 vertebrate species that previous analysis found to be sensitive to land-use change in the area.

Results: Scenarios enhancing biodiversity conservation also enhanced the production of ecosystem services. Scenarios based solely on market returns resulted in increased provisioning/value of agricultural products and timber while decreasing the level of biodiversity conservation and other ecosystem services. However, policy intervention for carbon sequestration might alleviate this trade-off, as payments for sequestered carbon may make it more profitable for landowners to favour conservation. Scenarios involving payments for water quality, soil conservation and storm peak mitigation also promoted maintenance of biodiversity.

Source: Nelson *et al.* (2009)

References

Berghöfer, A., Dudley, N. and Förster, J. (2012) 'Ecosystem Services and Protected Areas' in Wittmer and Gundimeda (eds) *The Economics of Ecosystems and Biodiversity in Local and Regional Policy and Management*, EarthScan/Routledge, London and Washington, DC

Bertzky, B., Corrigan, C., Kemsey, J., Kenney, S., Ravilious, C., Besançon, C. and Burgess, N. (2012) *Protected Planet Report 2012: Tracking progress towards global targets for protected areas*. IUCN, Gland, Switzerland and UNEP-WCMC, Cambridge

Bitariho, R. and Barigyira, R. (2009) *The potential supply of plant resources for local community use in Bwindi Impenetrable National Park beekeeping zones, S. W Uganda*, Institute of Tropical Forest Conservation, Uganda

Bitariho, R., McNeilage, A., Babaasa, D. and Barigyira, R. (2006) 'Plant harvest impacts and sustainability in Bwindi Impenetrable National Park, S. W. Uganda', *African Journal of Ecology*, 44(1): 14–21

Canales, C. and Jouravlev, A. (2012) Water and a Green Economy in Latin America and the Caribbean (LAC), UNECLAC Natural Resources and Infrastructure Division UN-Water Decade Programme on Advocacy and Communication (UNW-DPAC), Santiago, Chile

Corbera, E., Kosoy, N. and Martínez Tuna, M. (2006) *Marketing ecosystem services through protected areas and rural communities in Meso-America: Implications for economic efficiency, equity and political legitimacy*, Tyndall Centre for Climate Change Research Working Paper 94, http://www.tyndall.ac.uk/sites/default/files/wp94.pdf, accessed 19 December 2012

Emerton, L., Bishop, J. and Thomas, L. (2006) *Sustainable Financing of Protected Areas: A global review of challenges and options*, IUCN, Gland, Switzerland and Cambridge, http://app.iucn.org/dbtw-wpd/edocs/PAG-013.pdf, assessed 19 December 2012

Gachanja, M. and Kanyanya, E. (2004) *Conservation of Kenya's Coastal Forests: Social Economic Resources*, WWF East African Coastal Forest Ecoregion Programme, Nairobi, Kenya

Goldman, R. L., Benitez, S., Calvache, A. and Ramos, A. (2010) *Water funds: Protecting watersheds for nature and people*, The Nature Conservancy, Arlington, Virginia

Hunter, D., Maxted, N., Heywood, V., Kell, S. and Borelli, T. (2012) 'Protected areas and the challenge of conserving crop wild relatives', *PARKS*, 19(1)

Icimod (2011) *Protected Areas and Payment for Ecosystem Services, A feasibility study in Shivapuri-Nagarjun National Park, Nepal*) International Centre for Integrated Mountain Development, http://www.netif-nepal.org/PA%20and%20payment%20for%20Ecosystem%20Services_Shivapuri-Nagarjun%20National%20Park[1]-PES.pdf, accessed 9 June 2013

Kajala, L. (2013) 'TEEB Nordic case: Estimating local economic benefits of visitors' spending in protected areas', in M. Kettunen *et al.*. *Socio-economic importance of ecosystem services in the Nordic Countries – Scoping assessment in the context of The Economics of Ecosystems and Biodiversity* (TEEB), Nordic Council of Ministers, Copenhagen

Kettunen, M., Berghöfer, A., Brunner, A., Conner, N., Dudley, N., Ervin, J., Gidda, S. B., Mulongoy, K. J., Pabon, L. and Vakrou, A. (2011) 'Recognising the value of protected areas', in ten Brink P. (ed.) *The Economics of Ecosystems and Biodiversity in National and International Policy Making*, Earthscan, London and Washington, DC

Mhando Nyangila, J. (2006) *Museums and community involvement: A case study of community collaborative initiatives*, National Museum of Kenya, Intercomm 2006 Conference, Taiwan

Nelson, E., Mendoza, G., Regetz, J., Polasky, S., Tallis, H., Cameron, D. R., Chan, K. M. A., Daily, G. C., Goldstein, J., Kareiva, P. M., Lonsdorf, E., Naidoo, R., Ricketts, T. H. and Shaw, M. R. (2009) 'Modeling Multiple Ecosystem Services, Biodiversity Conservation, Commodity Production, and Tradeoffs at Landscape Scales', *Frontiers in Ecology and the Environment*, 7(1): 4–11

New Zealand Government (2010) *Arawai Kakariki wetland restoration programme – Whangamarino wetland*, Department of Conservation Waikato Conservancy http://www.doc.govt.nz/documents/conservation/land-and-freshwater/wetlands/ak-whangamarino-factsheet-web.pdf, accessed 19 December 2012

Pabon-Zamora, L., Bezaury, J., Leon, F., Gill, L., Stolton, S., Grover, A., Mitchell, S. and Dudley, N. (2008) *Nature's Value: Assessing protected area benefits*. Quick Guide Series, The Nature Conservancy, Arlington, Virginia

Stolton, S. and Higgins, L. (2009) *Protected Area – Benefits Assessment Tool. Kure Daglari Milli Parki. Turkey*, WWF

Stolton, S. and Dudley, N. (2009) *The Protected Areas Benefits Assessment Tool*, WWF, Gland, Switzerland

Stolton, S. and Dudley, N. (2010) *Arguments for Protected Areas: Multiple benefits for conservation and use*, Earthscan, London

ten Brink, P., Bassi, S., Bishop, J., Harvey, C. A., Ruhweza, A., Verma, M. and Wertz-Kanounnikoff, S. (2011) 'Rewarding Benefits through Payments and Markets', in ten Brink P. *The Economics of Ecosystems and Biodiversity in National and International Policy Making*, Earthscan, London and Washington, DC

TNC (undated), *Investing in Nature to Protect Water*, TNC, Washington, DC

Troëng, S. and Drews, C. (2004) *Money Talks: Economic Aspects of Marine Turtle Use and Conservation*, WWF-International, Gland, Switzerland, http://awsassets.panda.org/downloads/moneytalks.pdf, accessed 23 November 2012

UNEP (undated) *FONAG: Quito's water fund: A municipal commitment to protect the watersheds*, TNC, Washington, DC

UN-REDD Programme (2012) Information about REDD+, http://www.un-redd.org/aboutredd/tabid/582/default.aspx, accessed 28 December 2012

van Rensburg, T. M., Kelly, H. and Yadav, L. (2009) *Farming for Conservation of the Upland Landscape and Biodiversity in the Burren, Department of Economics*, National University of Ireland, Galway, http://aran.library.nuigalway.ie/xmlui/bitstream/handle/10379/1140/EconWP_0153.pdf?sequence=1, accessed 9 June 2013

Wunder, S., Engel, S. and Pagiola, S. (2008) 'Taking stock: A comparative analysis of payment for environmental services programs in developed and developing countries', *Ecological Economics*, 65(4):834-852

16

CONCLUSIONS AND WAY FORWARD

Kettunen, M. and ten Brink, P.

16.1 Towards a full appreciation and uptake of benefits provided by protected areas

This assessment guide has aimed to provide consolidated information and advice on how to identify multiple benefits associated with PAs, including how to estimate their socio-economic value. It has also shown how estimating the value(s) of PAs (monetary, quantitative and qualitative) can create a basis for both communicating their importance and developing concrete measures to support the management of sites. Finally, the guide has also synthesised a wealth of concrete examples around the world that demonstrate how the assessment of PA-related benefits and their socio-economic value can be carried out in practice.

While the appreciation of PAs' role in maintaining welfare is increasing, further efforts are needed to highlight their importance to different stakeholders and increase the evidence base on their socio-economic value. This assessment guide has aimed to show how the assessment of PA benefits can be an influential tool for both supporting human well-being, the economy and stepping up conservation efforts. It is hoped that the combination of evidence and advice provided by the guide will serve as a source of inspiration and practical guidance for its readers around the globe, supporting the realisation of biodiversity objectives at local to global levels.

The usefulness and efficacy of socio-economic assessments is ultimately determined by their uptake and implementation. As emphasised throughout the book, tailoring assessments to answer specific questions related to PA decision-making and management helps to guarantee uptake of results. Furthermore, carrying out balanced, inclusive and robust valuation studies, and interpreting and communicating the outcomes to the right stakeholders, will maximise the chance that the values of protected areas are taken account of in decision-making. Several tools and mechanisms are already available to integrate PA benefits into decision-making and management (Chapters 12 and 15 and see also ten Brink, 2011; Wittmer and Gundimeda, 2012). These include, for example, using environmental impact assessments and cost-benefit studies to integrate and mainstream information on the status and value of ecosystem services into decision-making. Furthermore, existing spatial planning tools, complemented with dedicated instruments for ecosystem services (PES-schemes, etc.), can be used to help on the ground integration. However, more work is needed to employ these tools to mainstream the value of PAs into concrete decision-making and implementation. This applies at the level of individual sites, in the context of local and regional land use and planning, and in relation to national and international policies and programmes.

Naturally, while socio-economic assessments of PAs can help to support effective PA management, they cannot automatically guarantee it. Successful establishment and management of PAs, and the delivery of associated benefits, requires multi-level policy support and effective institutional frameworks (Kettunen *et al.*, 2011; Berghöfer *et al.*, 2012). While monetary values can help to translate ecological concerns into economic arguments, these arguments must always be considered within the bigger picture of PA governance. General compliance with rules and regulations governing PAs (e.g. respecting their biodiversity goals) is critical and several preconditions, such as well-defined and stable property and/or use rights and capacity-building to support the uptake of new instruments, need to be met. Furthermore, good monitoring mechanisms are still needed to measure progress towards set objectives – biodiversity and ecosystem service objectives alike – and to assess efficiency of management actions and/or financial investment.

It is also of crucial importance that realising the identified benefits of PAs – and compensating for the associated costs – is done in an equitable manner. The benefits provides by PAs are often broadly disbursed, long-term and non-market, whereas the costs of protection are more immediate and the earning potential from conversion options more short-term and concentrated (Kettunen *et al.*, 2011). As highlighted in this guide, ensuring the participation of all relevant stakeholders, recognising and respecting local rights and creating mechanisms to make beneficiaries pay for or invest in what they receive form a starting point for integrating identified benefits successfully into PA management. Such actions also increase the perceived fairness of PAs and optimise their contribution to both biodiversity conservation and human well-being.

Finally, while socio-economic assessment can help to increase understanding on the value of PAs and guide decision-making, it needs to be kept in mind that valuation, whether based on monetary, quantitative or qualitative information, should never be considered a precondition for acting to conserve biodiversity. Assessing the socio-economic importance and value of PAs should not aim to undermine the intrinsic value of biodiversity.

16.2 An integrated future vision for protected areas: benefits for biodiversity and people

At the tenth Conference of the Parties to the UN Convention on Biological Diversity (CBD) in 2010 (Nagoya, Japan) the global community agreed on the Strategic Plan for Biodiversity for 2011–2020. As outlined in Chapter 1, one of the key objectives of the plan is to enhance the benefits from biodiversity and ecosystem services and to ensure that biodiversity values are integrated into all relevant strategies and decision-making processes, including being incorporated into national accounting and reporting systems (Aichi Biodiversity Target 2). In practice, this requires the implementation of national and/or regional assessments focused on the socio-economic importance of biodiversity and ecosystem services. A number of such assessments are already planned and/or taking place (e.g. UK NEA, 2011; Kettunen *et al.*, 2013; TEEB, 2013) and it is important that the socio-economic assessments of PAs are, depending on their scope, carried out as an integral part or with a due consideration of these broader assessments. PA networks are the backbone of well-functioning ecosystems and therefore it is crucial to ensure that they are integrated into wider ecosystem assessments and appropriately addressed in the context of future policy action.

Building on the above, there is a need to better align biodiversity objectives and PAs with other policies to ensure broad policy coherence and build on opportunities for synergies. PAs' role in sequestering and storing carbon and regulating natural hazards is recognised as key to both climate change mitigation and adaptation. PAs play also a crucial role in supporting global food security by, for example, maintaining pollination, protecting genetic variability of species used as food (crops, fruit, etc.), protecting breeding grounds and nurseries for fish, and functioning as a key resource of protein for the world's poorest populations. Furthermore, PAs are important in terms of global health: they help to secure clean water supply, control the spread of harmful diseases, provide alternative and/or novel medicinal resources and maintain recreation related health benefits. Consequently, well-designed coherent networks of appropriately managed and ecologically connected PAs provide an effective response not only to biodiversity conservation, but also to a range of other current global policy challenges.

As highlighted throughout this guide, PAs play an important role in underpinning the functioning of socio-economic systems, creating business opportunities and providing cost-effective solutions for different sectors. Consequently, sustainable use, protection and restoration of nature – with the help of PAs – can play a key role in the development of more sustainable and greener future economies (UNEP, 2011; UNCSD, 2012; ten Brink *et al.*, 2012). However, a suite of long-term measures is needed to fully account for the true benefits of safeguarding biodiversity, ecosystems and related services to socio-economic well-being. Investment in a comprehensive and well-managed global PA network and communicating its many biodiversity and socio-economic values is one such measure.

References

Berghöfer, A., Dudley, N. and Förster, J. (2012) 'Ecosystem Services and Protected Areas' in Wittmer and Gundimeda (eds) *The Economics of Ecosystems and Biodiversity in Local and Regional Policy and Management*, EarthScan/Routledge, London and Washington, DC

Kettunen, M., Berghöfer, A., Brunner, A., Conner, N., Dudley, N., Ervin, J., Gidda, S. B., Mulongoy, K. J., Pabon, L. and Vakrou, A. (2011) 'Recognising the value of protected areas', in ten Brink P. (ed.) *The Economics of Ecosystems and Biodiversity in National and International Policy Making*, Earthscan, London

Kettunen, M., Vihervaara, P., Kinnunen, S., D'Amato, D., Badura, T., Argimon, M. and ten Brink, P. (2013) *Socio-economic importance of ecosystem services in the Nordic Countries – Scoping assessment in the context of The Economics of Ecosystems and Biodiversity* (TEEB), Nordic Council of Ministers, Copenhagen

TEEB – The Economics of Ecosystems and Biodiversity (2013) Information on ongoing TEEB studies, http://www.teebweb.org/teeb-implementation, accessed 9 June 2013

ten Brink, P. (ed.) (2011) *The Economics of Ecosystems and Biodiversity in National and International Policy Making*, Earthscan, London

ten Brink, P., Mazza, T., Badura, T. and Kettunen, M. (2012) *Nature and its Role in the Transition to a Green Economy: A contribution to Rio + 20* (executive summary), http://www.teebweb.org/wp-content/uploads/2013/04/Nature-and-Green-Economy-Exec-Sum.pdf, accessed 9 June 2013

UK NEA – National Ecosystem Assessment (2011) *The UK National Ecosystem Assessment: Synthesis of the Key Findings*, UNEP-WCMC, Cambridge, http://uknea.unep-wcmc.org, accessed 18 June 2012

UNCSD (2012) *Rio+20 declaration – The Future We Want*, http://www.uncsd2012.org/thefuturewewant.html, accessed 19 July 2012

UNEP (2011) *Towards a Green Economy – Pathways to Sustainable Development and Poverty Eradication – Synthesis for Policy Makers*, http://www.unep.org/greeneconomy/Portals/88/documents/ger/GER_synthesis_en.pdf, accessed 30 December 2012

Wittmer, H. and Gundimeda, H. (eds) (2012) *The Economics of Ecosystems and Biodiversity in Local and Regional Policy and Management*, Earthscan, London

INDEX